Renewable Integration in Power Systems

Renewable Integration in Power Systems

Lokesh Pandey

ARCLER PRESS

www.arclerpress.com

Renewable Integration in Power Systems

Lokesh Pandey

Arcler Press
4164 Lakeshore Road
Burlington, ON L7L 1A4
Canada
www.arclerpress.com
Email: orders@arclereducation.com

ISBN: 978-1-77956-977-6 (Hardcover)

ABOUT THE AUTHOR

Lokesh Pandey is currently pursuing his PhD in Mechanical Engineeing from Uttarakhand Technical University, India, where he also completed his M.Tech in Thermal Engineeing. He has more than 3 years teaching experience as a Faculty for Mechanical Engineering. He has published articles in reputed Journals and has also chaired many conferences and workshops.

TABLE OF CONTENTS

LIST OF FIGURES

PREFACE

Energy has ever been the basis of human progress, and the ways societies produce, transport, and consume it have shaped civilizations historically. From the era of coal and the industrial revolution to the era of oil and gas of the 20th century, structures of energy have evolved along with economic development and technological progress. In our era, the 21st century faces challenges hitherto unimagined: rising demand for electricity power, environmental sustainability, and climate change mitigation on an urgent basis. Under all this, renewable energy has moved beyond alternative status—today it has become a future necessity for delivering clean, stable, and sustainable energy.

The integration of renewables into today's power grids is a technical, economic, and policy-making challenge that requires an interdisciplinary approach. Unlike conventional fossil-fuel-based plants, renewables such as solar, wind, hydro, and biomass bring variability, decentralization, and resource-dependence into the grid. Their handling of intermittency, matching generation with demand, and maintaining power quality are challenges that put existing infrastructure to new limits. But with advancing energy storage, smart grid appliances, and power electronics, the hope of a more resilient and cleaner grid is within reach. It is essential that engineers, policymakers, and researchers who shape the electricity future understand the basic principles of power systems and nature of different renewable resources.

Renewable Integration in Power Systems is written with the objective of bridging technical basics to real-world integration practices. The book shows a comprehensive view of how renewable resources can be seamlessly integrated into existing and future grids. It highlights not only the scientific and engineering basics but also the practical mechanisms that offer stability, efficiency, and reliability of electricity networks in the face of growing renewable penetration. Every chapter addresses a key topic of renewable energy integration, presenting readers with clarity and depth on conceptual as well as applied sides.

Chapter 1 offers an introduction to renewable energy and power systems with the reader, presenting them with information on the history of energy sources, global trends in renewable adoption, and drivers to force change. The chapter also establishes the reasons why switching to renewables is a priority, such as climate concerns, energy security, and competitiveness.

Chapter 2 is about power system basics. It discusses the design and functioning of power systems, from generation to transmission, distribution, and consumption. There is coverage of load flow, stability, and frequency management—all the essentials of getting an idea of how renewables fit into the larger picture of electricity grids.

Chapter 3 discusses the characteristics of the renewable sources. Solar radiation profiles, wind speed distribution, hydrology cycle, and biomass availability are discussed and their geographic and seasonal dependence examined. The chapter highlights the manner in which the inherent character of every renewable source affects the system planning and integration activities.

Chapter 4 is focused on solar power integration. The chapter touches on photovoltaic and concentrated solar technologies, grid connection systems, and variability and intermittency management issues. Distributed generation, forecast models, and hybrid solar-storage system solutions are analyzed, making this section applicable for utility-scale and decentralized solar implementation.

Chapter 5 is also dedicated to the integration of wind power. Wind turbine technologies, compliance with grid codes, and voltage control are all addressed topics. The chapter emphasizes topics such as frequency regulation and transient stability, graphically showing how wind farms react to the grid. The section also addresses the place of offshore wind schemes and the greater importance of such schemes in developing renewable capacity.

Chapter 6 addresses small-hydro and integration of hydropower. Large hydro is well established but run-of-the-river systems and small-hydro make ever-growing contributions to local self-sufficiency in energy. The chapter offers design issues, environmental aspects, and the role of pumped storage hydropower in renewable penetration integration.

Chapter 7 is dedicated to biomass and the other renewable sources. It looks at the transformation of organic material to electricity, and the lesser-known sources such as tidal and geothermal power. The chapter analyzes the viability of these resources, sustainability challenges, and technical integration obstacles.

Chapter 8 is concerned with power electronics for renewable integration. As renewables are typically interconnected through inverters and converters, familiarity with power electronics is important. The chapter covers converter design, harmonics, control methods, and fault-ride-through capabilities that enable renewables to be integrated into modern power systems.

Chapter 9 covers energy storage for renewable energy integration. Batteries, pumped hydroelectricity, compressed air energy, and other emerging storage technologies are defined by their capacity, efficiency, and stabilization of the grid. Particular focus is given on how storage supports load balancing, frequency regulation, and reliability with high renewable penetration.

Chapter 10 focuses on smart grids and demand response. The chapter highlights digitalization of power systems, sensor technology and communication networks, and active consumer involvement in electricity markets. Demand-side management, real-time pricing, and flexible consumption patterns are suggested as major tools to deal with renewable variability.

The book is appropriate for a wide audience of students, researchers, engineers, and energy professionals. For students, it provides systematic groundwork that traces a path from basic power system facts to advanced integration topics. For researchers, it serves as a platform to understand current challenges and innovation topics. For practitioners and policymakers, it will serve as a handbook to assist in guiding renewable integration measures that balance technical feasibility with economic and environmental requirements. Read from beginning to end or used as a reference for specific themes, the book aims to help support the common objective of maximizing the role of renewable energy as a foundation of electricity systems around the world.

CHAPTER 1

INTRODUCTION TO RENEWABLE ENERGY AND POWER SYSTEMS

Contents

INTRODUCTION

The transition to renewable energy from fossil fuel is one of the major characteristics of contemporary energy systems brought about by the strong imperative of managing the environment as well as enhancing energy security. Renewable energy from naturally renewable sources such as solar, wind, and hydroelectricity is a sustainable option towards meeting global energy requirements while minimizing carbon footprints. According to upcoming trends, countries like Ecuador have embraced this shift by implementing vast renewable projects, primarily in hydropower, which have contributed significantly to the energy autonomy of the nation. Not only are these projects environmentally friendly sources of energy in terms of comparison to fossil fuels, but they also drive economic growth through job creation and improved standards of living, particularly in remote regions.

Figure 1.1. *Renewable energy.*

Source: https://images.takeshape.io/86ce9525-f5f2-4e97-81ba-54e8ce933da7/dev/9d498117-3179-4718-81f4-f1c7fe8e892b/Pillar%201%20Energy%20Transition%20wind%20energy%20solar%20panels%20dreamstime.png?auto=compress%2Cformat&w=1440.

All of this comprehensive strategy focuses on the importance of integrating renewable power into power systems, paving the way toward a future that is sustainable (Chilán et al., 2018). At the root of all this knowledge of how renewable energy can be integrated into existing power systems stands the distributed generation concept. This new thinking allows for the deployment of low-power energy projects that generate electricity close to points of consumption, reducing transmission loss and grid unreliability. Though distributed generation integration can significantly improve voltage profiles and operating efficiency, it also introduces some technical challenges, including fault clearing problems and a need for advanced grid management technologies. The contrast between these advantages and disadvantages must be managed very carefully to facilitate orderly integration into established distribution networks. Lastly, with more countries designing renewable energy systems, understanding the dynamics between traditional and renewable sources becomes more vital to maintain healthy and efficient power systems (Afifi et al., 2014). Besides its technological features, the utilization of renewable energy has broader social and cultural implications. As people live through these technologies, they experience increased environmentalism and increased local engagement with energy issues. In regions like Loja, Ecuador, wind farm use, and solar photovoltaic systems have not only reduced pollution and its effects on health but have also raised people's interest in energy literacy. This cultural change enables broader adoption of sustainability programs and inspires innovative solutions to local energy challenges. In addition, as the renewable energy industry keeps on growing, it is bound to impact tourism by promoting sustainable technologies and practices that result in new economic opportunities and revenue streams. Therefore, the adoption of renewable energy is not just an environmental initiative but a far-reaching force within societies (Chilán et al., 2018) (Figure 1.1).

Definition and Importance of Renewable Energy

The definition of renewable energy is those sources of energy that are naturally replenished in a time frame that is acceptable to humans and serve as alternatives to traditional fossil fuels. Solar, wind, hydroelectric, and biomass energy are important among the renewable energy types. The use of these resources not only serves to reduce greenhouse gas emissions but also plays a crucial role in the battle against climate change. As more and more societies struggle with the adverse effects of the use of fossil fuels, moving towards renewable energy systems is more imperative than ever before. It is both a technological challenge as well as a social imperative, as changing consumer markets and pro-sustainability regulation policies make the transition inevitable. Therefore, the crux of knowledge regarding renewable energy is realizing its two-fold position as a clean source of energy and a force for social change and bringing it at the forefront of modern-day energy systems, policies, and community initiatives (Arrobbio et al., 2020). The value of renewable energy extends beyond the environmental dimension; it encompasses economic and social dimensions as well. With advancement in energy independence, clean resources offer nations an avenue of reducing their reliance on foreign fuels, enhancing the security of energy. Job creation locally in renewable industries such as solar panel installation and wind turbine maintenance offers great contribution to local economies while enhancing community participation in energy management. In addition, the use of renewable energy systems also tends to empower local communities by promoting decentralized energy generation and thus enabling individuals and small groups to contribute directly to energy production. The participatory nature can engage local economies and improve social cohesion as community members collaborate on shared goals towards sustainability. Such efforts have been referred to as Collective Action Initiatives (CAIs), which are increasingly being recognized as drivers of social innovation, as well as integrating the effects of traditional energy regimes (Arrobbio et al., 2020). Besides, renewable energy facilitates a paradigm shift in energy consumption and production trends, as required for sustainable development to be achieved. Notably, technologies such as anaerobic digestion demonstrate how the combination of renewable methods can provide energy security while addressing waste management issues. In industries such as food processing, energy management strategies are increasingly focused on aligning environmental, economic, and social factors. Such approaches contribute not only to optimizing operation efficiencies but also to broader sustainability goals by minimizing waste and optimizing resource utilization. Thus, the applicability of renewable energy is enhanced in cross-cutting interactions among environmental policies, economic interventions, and social commitments. With these systems evolving, there is a continued need for integrated strategies that connect energy and power management, ultimately designing more sustainable energy futures (Figure 1.2).

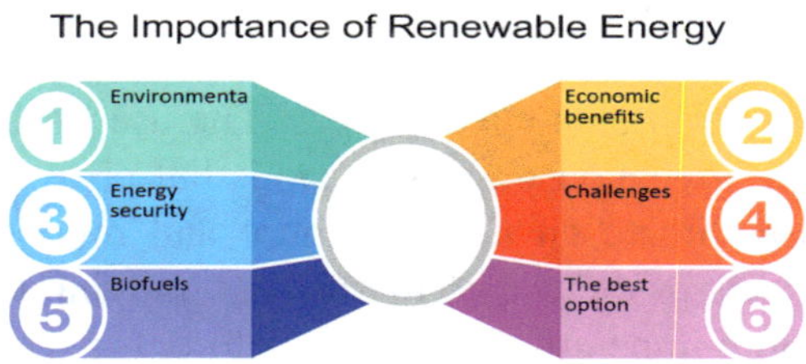

Figure 1.2. *Importance of renewable energy.*

Source: https://fastercapital.com/i/Renewable-Energy--Revolutionizing-Renewable-Energy--The-Role-of-Biofuels--The-Importance-of-Renewable-Energy.webp.

Overview of Different Types of Renewable Energy Sources

The landscape of renewable energy resources is vast and diverse, ranging over resources that utilize natural processes to generate power. Amongst these, solar power is one of the most popular ones, which utilizes sunlight in photovoltaic cells or solar thermal collectors to generate electricity or heat. Wind energy ranks second in popularity, which harnesses the kinetic energy of wind through turbines to produce electrical power. The complementarity of these two sources frequently raises their efficiency; for example, when solar production reaches its peak on sunny days, wind patterns usually result in energy production at night or on bad weather days. As has been emphasized by current research, combining various renewable sources can alleviate the intermittency issue related to their individual outputs, ensuring a more even energy supply to address fluctuating levels of demand (Beluco et al., 2019). Other than solar and wind, hydropower is an ancient yet highly efficient source of renewable energy. By harnessing the energy of flowing water through the utilization of dams or run-of-river plants, hydropower contributes a significant portion of the global energy mix. Hydropower offers immense advantages, one of which is the storage of energy, which can be vital in balancing supply and demand. However, the environmental impact of mega hydropower schemes usually raises concerns, and there is thus a need for cleaner options, where micro-hydropower schemes with lower ecological disruption are some of the alternatives.

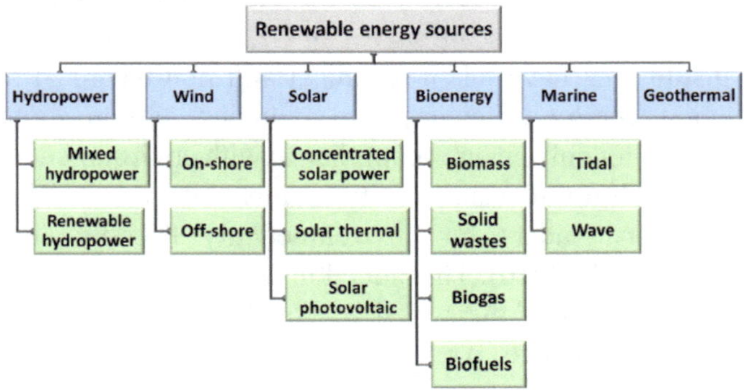

Figure 1.3. *Different types of renewable energy.*

Source: https://www.researchgate.net/profile/Ahmed-Osman-43/publication/369521612/figure/fig1/AS:11431281130182396@1679758371576/Different-types-of-renewable-energy-The-field-of-renewable-energy-technology-encompasses.png.

Advanced technologies in energy storage, for example, the vanadium redox flow batteries, have promise to maximize energy storage from the available renewable resources, thereby enhancing the efficiency and reliability of power systems affected by hydropower (Guarnieri et al., 2016). Other recently discovered renewable energy sources add to this system's strength. Biomass energy exploits organic matter, converting waste products into exploitable forms of energy through burning or biochemical processes. Geothermal energy exploits heat within the earth's interior and is especially uniform where geology permits. Ocean energy, dependent upon tides and waves, also promises to be a huge as-yet-unexploited resource. As the energy scenario globally keeps improving, integration of these heterogeneous renewable sources—aided by sustained research and technology innovation—holds tremendous prospects for sustainable energy supply. Notably, hybridization of the dual or multiple sources can overcome inherent limitations from any one technology to deliver a stable and resilient energy supply to suit contemporary pattern of demand (Beluco et al., 2019) (Figure 1.3).

Current Trends in Renewable Energy Adoption

The contemporary landscape of renewable energy adoption has witnessed a monumental shift towards sustainability across numerous sectors. The momentum is driven by increasing concern about climate change, technological innovation, and policies promoting the same. There has been a general transition in most countries toward utilizing renewable energy resources such as solar, wind, and hydroelectric power in a bid to end their reliance on fossil fuels. For instance, according to a large-scale survey of UK higher education institutions, 83% of the institutions have taken technical and non-technical steps geared toward achieving energy efficiency and reducing carbon footprints. This collaborative effort not only ensures compliance with environmental regulations but also creates a culture of sustainability in universities and colleges, thereby positioning them as leaders for environmentally friendly change in the community (Altan et al., 2010). This added investment in renewables indicates a heightened recognition of their vital role in lessening climate change and ensuring economic resilience. As the world continues to embrace renewable energy, legal frameworks that facilitate this transition are rapidly evolving. The Rio 20 Declaration serves as a watershed in resetting legal preparedness in meeting the opportunities and threats that come with a green economy transition. This global movement challenges legal professionals to think beyond the box and propose innovative solutions to facilitate adoption of renewable practices and technologies. In Europe, many member states are already implementing national and regional legal reforms that would spur renewable energy investment and lower regulatory barriers. These efforts not only enhance energy security but also attract needed investment to assist in developing new technologies and infrastructures. The ongoing development of regulatory frameworks is the major challenge for governments to create an enabling framework for renewable energy and tackle methodically the transition challenges (Gehring et al., 2013). Second, consumer behavior is the major stimulus in hastening the adoption of renewable energy. Individuals and institutions are increasingly opting to make sustainability a priority in purchasing choices and, in doing so, create massive demand for renewable energy products. Utilities and energy providers are responding by expanding their product offerings to include renewable sources of energy, hence making it accessible for consumers. For example, the dissemination of community solar projects enables households, as well as those who cannot provide roof space for solar panels, to gain access to the market of renewable energy. This liberalization of energy supplies not only enhances participation but also facilitates the development of grassroots organizations aimed at financing local renewable energy schemes. The greater the involvement of consumers, the faster it fuels the importance of sustainable practices and policy, justifying constant investments in green infrastructure to meet the growing demand from eco-conscious consumers.

1.1. SOLAR ENERGY

Utilization of solar power in energy systems is a historic step towards green energy technologies. With increasing concerns over dependence on fossil fuels and destruction of the environment, the capability of solar power has been under increasing focus. Solar power uses sunlight through photovoltaic cells or concentrated solar power systems to provide a multi-purpose source of energy for various applications.

The ability to use electricity from a ubiquitous renewable source reduces greenhouse emissions and results in cleaner air. More significantly, the deployment of solar energy plays a role in furthering energy independence, creating a more secure energy system that can hedge against the volatility of global oil markets. Microgrids have been shown to be an exciting way of maximizing the performance and reliability of solar power systems. These local, smaller energy networks allow for synchronization of solar energy output with conventional energy sources, maximizing supply and demand balance. Among the innovations that come under this are thermal storage systems that, in practice, address the intermittency that has long been inherent with solar energy production.

The integration of solar thermal power generation within microgrids allows for strategic dispatching, in which thermal energy is stored and used to supply a stable power output, thereby reducing the reliance on conventional grid power. As shown, the presented dispatching strategy can reduce conventional energy dependency by 10%, showcasing the capability of such systems in revolutionizing energy production and consumption patterns (Salim et al., 2015). Moreover, the use of power electronics technology is at the core of enhancing the performance of solar power systems. As there is more development in electronics, technologies like photovoltaic inverters and Maximum Power Point Tracking (MPPT) technology play important roles in maximizing the use of solar energy. These technologies convert and maximize solar energy efficiently, making solar panels provide maximum power in every condition. In addition, power electronics facilitates seamless integration into the vast energy networks, making the grid more stable and reliable. Its capacity to manage the energy flow and enhance the efficiency of operations not only boosts the efficiency of solar systems but also makes the overall operation of renewable energy systems in the broader energy industry even more efficient (Xiao et al., 2024) (Figure 1.4).

Figure 1.4. *An array of solar panels converts sunlight into electricity.*

Source: https://cdn.britannica.com/94/192794-050-3F3F3DDD/panels-electricity-order-sunlight.jpg.

1.1.1. Mechanisms of Solar Power Generation

The working principles of solar power generation are on the basis of the conversion of sunlight to workable electrical energy through various technologies that employ the photovoltaic (PV) effects. Basically, photovoltaic systems rely on the employment of semiconductor material, typically silicon, to absorb sunlight and generate direct current (DC) electricity through electron movement. This process is triggered when sunlight photons strike the semiconductor, resulting in electrons being energized and causing an electric flow of charge. The generated electricity is then typically converted to alternating current (AC) using inverters in an effort to render it grid-friendly. Material technology advances have also led to high-efficiency solar panels that enable additional energy capture and maximize the yield of solar installations. As a fundamental component of solar energy devices, an understanding of these simple mechanisms is essential to achieving optimum power output and integration into large-scale energy schemes. In addition to traditional PV systems, concentrated solar power (CSP) is another sophisticated method for the capture and conversion of sunlight. CSP technologies use mirrors or lenses to concentrate sunlight onto a small surface, generating heat that can be utilized to produce steam. This steam energizes turbines that are connected with generators and thus convert thermal energy into mechanical energy and subsequently into electricity. This method offers a convenient way for giant solar power plants to generate as well as store energy and has the additional advantage of thermal storage systems giving a constant supply even in darkness. Along with this, the integration of CSP within the smart grid infrastructure is becoming increasingly relevant, maximizing efficiency in the management and distribution of energy. The cooperation between these mechanisms also highlights the variability of the generation of solar power and highlights its contribution to the transition to renewable energy sources within a decarbonizing world economy. The application of smart grid technologies also maximizes the efficiency of solar power generation systems. The energy grids of today that are capable of accommodating more than one energy source rely heavily on Information and Communication Technologies (ICT) to promote grid advancements. Equipped with the ability to process real-time data as well as communication, smart grids allow for simple integration of solar energy, responding to variations in energy supply as well as demand. The incorporation of big data analytics is at the center of this in the sense that it allows for stakeholders to handle volumes of data produced from solar installations and grid operations. This analysis capability enhances decision-making, enhances energy distribution reliability, and reduces the cost of operations. As noted, smart grid modernization of power networks plays a central role in addressing issues in contemporary energy markets, with a focus on recent technological advancements in solar power generation (Borghetti et al., 2013; Gharehpetian et al., 2019) (Figure 1.5).

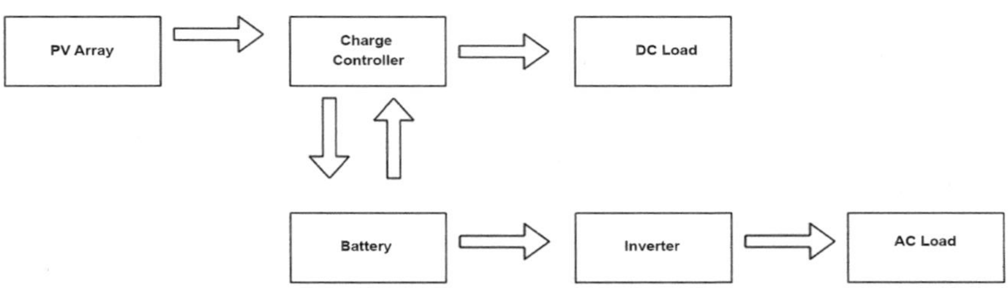

Figure 1.5. *Block diagram of solar energy.*

Source: https://media.geeksforgeeks.org/wp-content/uploads/20240206170916/pp-drawio.png.

1.1.2. Advantages and Disadvantages of Solar Energy

The integration of solar power into the structure of power systems is associated with numerous advantages, most significantly the prospect for sustainability and environmental benefit. Solar power utilizes sunlight, an unlimited and recyclable resource, which reduces considerable reliance on fossil fuels. The transition is necessary as a response to climate change since solar power does not produce any short-term greenhouse gas emissions during use. Moreover, the utilization of solar power plants allows decentralized power generation, and as such, it fosters energy security and reduces transmission losses characteristic of traditional centralized power supply. As noted in recent research, the utility sector has emerged as a principal source of finance for green bonds' renewable energy projects, and this can invite the expansion of solar energy schemes such that this transition is not only a feasible but also a preferred choice for investors who want to support sustainable strategies (Fox-Penner et al., 2019).

Figure 1.6. *Advantages and disadvantages of solar energy.*

Source: https://www.renovablesverdes.com/wp-content/uploads/2021/03/energia-solar-en-las-viviendas.jpg.webp.

1.1.3. Innovations in Solar Technology

Technological innovation in solar technology has become the core element spearheading the transition to sustainable energy systems, with substantial impacts on energy generation as well as consumption patterns. The transition is necessitated by research into photovoltaic (PV) materials, storage technologies, and smart grid integration techniques. Recent discoveries in thin-film solar cells as well as perovskite materials have enhanced the efficiency of energy conversion by considerable factors while reducing costs of production. These advancements not only increase the access to solar energy but also increase the adaptability of solar power systems to different environments and applications. Whereas governments and institutions prioritize the deployment of renewable energy sources (RES), the continuous advancement in solar technologies positions them as key agents of the cleaner energy future to finally achieve the global goals of mitigating climate change and achieving energy security, as argued by Britchenko et al. (2018). Solar technology development goes hand in hand with changing energy policy and market trends. The emergence of distributed energy production allows users to harness solar energy and become active participants in their own energy supply by sharing solar energy via peer-to-peer (P2P) energy trading.

Figure 1.7. *Innovations in solar technology.*

Source: https://www.taradigm.com/wp-content/uploads/2023/05/AdobeStock_350406348-scaled.jpeg.

Energy management systems and smart inverters are under development for optimum self-consumption of locally generated power. The integration of solar energy into microgrid systems is a classic example of how innovation can increase energy autonomy and resilience. Also, with countries seeking to lower their fossil fuel dependency, the deployment of solar technology is critical in raising energy security, particularly in the European Union, where a shift toward low-carbon energy production is becoming increasingly important, as discussed by Abrhám et al. (2018). Furthermore, the development of battery technology for power storage from solar energy is critical in ensuring the optimal use of solar energy despite its inherently intermittent form. Energy storage equipment such as the lithium-ion battery and the new-generation solid-state battery allow excess solar power generated under high-sun conditions to be stored and used during low generation hours. This capability not only enhances the supply of solar energy but also facilitates grid stability through the provision of ancillary services such as load balancing and frequency regulation. As technologies advance, the convergence of artificial intelligence and machine learning in solar power systems is encouraging greater energy management and predictive maintenance, culminating in the path of smarter and stronger renewable energy infrastructures. Solar technology is thus at the forefront of renewable energy technology, fusing economic viability with environmental responsibility (Figure 1.7).

1.2. WIND ENERGY

As the world boosts demand for clean energy, wind power becomes a leading impetus for the transition towards sustainable energy systems. Wind turbines harness the kinetic energy of the wind and transform it into electricity, thereby offering a cheap and environmentally friendly alternative to fossil fuels. This form of renewable energy is particularly useful with its ability to come in large amounts as well as having an incredibly low carbon footprint when used. However, the integration of wind energy into present power systems poses a definite challenge, particularly for reliability and grid stability. This requires the need for design and optimization of modular power converters that are essential in fostering grid compatibility and facilitating flexible energy services. Technological breakthroughs such as these are crucial in managing dependence on traditional energy sources and ensuring easy integration of renewable technologies into the energy matrix (Hahn et al., 2022). Intermittency of wind energy caused by varying wind patterns is another hindrance to its large-scale embracement. Volatility impacts the consistency of the energy supply, complicating the management of the electrical grid. To fight this challenge, innovative technology such as hybrid solar-wind power plants is rapidly gaining popularity.

Due to the complementary nature of wind and solar resources, these systems are capable of eradicating the effect of intermittency. By efficiently combining the two renewable sources, hybrid plants can generate a more optimum and stable energy output, finally stabilizing the grid. Such advancements not only increase the reliability of renewable energy systems but also render integration of such technologies in more energy-consuming cities economically feasible (Baquedano et al., 2021). Additionally, design and operation of wind energy systems involve profound knowledge of complex variables like environmental parameters, policy regulations, and technology capabilities. As cities and states strive to meet their renewable energy goals, the coming of decentralized power generation, spearheaded by wind power, is still central to breaching the limitations of the conventional sources of power.

The future of wind power is not only in its capacity to reduce greenhouse gas emissions but also in its ability to usher in economic growth and energy independence. Thus, continued research and innovation in hybrid systems and modular power converters will be vital in ensuring a green energy transition, in accordance with global agreements to mitigate climate change while meeting the needs of a growing population (Baquedano et al., 2021; Hahn et al., 2022) (Figure 1.8).

Figure 1.8. *Wind farm.*

Source: https://images.nationalgeographic.org/image/upload/t_edhub_resource_key_image/v1652340973/EducationHub/photos/wind-farm.jpg.

1.2.1. Principles of Wind Power Generation

The theory behind harnessing power from wind involves the conversion of blowing air's kinetic energy into electrical energy, largely using wind turbines. With the wind causing a turbine's blades to turn, this turning is converted into mechanical energy, which is, in turn, translated into electrical energy by an attached generator. This is a highly successful process since wind power is renewable and abundant, aiming to stimulate the global energy mix in pursuit of renewable sources of power. However, the use of wind power also needs to overcome location and intermittency challenges, with tactical planning and incorporation within broader power systems. Additionally, the turbine technology scale economies significantly affect the feasibility and efficiency of wind power such that constant design and material improvements are required in order to achieve maximum energy harvesting (Robert, 2008). Incorporating wind power into the current power infrastructure means there are fundamental issues about grid stability and reliability. Wind generations variability also requires backup sources that will supplement this in order to ensure stable electricity supply, particularly during periods of low winds. This dependence on conventional generation not only makes it more difficult to incorporate renewables into the grid but also questions the overall carbon reduction potential of wind power as a standalone source, since it has a tendency to be fueled by fossil-fired generation in peak load situations. Moreover, the technical constraints of the distribution network must be taken into account as the integration of distributed energy resources like wind turbines can add value to voltage profiles but also poses challenges in fault response and conflicts of operation (Afifi et al., 2014). Hence, good policy design and technological advance play a crucial role in addressing these challenges in order to make the most of wind powers contribution towards renewable energy targets. Public opinion and policy climate towards wind power have leading roles to play in shaping its future development. With public awareness of climate change and energy sustainability on the rise, many locations have implemented renewable portfolio standards (RPS) that mandate a percentage of electricity to be derived from renewable resources, including wind. Although such initiatives demonstrate sincere intent at avoiding greenhouse gas emissions, they are generally condemned on the basis of their economic feasibility and effectiveness. Critics have argued that the majority of RPS programs have not been able to realize the expected increase in renewable energy output, as evidenced by Californias flat line renewable percentage since 1998 (Robert, 2008). This underscores the need for advanced solutions to energy policy based on economic efficiencies and technological advances to bring about significant change. Along with incentives for technological advancement in wind turbines and maintenance practices, strategic policy can go a long way in enhancing the role of wind power in an energy future that is sustainable and resilient (Figure 1.9).

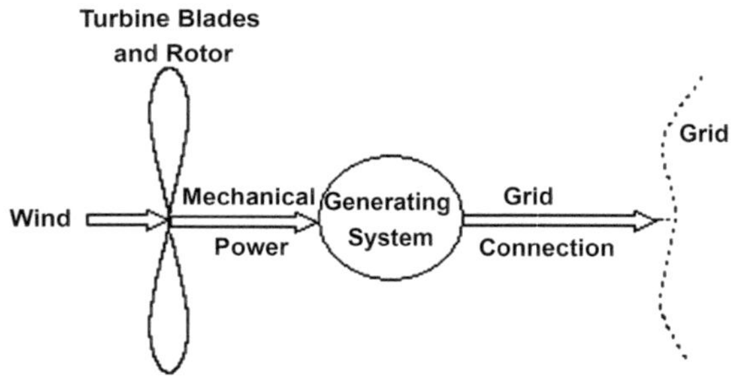

Figure 1.9. *Basic principle of wind power generation.*

Source: https://www.researchgate.net/profile/Mohamed-El-Shimy-3/publication/4346079/figure/fig1/AS:649633881669653@1531896195507/Basic-Principle-of-Wind-Power-Generation.png.

1.2.2. Environmental Impact of Wind Farms

The development of wind farms as a major component of renewable energy systems has attracted mixed reactions because of their implications for the environment. Wind farms are on the one hand a source of clean, sustainable energy whose use cuts greenhouse gas emissions and dependency on fossil fuels. This shift is vital as the world struggles to deal with the consequences of climate change. Nonetheless, wind installations' placement and functioning have undesirable impacts, mainly on the surrounding fauna. Birds and bats, for example, are reported to suffer mortality due to impacts from turbines, which bothers green activists. Therefore, there are constant arguments about the environmental effects of wind power, pushing stakeholders to harmonize energy generation and preservation of ecosystems, as summarized in the planning and siting methodologies outlined on the WePWEP website (Simao et al., 2006). Land utilization and habitat disturbance is yet another environmental effect factor. Wind farms cover big areas of land, with a tendency to change land utilization patterns. In certain regions, such changes will overlap with significant ecosystems, and hence the importance of careful site selection and planning. Topography, vegetation cover, and wildlife habitats are critical factors to consider in minimizing ecological disturbance. Methods highlighted for involving the public in planning, as in wind power projects, can bring maximum conservation through the inclusion of community involvement. This collaborative method can raise more awareness and encouragement of environmentally friendly measures that avoid adverse consequences while encouraging the development of renewable energy sources (Simao et al., 2006). Together, the discussion regarding wind farm location should incorporate environmental consideration to ascertain that renewable energy systems are not paid for by an unexpected cost to the environment. Also, studies show that there may be synergies between wind energy and existing installed thermal power plants, particularly in the developing world with heavy energy consumption requirements. For example, in Mexico, modeling shows that integrating wind farms with thermal power plants is able to effectively limit peak electricity consumption by aligning wind production with consumption patterns. This cogeneration policy can take advantage of the complementarity of the two energy sources, ensuring stability of the power grid while reinforcing clean energy objectives. The Energetic Transition Law emphasizes the increase in clean energy production by countries as a way of highlighting the suitability of such models. By using wind power in strategic complementarity with existing power infrastructure, not only is energy reliability enhanced but also the environmental price of fossil fuel consumption is lowered, thus encompassing a holistic vision of renewable energy systems (Figure 1.10).

Figure 1.10. *Impact of wind energy on flora and fauna.*

Source: https://blablanegocios.com/wp-content/uploads/sites/2/2017/02/IMPAC-TO-AMBIENTAL-DE-LA-ENERG%C3%8DA-EOLICA-3-1024x768.jpg.

1.2.3. Economic Considerations in Wind Energy Development

The economic implications in wind power development are core to determining its viability as a renewable energy source within renewable energy systems in general. One of the primary economic implications is the level of capital required to construct and install wind farms. The venture entails huge financing to acquire land, turbines, and infrastructure development. Whereas these upfront expenses, though, have deterred investment so far, wind power has shown a decline in levelized cost because of technological progress and higher manufacturing efficiencies. A judiciously constructed policy framework complemented by incentives like tax credits and feed-in tariffs can further ensure breaking through these upfront expenses by making the investment environment more favorable. Aside from this, effective implementation can also create jobs in local economies, addressing environmental and socioeconomic concerns relevant to sustainable development (Ramamurthi, 2016). Yet another key economic consideration is the operational and maintenance cost of wind power plants. The periodic costs, although normally lower compared to fossil fuel power plants, are required to keep wind systems running efficiently and still contribute to energy supply. Correct forecasting of these expenses is vital to long-term finance planning and attracting investors. Additionally, volatile energy prices may influence the profitability of wind power ventures. While reducing fossil fuel prices have the capacity to challenge the competitiveness of wind power in the short run, prudent economic control and diversification of energy sources will facilitate finance returns within a diversified mix of renewable energy. The analysis of the relationship between energy market dynamics and operating efficiency is crucial in optimizing the economic system that supports wind energy development (Ault et al., 2005). Lastly, coupling energy storage systems with wind energy generation is gaining growing economic significance. Energy storage technologies possess the ability to significantly enhance the reliability and stability of wind energy by offsetting the intermittency character of wind speeds. Although initial capital expenses on energy storage systems are costly, long-term economic benefits from improved grid stability and reduced dependence on fossil fuels are significant. An in-depth cost-benefit analysis of energy storage can demonstrate the economic viability of incorporating such systems on wind farm projects. Also, with a change in the role of storage, regulatory frameworks must adapt to enable new business models capable of making economic efficiency in renewable energy systems feasible. This will be essential to maximize the overall performance and attractiveness of wind energy investments in the long term (Ault et al., 2005).

1.3. HYDROPOWER

Hydropower is one of the most established and reliable renewable energy sources, and it harnesses the kinetic energy of flowing water to produce electricity. As energy demands continue to rise across the globe, this renewable resource is a large part of the solution to reducing the demand for fossil fuels and maintaining low greenhouse gas emissions. The adaptability of hydropower does not stop at the generation of simple electricity; it can even provide ancillary services such as flood control, support for irrigation, and recreational purposes. Being enabled by hydropower are facilities such as dams and run-of-the-river systems, hydropower can be utilized for various geography-dependent needs, making it a varied choice amongst the renewable sources of power. The integration of hydropower into a diversified energy mix can enhance energy security and stability, facilitating a transition to more sustainable power systems. Despite the advantages, hydropower is not free of issues since it has environmental and social problems that require mitigation. While constructing dams can facilitate energy generation, it can disrupt surrounding areas, lead to habitat loss, and displace individuals. Proper planning and evaluation need to be done to mitigate these impacts, and innovations of technology now available over the past few decades have opened up opportunities for improved fish passage and ecological integrity. In addition, hydropower production also depends very much on climate variability, and thus producing hydropower needs to be complemented with other renewable energy sources. This diversified model forms a more robust energy system, such that the energy can be maximized in times of drought or climatic conditions. Thus, synergizing hydropower with the most recent management techniques warrants a green route towards renewable energy expansion. In the years to come, the prospects for the role of hydropower as an input to modern energy systems are great, particularly since the world's need for clean electricity increases at an increased pace. Emerging technologies, such as small modular hydropower plants and pumped-storage hydropower, can potentially improve the efficiency and availability of hydropower plants. Furthermore, studies on integrating solar and wind power with hydropower in hybrid systems hold the promise of building robust and flexible energy platforms to meet variable demand. Educational programs that enhance awareness of alternative modes of transport, such as those highlighted in the 2008 symposia, are extremely vital in informing the future generation of engineers and policymakers who will advance hydropower in wider renewable energy frameworks (N/A, 2008). It is through such collective push that hydropower can continue to evolve, adopting best practices that not only meet energy needs but also for the integrity of the environment (Gelil et al., 2012) (Figure 1.11).

Figure 1.11. *Representation of hydropower generation.*

Source: https://www.thestatesman.com/wp-content/uploads/2024/08/IMG_3603-jpeg.webp.

1.3.1. Types of Hydropower Systems

In renewable energy, hydropower facilities are the key to harnessing the energy of running water for generating electricity. This type encompasses primarily three major types: run-of-river, reservoir, and pumped storage hydropower facilities. Run-of-river facilities alter the natural flow of water slightly, generating power as river currents make turbines rotate. These systems are likely to be less disruptive to ecosystems than traditional dams but may have limited power capacities depending on seasonal water availability. Reservoir systems, however, have enormous volumes of water impounded behind dams and provide high energy flexibility in generation, particularly during peak demand hours. The advantage of water release control leads to stable and predictable power supply, an advantage essential to grid stability. Pumped storage hydropower plants are a unique way of storing and managing energy in the hydropower industry.

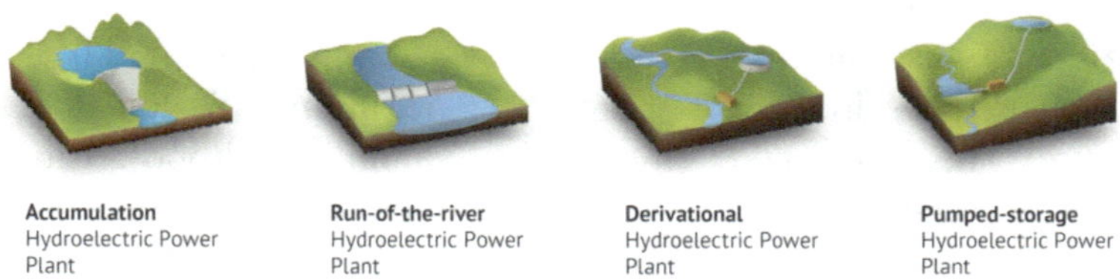

Figure 1.12. *Type of hydroelectric power plants.*

Source: https://www.energyencyclopedia.com/cache/img/f-3/data-web-free-downloads-water-png-ee-hydro-types-4k.fitbox.x1280.y1280.r0.q80.nr0.me2.png.webp.

Pumped storage plants operate by pumping surplus electricity to shift water around from the lower reservoir to the higher one when there is low demand. When there is high energy demand, the water is pumped downwards through turbines to generate electricity. This strategy not only increases energy supply reliability but also achieves demand variation balancing more efficiently. Hybridization of energy sources, as studied, enhances the efficiency and stability of renewable energy systems as a whole because it facilitates integration of variable generation resources such as solar energy and wind energy, which are highly weather-dependent (Beluco et al., 2019). The evolving energy demand calls for innovative measures, particularly with the incorporation of hydropower plants into larger renewable energy systems. The character of hydropower systems can also be improved with technology and economic factors. The increasing effectiveness of turbines and generation technologies enables more energy to be extracted from the available water sources, thus enhancing the overall role of hydropower in energy production. Despite being currently only providing a percentage of world energy, the potential of future growth of hydropower alongside other renewable options like solar and wind power is significant. Knowledge of how these systems interact can unlock the door to more sustainable, resilient energy infrastructures, which are critical in managing the impacts of climate change as well as future energy requirements (Aheicheva et al., 2018) (Figure 1.12).

1.3.2. Benefits of Hydropower as a Renewable Resource

Hydropower is among the most efficient and oldest of renewable energy resources, whose advantage in terms of sustainability as well as environmental sustainability is significant. It harnesses the kinetic energy in flowing water, primarily rivers, to generate electricity and thereby reduce our reliance on fossil fuels. Not only does it lower greenhouse gas emissions, but it also reduces the ill effects of climate change. Through focusing on hydropower, countries like Brazil can transition to cleaner energy systems. As suggested by Lucon et al. (2015), Brazil can invest approximately 0.5 trillion U.S. dollars in energy infrastructure. This can provide room for innovation in hydropower to position Brazil as an inspiration in the low-carbon economy and safeguard the health of its citizens as well as international climate requirements. One of the most significant benefits of hydropower is its ability to offer a reliable and constant supply of power, usually referred to as base-load electricity. As opposed to solar and wind power, whose production is variable and weather-dependent, hydropower schemes can operate constantly, thus providing a steady power supply. This reliability is greatly significant in grid balancing, especially as more variable renewable sources are fed into energy grids. Furthermore, the possibility to quickly ramp up or down power production makes hydropower respond favorably to demand variations, thus enhancing overall grid stability. The strategic employment of hydropower can help reduce dependence on fossil fuels, producing an energy system with greater resilience and multiple sources, as stated by Kammen et al. (2019), who call for a mix of renewable technologies at river basins. Additionally, hydropower has economic benefits in the form of job opportunities and local economic growth. The construction and operation of hydroelectric plants create economic activity and jobs in urban and rural areas. Furthermore, the presence of hydropower plants can lead to ancillary industries and businesses, such as transport and tourism, contributing to the local economy. These projects are also rendered financially viable in the long term by the negligible cost of running once infrastructure is built. Thus, the integration of hydropower into a country's energy strategy ensures not only better energy security but also the promotion of economic development. The capacity of hydropower to be married with general environmental goals and regional economic development speaks volumes about its multi-faceted benefits as a renewable energy source in the overall renewable power and energy systems (Figure 1.13).

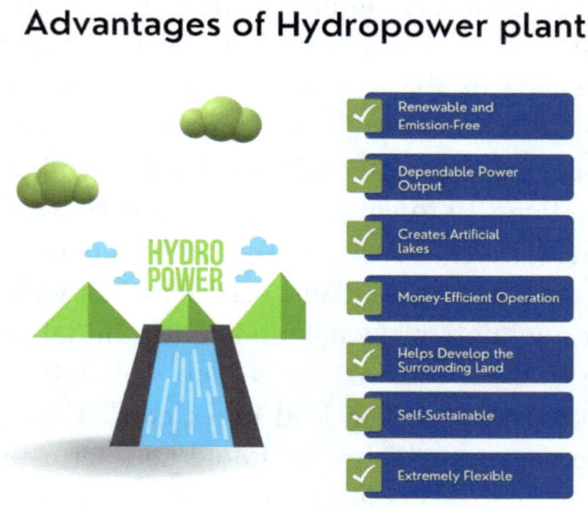

Figure 1.13. *Advantages of hydropower plant.*

Source: https://i0.wp.com/www.techquintal.com/wp-content/uploads/2022/09/Advantages-of-Hydropower-Plant.jpg?w=1080&ssl=1.

1.3.3. Challenges and Limitations of Hydropower

Its expansion as a form of renewable energy is accompanied by a myriad of challenges and limitations that must be adequately addressed. One of the most significant issues may be the environmental impact of dam construction and reservoir creation. The plants can tend to significantly disrupt local ecosystems, particularly aquatic ones, leading to loss of biodiversity. Damming of rivers alters natural flow regimes that could risk fish populations and other fauna accustomed to particular environments. In addition, land inundation for reservoir construction has a tendency to dislocate human settlements and break local economies. With nations seeking to meet renewable energy targets, the balance between hydropower development and environmental preservation remains a critical challenge that influences public perception and policy toward the construction of hydropower facilities (Huang et al., 2019). As such, environmental trade-offs for hydropower projects need to be seriously taken into account in the planning of energy. Besides, geographical and climatic limitations further limit the use of hydropower. Hydroelectric generation relies heavily on consistent flow of water, which is likely to be affected by variations in precipitation during seasons and global climate change. Such variability makes it impossible to commit power supply in advance, a critical problem with grid operators seeking stability and reliability in energy systems. The variable nature of hydropower, particularly in regions where rain is variable, poses a challenge that is closely similar to what other intermittent sources of renewable energy like the sun and the wind energy encounter. Lower water levels can lower the capacity for generating energy, and that makes integration into existing power grids challenging. The need for enhanced grid flexibility to address such issues becomes the order of the day, with alternatives being pursued by interested parties to enhance system robustness during periods of hydropower volatility (Botterud et al., 2020). Lastly, economic sustainability of hydropower plans represents a formidable barrier to their widespread use. Even though hydropower might offer low operating costs, initial capital expenses in dam and supporting facilities construction are substantial. Economic challenges are further exacerbated by the long lead times needed for project planning, permitting, and construction, which would hold up returns on investment and lead to uncertainty among stakeholders. The cost-effectiveness of hydropower might also come into question as competition intensifies from other sources of renewable power. Relatively affordable natural gas prices and technological advance in solar and wind technologies offer alternative investment avenues, potentially sidestepping hydropower for economically more viable actions. Resolution of these economic constraints is thus the answer to the future potential of hydropower in the grand scheme of renewable energy measures (Botterud et al., 2020; Huang et al., 2019) (Figure 1.14).

Figure 1.14. *The reliability of hydropower makes it a popular choice worldwide, extending its reach to regions with unique challenges.*

Source: https://www.power-technology.com/wp-content/uploads/sites/21/2021/11/american-public-power-association-FUeb2npsblQ-unsplash-3.jpg.

1.4. BIOMASS ENERGY

Biomass energy, which is derived from organic material such as plant material, agricultural waste, and animal waste, is a suitable alternative to fossil fuels when it comes to renewable power and energy supplies. Since biomass is a renewable energy source, it can serve a critical role in mitigating greenhouse gas emissions as well as attaining energy autonomy. Biomass energy, electricity, and heat conversion can efficiently use waste materials, thus lowering the environmental impact of disposal. Furthermore, increased demand for biomass energy promotes new rural economic opportunities that would stimulate stagnant agricultural markets. Transition to biomass aids not only the environment but also economic growth in the form of job creation in the production, processing, and distribution of bioenergy. As states like Iowa make investments in building their biomass capacity, the hope for a sustainable energy future becomes increasingly tangible (N/A, 2008). Incorporation of biomass power into existing power infrastructure is still the main challenge, and quantum technological and infrastructural development is required.

Figure 1.15. *Biomass energy.*

Source: https://cdn.educba.com/academy/wp-content/uploads/2023/02/Advantages-and-Disadvantages-of-Biomass-Energy.png.

Maximizing the conversion technology efficiency, such as anaerobic digestion, combustion, and gasification, will be crucial to attain maximum energy with decreased emission. It is also crucial that there exists a safe supply chain of biomass feedstocks, including for sustainable harvesting and logistics of transportation. Once policymakers begin to work on and execute energy plans, land use, agriculture practices, and food production impacts are the considerations that need to be balanced to avoid undesirable impacts on the agricultural sector. Engaging farmers in the biomass energy debate could stimulate synergies between agricultural activities and the use of renewable energy as many farm crops can be used in energy generation (Werner et al., 2001). The creation of biomass energy systems could, in fact, stimulate a circular economy, where products wasted in the first place would contribute to the generation of energy instead of polluting the environment. It aims at optimizing resources and innovative uses of by-products, thus creating avenues towards enhanced sustainability in the energy industry. As the urgency to leave the use of fossil fuels behind continues to grow, biomass energy presents a powerful path to achieving a green future. Thus, it is essential to create a dialogue between the stakeholders, i.e., policymakers, farmers, and energy companies, in order to align objectives and make certain that the growth of biomass energy is concurrent with larger environmental goals. By taking advantage of the proprietary nature of biomass, society can help create a diversified and secure energy mix that not only helps reduce climate change but also supports rural economies (N/A, 2008) (Figure 1.15).

1.4.1. Sources and Types of Biomass Energy

Biomass energy, derived from organic material, is a natural part of renewable energy solutions. Plant products such as agricultural waste, energy crops harvested for energy, and woodlot residues are some of the numerous sources of biomass, as well as organic industrial and municipal waste. They can be put to use in a diverse array of formats, such as direct combustion, gasification, or anaerobic digestion, to generate electricity, heat, or biofuels. For instance, crop waste plays a dual purpose for energy production and waste treatment, while bioenergy crops specifically grown for cultivation provide a renewable material for use in bioenergy processes. Biomass multi-purpose uses guarantee its integration into traditional energy structures, mainly through co-generation systems, where both heat and power production are simultaneously utilized, enhancing overall efficiency of energy production (Lima et al., 2012). The energy efficiency of biomass is also increased by co-generation, a process that produces electricity and heat from the same energy source. This technique is particularly valuable when using biomass due to its potential for higher energy efficiency by utilizing waste heat for space heating or producing hot water. By making use of and tapping this by-product, distributed generation systems benefit from higher fuel input-to-energy output ratios compared to conventional power plants. This concept is vital in reducing greenhouse emissions as well as meeting energy needs in a sustainable way (Lima et al., 2012). Moreover, development of local biomass energy systems can be a valuable contribution to energy security by reducing reliance on fossil fuels and stimulating rural economy through employment opportunities in the production of biomass and processing. In addition to the economic and environmental benefits, social aspects of biomass production of energy are very important. Such implementation projects for biomass energy systems can be attached to larger community projects, involving local people and promoting the use of renewable power. For example, most rural communities lack stable access to energy resources since distribution networks are in poor condition. Promoting local production of biomass energy technology serves not only to address energy requirements but also to equip communities with the infrastructure to operate their systems. With communities independent in their power generation, they stand a better chance of being accessed by sustainable practice more intensely, confirming the harmony of energy policy, community development, and environmental management under the broader platform of renewable energy networks (Figure 1.16).

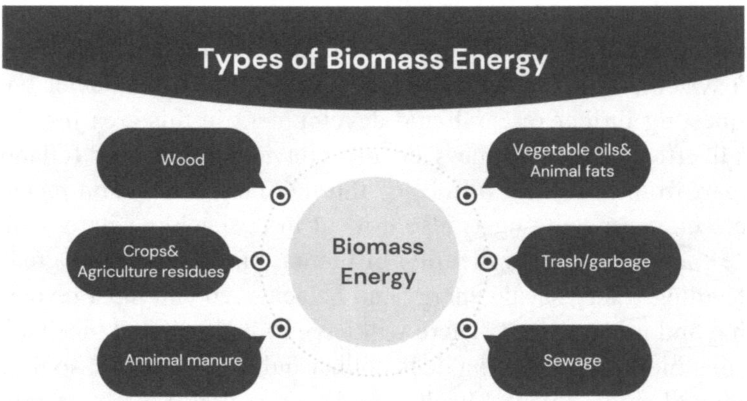

Figure 1.16. *Types of biomass energy.*

Source: https://www.wtsenergy.com/wp-content/uploads/2023/04/Glossary-images-44.webp.

1.4.2. Conversion Processes for Biomass Energy

Biomass energy conversion processes form the core of renewable energy systems development. Conversion processes like combustion, gasification, and Fischer-Tropsch (FT) synthesis are crucial for converting biomass into usable forms of energy. Of these, FT technology has also been re-spurred by increased global demand for sustainable transportation fuels amid declining fossil fuel reserves. Biomass-to-Liquid (BTL) processes are a carbon-neutral way of producing liquid fuels from biomass feedstock. While promising, traditional biomass gasification is far from trouble-free, not least of which is carbon utilization rates, typically between 25% and 40%. The resulting inefficiency is highly damaging to the economic feasibility of the BTL process. However, achievements like CSP coupling with BTL can optimize carbon efficiency and in the future make possible storing solar energy, thus creating a more viable and sustainable route towards biomass conversion and utilization (James et al., 2013). Gas turbines are also a promising line of development towards realizing biomass energy due to their suitability with a wide array of biofuels. Employing biofuels such as biogas, ethanol, and bio-syngas in combustion technology can lead to significant reductions in CO_2 emissions while utilizing established turbine systems. Despite the operational setup seeming feasible, the realization remains that there is immense technological development necessary to optimize performance and efficiency. Energy and exergy analysis confirms the feasibility of these biofuel alternatives in gas turbines, particularly when compared to conventional natural gas systems. The prompt indicates towards the overall impact of using a Biomass Integrated Gasification Combined Cycle (BIGCC) system, which would facilitate easy generation of power on a large scale, requesting further research and development in this area for enhancing the overall efficiency of biomass energy conversion systems (Olano et al., 2011). Apart from providing efficiency through new conversion mechanisms, sustainable biomass sourcing is also critical in unshackling its potential as a renewable energy source. Availability of biomass feedstock has a great role to play in deciding if large-scale energy can be achieved and must be considered in planning and management. There will have to be instituted mechanisms for ensuring that biomass harvesting does not harm the environment so that energy production and environmental health can be promoted. Moreover, integrating biomass conversion technologies into existing energy systems requires a cross-cutting policy of energy, technology, and sustainable practice. As the global renewable energy movement gains momentum, the advancement of biomass conversion technologies must advance hand in hand with world climate goals to maximize energy production while having as little impact on the environment as possible so that biomass can become a key sector for moving the world towards sustainable energy systems.

1.4.3. Sustainability and Environmental Concerns

Greater integration of sustainability and environmental concerns into renewable energy systems is essential for the development of a sustainable and eco-friendly future. Technology innovation in solar, wind, and hydropower holds enormous potential to mitigate the ill effects of climate change. Shifting away from fossil fuel dependence and towards renewable forms of energy not only curbs greenhouse gas emissions but also enhances energy security and economic stability. However, to utilize these renewable systems efficiently, it is crucial to consider their environmental impact, e.g., land use, resource extraction, and biodiversity. Therefore, sustainability should be a guiding principle in renewable energy technologies design and deployment so that they promote ecological well-being and social justice alongside addressing the issues of climate. Hence, community planning and consultation will be essential in achieving this balance and gaining broad public acceptability for renewables. In addition to technological advancement of renewable energy, public education and awareness are fundamental drivers of promoting sustainable practices. The FLOWS 2B Information and Symbols Project is a good example that emphasizes dissemination of information on flood hazards and sustainability as part of environmental learning. By applying interactive ways of engaging people, sustainability programs can be communicated and carried out in the best possible manner. Not only does this approach inform people about ecological risks but also allows them to actively participate in climate change mitigation actions and adaptive measures. People-centered initiatives, which combine awareness with education, can make people more aware of local community environmental concerns and build collective solutions. As in other models of sustainability, these efforts enhance impacts of clean systems of energy and make the community more resistant to environmental swings, particularly in flood-prone areas (Abeyasekera et al., 2007). Lastly, the need to incorporate historic preservation into the sustainable development effort cannot be overlooked. For example, the District of Columbia has shown the world how historic districts offer environmental sustainability while enhancing aesthetic and cultural heritage. Preservation efforts themselves promote energy-efficient rebuilds, wisely utilize existing assets, and reduce the carbon footprint of new structures. These types of programs benefit broader sustainability aims by preserving historic neighborhood character and integrity and promoting community investment. Moreover, partnerships with groups like the National Trust for Historic Preservation bring historic preservation together with modern-day sustainability initiatives and further solidify its role as an impetus for comprehensive environmental schemes. Therefore, by recognizing the significance of preserving history, not only is sustainability maintained, but also the discussion of renewable energy and power grids is enriched since both areas can enhance each other's objectives (Stein et al., 2009).

1.5. GEOTHERMAL ENERGY

Geothermal power is generally regarded as a key component in the set of renewable energy resources, and it plays a central part in the transition towards sustainable power systems. As opposed to solar and wind energy, which rely on specific weather patterns, geothermal power taps into the internal heat of the Earth to provide a consistent and reliable source of energy. This reliability is particularly valuable for baseload power generation, maintaining electricity supply as a constant regardless of external variables. Government regulations worldwide have been changing to help develop geothermal resources, with at least 48 countries implementing one form of renewable energy incentive policy. This government support is integral in accelerating readiness in the markets and increasing the production of geothermal power, which is integral in the achievement of long-term energy goals and reducing dependence on fossil fuels (N/A, 2005).

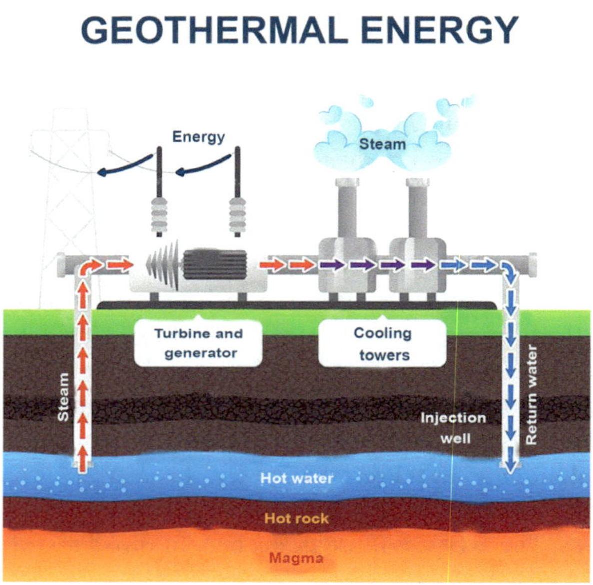

Figure 1.17. *Geothermal energy.*

Source: https://i0.wp.com/i.pinimg.com/736x/a2/c2/7e/a2c27ef3d8cf72b211a428e4b248b7cf.jpg?w=640&ssl=1.

The growth potential of geothermal energy is significantly highlighted by its potential capacity growth, especially where the resource is predominant. For instance, studies indicate that New Zealand can experience the expansion of geothermal capacity to a large extent, which is poised to increase fourfold to meet increasing energy demand with decreasing carbon footprint to levels comparable to 1990 levels emissions. The expansion is crucial as the nation continues to work towards maintaining a renewable generation contribution that remains impressively at around 67%. However, the maximization of geothermal energy production is never a simple task; site-specific greenhouse gas emissions along with resource management must be considered assiduously. It is essential to gain knowledge of the geothermal fields so that they may be managed and their environmental effects minimized, and this development in the renewable source of energy may be made compatible with sustainability (Atkins et al., 2010). In brief, though geothermal power holds huge potential to enhance the resilience and sustainability of power grids, its success is dependent upon wide policy frameworks, technological breakthrough, and prudent environmental stewardship. As countries move toward the use of renewable power solutions in greater numbers, geothermal power is a highly competitive candidate in the race for cleaner energy. Emphasis on governmental leadership and support cannot be exaggerated either, as witnessed in the varying degrees of success in nations advancing with renewable energy technology. By emphasis on research, investment, and public-private collaboration, the resource of geothermal power can be maximally exploited, hence effectively contributing to the greater objective of decarbonizing energy systems globally (N/A, 2005) (Figure 1.17).

1.5.1. How Geothermal Energy Works?

Geothermal energy employs the heat that is trapped beneath the Earth's surface as a constant and sustainable source of power. The method depends on natural heat from the Earth's interior, generated by remaining heat from planetary formation and radioactive isotope breakdown. Where there is intense geothermal activity, such as along the edges of tectonic plates, for instance, hot water and steam at comparatively low depths can be found. These thermal energies can be harnessed by wells drilled into geothermal reservoirs. If steam is generated and extracted to the surface, it is utilized to power turbines connected to electricity generators, transferring thermal energy into electrical energy. Geothermal systems are sources of clean energy that emit minimal greenhouse gas emissions in comparison to fossil fuels, making them relevant in the pursuit of renewable energy projects to reverse the effects of climate change (N/A, 2005). Geothermal energy has one of the most significant benefits, and that is being able to provide a steady and consistent source of energy, usually termed as a baseload power. Unlike solar or wind energy, which relies on the situation outside, geothermal power plants are able to operate continuously, generating electricity day and night. The reason for this steadiness is the ever-constant temperature of the Earths subterranean environment, unchanging with regard to seasonal or weather conditions. Furthermore, advancements in geothermal technology have broadened the application of the energy source beyond traditional hydrothermal deposits. Enhanced geothermal systems (EGS) facilitate the exploitation of energy from hot and dry rock formations via the injection of water into the areas, essentially creating artificial reservoirs. The technology expands the potential area for the production of geothermal energy, hence becoming more plausible in various geographical environments (Gelil et al., 2012). Geothermal energy integration into the broader context of renewable energy is also supported by increasing government and institutional awareness of what it can offer. Current trends indicate an increasing interest in renewable technology internationally, with policies in development to enable sustainability (N/A, 2005). Plans for incorporating geothermal systems into national energy planning demonstrate an increasing realization of the need for diversified energy portfolios. Countries with geothermal potential are exploring and investing in infrastructure to develop the potential to its best, enhancing local economies and energy independence. As technology continues to advance, the role of geothermal power to diversify, stabilize, and decarbonize energy sources will ultimately prove to be a success formula in meeting climatic aspirations all over the world, cementing its place further in the renewable energy matrix (Figure 1.18).

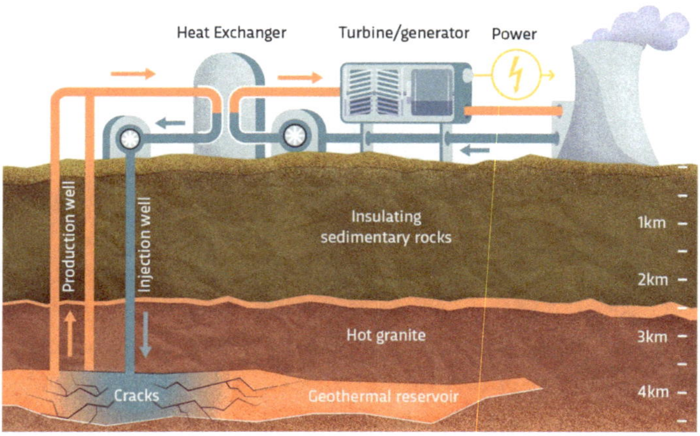

Figure 1.18. *How does geothermal energy work to produce electricity?*

Source: https://c02.purpledshub.com/uploads/sites/41/2021/12/geothermal-00eef72.jpg?webp=1&w=1200.

1.5.2. Applications of Geothermal Energy

The applications of geothermal energy are multifarious, particularly in terms of heating and electricity generation. Geothermal systems can access the internal heat of the Earth, making them extremely efficient in terms of heating homes and businesses. When geothermal resources are plentiful, as they would be in volcanic or tectonic areas, ground-source heat pumps can be employed to generate steady heating and cooling, reducing by a great extent the consumption of fossil fuels. As the Waitakere City case attests, community energy initiatives are increasingly becoming alert to the potential of geothermal energy in meeting growing energy demands sustainably. Indeed, the city has identified geothermal and solar applications as central strategies for achieving energy self-sufficiency and avoiding temporary blackouts, underscoring geothermal energy's role as the foundation of alternative energy initiatives (Grooby et al., 2005). Besides direct heating, geothermal energy also has widespread application in the production of electricity. High-temperature geothermal resources can be utilized in power generation facilities that directly produce electricity from the steam, thus serving as a steady base-load power resource. Unlike some of the other forms of renewable energy such as wind and solar, which are intermittent in nature, geothermal power plants can operate round the clock each year, providing a constant source of power.

Figure 1.19. *Example of an aquaculture farm.*

Source: https://thermtest.com/wp-content/uploads/fishfarm-1024x682.jpg.

The resilience of geothermal systems is further enhanced by ongoing research and development activities on sorption cooling and heating systems, which are hybrids of geothermal energy and cutting-edge thermal energy storage technology. Such novel applications have the potential to expand the application of geothermal energy in a wide range of sectors, hence further incorporating it with the remainder of the energy mix (Cabeza et al., 2017). The diversified applications of geothermal energy also extend into industrial processes. These include industries that need huge amounts of thermal energy like agriculture, aquaculture, and food processing, which can use geothermal resources to provide heat, reducing operational costs as well as carbon footprints. Through the utilization of geothermal systems for operations like drying or pasteurizing, companies are able to increase their sustainability standing as well as achieve energy efficiency. The labeling of flagship schemes in cities like Waitakere City indicates an emerging trend of accessing geothermal potential within regional economies. Geothermal power becomes a viable option as regions seek to transition into more sustainable practices because it provides twin solutions for both energy requirements and abatement of climate change, making it more significant in the matrix of renewable energy alternatives (Grooby et al., 2005) (Figure 1.19).

1.5.3. Potential and Limitations of Geothermal Resources

The potential for geothermal energy as a renewable source of fuel is enormous, especially as a baseload power source. Compared to most of the renewable resources like wind and solar, which are naturally intermittent, geothermal energy is based on a consistent output built upon the continuous heat transferred from the Earth's core. This inherent capacity for reliability positions geothermal sources as a valuable resource within the greater portfolio of renewable energy technologies. However, to be integrated into current power grids and operate successfully, there needs to be knowledge of local conditions and geology, coupled with resource management, coupled with good infrastructure capable of conducting long-term operations. While scientists continue to investigate curtailment issues at geothermal power facilities, such as those noted at The Geysers, the necessity of flexibility and responsiveness in optimizing the benefits of geothermal energy and addressing its deficits in a dynamic energy climate emerges (Dobson et al., 2020). Despite the advantages, natural geothermal power usage constraints create challenges that must be taken seriously. One of the major disadvantages lies in geographical limitations; geothermal resources are neither equally distributed nor they are found in large numbers in tectonically active areas. This geographical dependency also creates issues of logistics in transporting energy and high investment in infrastructure in order to connect the resources with inhabited settlements. Moreover, the environmental impacts of geothermal activities such as land use and potential induced seismicity are of concern among communities and stakeholders. The balance of these factors must be well managed in order to ensure the highest exploitation of geothermal resources and their integration into renewable energy applications. To this end, intelligence in local resources potential and effective alleviation of community issues is crucial in the successful operation of geothermal power systems (Driver et al., 2004). The potential of resources for enhancing geothermal energy provision of renewable power systems can be achieved through technological advances and integrated management strategies. New technologies, including enhanced geothermal systems (EGS) and advanced drilling, are capable of unlocking new resources as well as improving efficiencies of extraction. Additionally, as well as combining geothermal power production with energy storage technologies, curtailment can be treated through balancing supply with demand while aiding grid stability. Ongoing research on flexible generation opportunities in plants like The Geysers confirms the necessity of adaptive approaches in managing geothermal resources to help address the effects of economic and transmission-related curtailment (Dobson et al., 2020). More broadly, developing a well-balanced energy plan—such as that advocated for in the Balanced Energy Plan for the Interior West—would solidify geothermal resources in combination with other renewable resources as part of a more solid and diversified energy mix (Driver et al., 2004).

1.6. INTEGRATION OF RENEWABLE ENERGY INTO POWER SYSTEMS

The integration of renewable energy in power infrastructure is paramount towards building sustainable energy infrastructures that are capable of meeting modern requirements. With communities grappling with climate change and energy security, moving away from fossil fuels and towards renewable sources is important. The positive growth of renewable technologies such as solar and wind proves that they have the capability to generate energy at scale. Their incorporation into existing power grids demands advanced grid management strategies and storage technologies that can accommodate the volatility of renewables. Furthermore, new ideas such as sorption-enhanced gasification processes also have a potential of capturing electricity and hydrogen from biomass, essentially complementing traditional energy sources. Evaluating the benefit of such systems, like reduced emission and energy resilience, can be included in a more sustainable energy future (Arabkoohsar et al., 2024). The impact of effective incorporation of renewable energy extends beyond environmental benefit to economic implications and energy policy. Countries are offered profound decisions about their future energy futures amid global endeavors at reducing greenhouse gas emissions. In Brazil's case, investments in renewable energy infrastructure can either lock in carbon-intensive tracks or make the country a leader in low-carbon technology. With over $0.5 trillion earmarked for energy infrastructure in the next decade, coordinating funding towards green practices is essential. By analyzing energy-related greenhouse gas emissions and their causes, policymakers are able to design plans that reconcile economic development and climate goals. For instance, prioritizing decarbonization in the transport, industry, and power sectors presents tremendous opportunities for emissions reduction (Lucon et al., 2015). Furthermore, the transition to renewable resources can be achieved only if it is complemented by technological developments and effective policy frameworks that fortify energy resilience. Utilizing smart energy management systems has the potential to optimize electricity production and supply from renewables while maintaining resilience of the grid. This requires not just investment in smart grid infrastructure, but regulatory policies that promote energy efficiency and storage. By using response surface methodology, stakeholders can define and optimize system performance towards better energy and exergy efficiencies, and lead to more sustainable power generation processes. The integration of various renewable technologies can reduce the dependency on non-renewable inputs, as is evident from the advances in sorption-enhanced biomass gasification, which achieves remarkable energy and exergy efficiencies (Arabkoohsar et al., 2024). Such developments have the potential to lay the groundwork for more integrated, lower-carbon power system futures worldwide.

1.6.1. Smart Grids and Their Role in Energy Management

The development of smart grids represents a significant leap in energy management, particularly the use of renewable energy. Smart systems utilize upgraded digital infrastructure to enhance power distribution efficiency, reliability, and sustainability. The central aspect of smart grids is their ability for two-way communication between consumers and producers of energy, facilitating real-time information exchange and dynamic load balancing. Such capability is crucial when handling volatile energy supply from renewable technologies such as solar and wind, which have intermittent supply characteristics. By utilizing smart technologies, grid operators can optimize the production and supply of electricity most efficiently, reducing wastage and greenhouse gas emission, and ultimately opening the doors to a sustainable future for energy. Besides, the role played by multi-agent systems (MAS) in the development of smart grids cannot be overstated because these systems effectively manage distributed energy resources. MAS can enable numerous independent agents to coordinate and act as a group to perform complex tasks, thereby enhancing energy management processes. In smart grids, the systems identify consumer energy patterns, predict demand changes, and offer decentralized decision-making for grid operations. These are the features through which integration of renewable energy sources is facilitated, allowing for an adaptable response to supply and demand fluctuations. Application of MAS in energy management not only aids in optimizing resource planning but also aids in the uptake of renewable technologies. Through this mechanism, future smart grids can be more resistance-proof towards interruptions and be able to accommodate renewable energy goals (Wang et al., 2024). The integration of renewable energy and smart grid technology also reflects in its own opportunities and challenges, which were predominantly addressed at the National Conference on Renewable Energy, Smart Grid, and Telecommunication in 2023. Experts in the industry highlighted that strong policies and regulatory frameworks need to be developed that promote the implementation of a smart grid while taking environmental concerns into account. The key areas of discussion were integrating innovative communication technologies and telecommunication in improving renewable energy solutions. The subject of discussion highlighted the importance of effective data management techniques that can assist in optimizing energy systems more efficiently. Moreover, the conference also addressed new technologies such as battery storage and IoT innovation that play an important role in efficient energy management within the systems. Lastly, the formation of alliances among stakeholders will be crucial in bridging the challenges to smart grid installation, towards a sustainable future (Mishra et al., 2023).

1.6.2. Challenges of Integrating Renewables into Existing Infrastructure

Addition of renewable power sources into the existing infrastructure is attended by a number of challenges, largely due to the natural uncertainty and randomness of these power sources. Unlike all other types of energy generation, solar and wind are renewable power sources that respond to weather and time. This unpredictability strains the dependability of existing power grid systems, which were designed to handle constant and regulated electricity flow. Therefore, grid operators must become innovative in ensuring a stable service while accommodating such renewable power sources. An envisioned solution to this inherent dilemma is the utilization of Smart Grids (SGs) (Copeland et al., 2019). SGs allow for bi-directional power exchange and can adapt dynamically to changing energy supply and demand. This aspect is especially important in addressing the problems caused by incorporating high levels of renewable energy into legacy infrastructure. But another fantastic challenge is the centralized generation and one-way energy flow dependence of current infrastructures. Most power grids that currently exist cannot accommodate decentralized generation systems characteristic of renewable energy technologies. This requires extensive upgrades and restructuring, which sometimes require tremendous capital investments and long planning horizons. Further, the move towards a more responsive and adaptive grid system involves the implementation of advanced monitoring and control technologies, which form the foundation of Smart Grids. These technologies utilize the Internet of

Things (IoT) to collect real-time data and analyze the system, enabling grid operators to optimize performance and efficiency (Copeland et al., 2019). However, the introduction of new technologies can create resistance from stakeholders accustomed to traditional systems, further complicating the process of transformation of electrical grid infrastructure. Lastly, regional coordination also plays a crucial role in addressing the issues of integrating renewables into the existing energy systems, particularly in nations such as ASEAN that has aggressive targets for renewable energy. Grid flexibility evaluation shows that nations must focus on cooperative networks in order to establish their energy infrastructure and optimize resource utilization. Weak forecasting systems and restricted access to electricity markets are limiting factors for deploying renewable energy in large quantities. A holistic method of advancing regional cooperation is needed to optimize grid flexibility capacity and overcome these shortages. Collectively, countries can share resources, technologies, and best practices, ultimately leading to a more efficient and reliable electricity grid that can support greater penetrations of renewable energy at lower costs.

1.6.3. Policy and Regulatory Frameworks Supporting Integration

The transition to renewable energy needs firm policy and regulatory frameworks that promote the assimilation of such systems into conventional grids. Such frameworks are essential to create a healthy investment, innovation, and adoption environment for renewable technologies. Recent studies say there are changes taking place in legislation in various regions of the world and particularly in Europe towards supporting the transition towards cleaner economies. These reforms aim to transcend systemic hurdles in the context of efforts by domestic governments to incorporate renewable energy into power systems. By observing successful institutional and legal reforms, actors can adopt promising strategies to boost green economy interventions comprehensively. Thus, the strategic transformation of policy frameworks is critical in spearheading an unprecedented revolution in energy practices and infrastructure (Gehring et al., 2013). Furthermore, the international dimension establishes the quality of the regulatory framework for renewable integration. The European Union model of regulation is the blueprint for reforming the energy sector across the region's southern Mediterranean countries such as Algeria, Egypt, Türkiye, and Jordan. Through the promotion of a top-down paradigm of energy regulatory agencies, the EU aims to enhance energy exchanges and achieve maximum collaboration between regional states. Comparative mode of governance analysis is a pointer to the effectiveness of external governance in achieving energy regulatory goals. Such systematic frameworks not only prescribe standard compliance but also adopt best practice in energy integration, which can be emulated by other nations. It is this approach that is pivotal in rebalancing regional disparities and encouraging an integrated market for renewable energy (Cambini et al., 2014). Generally, adoption of renewable power into the power system hinges on robust policy and regulatory frameworks that ensure collaboration, standardization, and creativity. As nations combat climate change, the necessity for such frameworks is underscored by their ability to ensure the shift to green energy consumption. Legal preparedness for the green economy, as stipulated in recent studies, requires governments to be creative with the regulatory framework. The interlinkage between local, national, and international initiatives can be one step towards a more integrated power network embracing renewable energy. Lastly, building these frameworks is crucial not just to attend to energy needs but also to create economic resilience and environmental sustainability as an answer to global energy transitions (Gehring et al., 2013).

SUMMARY

Summarily, the transition towards the use of renewable energy sources is a radical paradigm change in terms of power system design and operation. With the world attempting to grasp the problem of climate change coupled with rising energy demands, there is the need for the integration of sustainable technology. Enhanced efficiency of energy, especially in existing buildings' infrastructure, is an example of the ability to reduce carbon emissions while preserving cultural heritage. Especially, technologies such as hydrogen-rich natural gas produced using excess renewable electricity demonstrate the possibility of transforming energy supply without a dimension of changes to existing building envelopes. This approach not only addresses energy efficiency but also contributes to sustainability in urban living. These breakthroughs form a realistic path to future reconciliation between heritage preservation and modern energy requirements, defending the argument that development never has to trade cultural values, as stressed by contemporary studies (Matteo et al., 2017). Further, the integration of distributed generation systems in current power grids is essential towards optimizing power system reliability and resilience. As renewable energy sources like wind and sunlight become more popular, the dynamics of electrical distribution need to adapt. Such technologies have the potential to improve the voltage profiles and system efficiency. The process, nevertheless, comes with a few challenges, among them coping with potential conflict during fault clearing and the right interference with protective relaying operations. It is essential to harmonize these new solutions with the current grid infrastructure, especially as concerns the continuation of supply. The two-pronged approach is key to the evolution of energy systems into resilient networks that can manage future energy demands and permit increased utilization of renewable resources (Afifi et al., 2014). Ultimately, the establishment of renewable energy and power infrastructures is a multi-faceted initiative that requires collaborative actions from various industries. Policymakers, engineers, and conservationists must synergize efforts to develop strategies ensuring energy sustainability and historic preservation. With the challenges imposed by distributed generation and other emerging technologies, continuous assessment of their impacts on current infrastructures will be required. The integration of hybrid renewable energy systems can enhance power system performance as a response to the booming demand for energy efficiency and environmental responsibility. Such convergence of activities, therefore, contributes to making not only a cleaner energy future but also a reliable power grid with the ability to respond to tomorrow's energy needs through an increasingly sustainable system.

The debate today is all about the revolutionary contribution of renewable energy to global energy systems. The underlying assumption of this debate is the increasing reliance on wind power, as it has emerged as a cornerstone of clean power generation. The description of how wind turbines operate, particularly in terms of the explanation using power curves, identifies the intricate relationship between wind speed and energy output. Later advances in statistical techniques, such as the application of empirical copulas, have taken comprehension of such a relationship even deeper. By facilitating a better understanding of the relationship between active power and wind speed, such techniques facilitate earlier detection of turbine inefficiency or malfunction, thus optimizing working efficiency and ensuring sustainable development of wind power in power systems (Galloway et al., 2012). With wind power, the greater sense of renewable energy policy is increasingly evident in urban development plans. Urban planning's integration of sustainable development agendas is a characteristic of a profound shift in governance where technological innovation is central to public policy deliberation. London Plan is a pioneering instance that illustrates how city planning not only reacts to but also shapes technological discourses of renewable energy deployment. This policy-technology interface constitutes a critical paradigmatic shift in the manner in which cities envision their futures as regards energy, and it underscores the imperative for stakeholders to critically interact with technicalities and the evidence base that informs strategic decisions (Rydin et al., 2010). With the debate on renewable energy gaining momentum, an understanding of how technology, policy, and performance assessment interact becomes ever more crucial. The key issues raised press home the importance of continuous enhancement in the energy infrastructure, with a perspective to making energy systems continually monitored and assessed for every changing technology and methodology. Inclusion of new statistical tools such as copula analysis is necessary in advancing wind energy reliability. Concurrently, the relationship between urban planning and renewable energy reiterates the manner in which spatial development policies can harmonize with sustainable strategies. Together, these discussions serve to define a holistic perspective of the future of power systems, further emphasizing the central role that innovation plays in realizing environmental sustainability as well as energy security.

REFERENCES

1. Abdel Gelil, I., Aguayo, F., Boyle, G., Breukers, S., et al. (2012). Policies for capacity development. https://core.ac.uk/download/9836421.pdf (accessed on 30th September 2025).
2. Abeyasekera, K., & Matthews, G. (2007). Sustainable exhibit design: Guidelines for designers of small scale interactive and travelling exhibits. https://core.ac.uk/download/56852.pdf (accessed on 30th September 2025).
3. Abrhám, J., Britchenko, I., Garškaitė-Milvydienė, K., Jankovic, A., et al. (2018). Energy security issues in contemporary Europe. https://core.ac.uk/download/322492552.pdf (accessed on 30th September 2025).
4. Adam, D., Giddings, B., Bunce, M., Underwood, C., Cloke, P., et al. (2007). Renewable energy in remote communities. https://core.ac.uk/download/4149082.pdf (accessed on 30th September 2025).
5. Afifi, S. N., & Darwish, M. K. (2014). Impact of hybrid renewable energy systems on short circuit levels in distribution networks. https://core.ac.uk/download/29140010.pdf (accessed on 30th September 2025).
6. Aheicheva, O., Bukhkalo, S. I., Byalskiy, Y., Derkunska, O., et al. (2018). Distance education possibilities analysis for integrated innovative projects. https://core.ac.uk/download/187298594.pdf (accessed on 30th September 2025).
7. Alcalá, G., Garrido, J., Hernandez-Escobedo, Q., Perea Moreno, M. A., et al. (2019). Wind power cogeneration to reduce peak electricity demand in Mexican states along the Gulf of Mexico. https://core.ac.uk/download/222767697.pdf (accessed on 30th September 2025).
8. Altan, H. (2010). Energy efficiency interventions in UK higher education institutions.
9. Arabkoohsar, A., Khoshbakhti Saray, R., Khosravi, S., Neshat, M., et al. (2024). Techno-economic analysis, energy assessment, and optimization using response surface methodology of a solar and biomass-based co-generation system. https://core.ac.uk/download/604505224.pdf (accessed on 30th September 2025).
10. Arrobbio, O., Candelise, C., Delvaux, S., Gilcrease, W., et al. (2020). Collective action and social innovation in the energy sector: A mobilization model perspective. https://core.ac.uk/download/289098471.pdf (accessed on 30th September 2025).
11. Atkins, M. J., Morrison, A. S., & Walmsley, M. R. W. (2010). Carbon emissions pinch analysis (CEPA) for emissions reduction in the New

Zealand electricity sector. https://core.ac.uk/download/29197470.pdf (accessed on 30th September 2025).

12. Ault, G. W., Currie, R. A. F., Foote, C. E. T., McDonald, J., et al. (2005). Ubiquitous energy storage. https://core.ac.uk/download/9034038.pdf (accessed on 30th September 2025).
13. Baquedano Aísa, M. (2021). Analysis of the techno-economic viability of a solar wind hybrid power plant. https://core.ac.uk/download/574567056.pdf (accessed on 30th September 2025).
14. Beluco, A., Canales, F. A., Guezgouz, M., Jurasz, J., et al. (2019). A review on the complementarity of renewable energy sources: Concept, metrics, application and future research directions. https://core.ac.uk/download/286642706.pdf (accessed on 30th September 2025).
15. Benoit, P., Boustead, I., Callaghan, C., Finkbeiner, M., Hall, G. M., et al. (2012). Energy from waste and the food processing industry. https://core.ac.uk/download/6112944.pdf (accessed on 30th September 2025).
16. Borghetti, G., Menga, G., Procaccianti, G., Kok, K., Ardito, L., Morisio, M., Mo, B., et al. (2013). Smart grid technologies in Europe: An overview. https://core.ac.uk/download/pdf/11431923.pdf (accessed on 30th September 2025).
17. Botterud, A., Levin, T., Mills, A. D., Seel, J., et al. (2020). Impacts of variable renewable energy on wholesale markets and generating assets in the United States: A review of expectations and evidence. https://core.ac.uk/download/305124749.pdf (accessed on 30th September 2025).
18. Britchenko, I., Jarosz, P., & Zielińska, A. (2018). Leading innovations and investments into the new energy technologies. https://core.ac.uk/download/322492939.pdf (accessed on 30th September 2025).
19. Bruce Driver, D., Berry, J., Nielsen, J., Lehr, R., & Innis, S. (2004). A balanced energy plan for the Interior West. https://core.ac.uk/download/71349464.pdf (accessed on 30th September 2025).
20. Cabeza, L. F., & Schossig, P. (2017). Advances in sorption systems for energy efficient heating and cooling. https://core.ac.uk/download/222669221.pdf (accessed on 30th September 2025).
21. Cambini, C., Franzi, D., Eberlein, B., Levy, B., Schimmelfennig, F., et al. (2014). Assessing the EU pressure for rules change: Perception by southern Mediterranean energy regulators. https://core.ac.uk/download/pdf/11432127.pdf (accessed on 30th September 2025).
22. Chilán, J. C., Cordova, A. J., Gamez, M. R., Machuca, A., et al. (2018). Social impact of renewable energy sources in the Province of Loja.

https://media.neliti.com/media/publications/268619-social-impact-of-renewable-energy-source-a9cad46b.pdf (accessed on 30th September 2025).

23. Copeland, R., Crespi, N., Rehmani, M. H., Saleem, Y., et al. (2019). Internet of Things-aided smart grid: Technologies, architectures, applications, prototypes, and future research directions. http://arxiv.org/abs/1704.08977 (accessed on 30th September 2025).
24. Di Matteo, U., & Nastasi, B. (2017). Innovative use of hydrogen in energy retrofitting of listed buildings. https://core.ac.uk/download/141694922.pdf (accessed on 30th September 2025).
25. Dobson, P., Dwivedi, D., Garcia, J., Kiran, M., et al. (2020). Analysis of curtailment at The Geysers geothermal field, California. https://core.ac.uk/download/323310904.pdf (accessed on 30th September 2025).
26. Escudero Olano, M., González Fernández, M. C., & Jiménez Alvaro, Á. (2011). Analysis of the behaviour of biofuel-fired gas turbine power plants. https://core.ac.uk/download/148661826.pdf (accessed on 30th September 2025).
27. Fox-Penner, P., Hatch, J., & Lam, T. (2019). Assessing the potential for U.S. utility green bonds. https://open.bu.edu/bitstream/2144/39183/1/Green-Bond-Final-Paper-5_16-2019.pdf (accessed on 30th September 2025).
28. Galloway, S., Gill, S., & Stephen, B. (2012). Wind turbine condition assessment through power curve copula modelling. https://core.ac.uk/download/9036228.pdf (accessed on 30th September 2025).
29. Gehring, M., & Kent, A. (2013). Innovative regulatory frameworks promoting green economy for sustainable development and poverty eradication in Europe. https://core.ac.uk/download/41991363.pdf (accessed on 30th September 2025).
30. Gharehpetian, G. B., Marzband, M., Moradi, J., Nafisi, M., et al. (2019). Attributes of big data analytics for data-driven decision making in cyber-physical power systems. https://core.ac.uk/download/305121394.pdf (accessed on 30th September 2025).
31. Grooby, L. D. (2005). *The Renewable Energy and Energy Efficiency Potential of Waitakere City: A Thesis Presented in Partial Fulfilment of The Requirements for The Degree of Masters of Technology in Energy Management at Massey University*. https://core.ac.uk/download/148644048.pdf (accessed on 30th September 2025).
32. Guarnieri, M., Mattavelli, P., Petrone, G., Spagnuolo, G., et al. (2016). Vanadium redox flow batteries: Potentials and challenges of an emerging

storage technology. https://core.ac.uk/download/80155277.pdf (accessed on 30th September 2025).

33. Hahn, F. (2022). Optimized modulation and thermal management for modular power converters. https://core.ac.uk/download/541038916.pdf (accessed on 30th September 2025).
34. Huang, Y. W., Kammen, D. M., & Kittner, N. (2019). ASEAN grid flexibility: Preparedness for grid integration of renewable energy. https://core.ac.uk/download/323310932.pdf (accessed on 30th September 2025).
35. James, O. O. (2013). Improving carbon and process utilization for biomass-to-liquid fuel operations. https://core.ac.uk/download/16436077.pdf (accessed on 30th September 2025).
36. Kammen, D. M., Kittner, N., Kondolf, G. M., Schmitt, R., et al. (2019). Deploy diverse renewables to save tropical rivers. https://core.ac.uk/download/323311068.pdf (accessed on 30th September 2025).
37. Lima, M., Machado, J. M., & Seabra, E. (2012). Conceptual design and development of an automated co-generation system. https://core.ac.uk/download/55623114.pdf (accessed on 30th September 2025).
38. Lucon, O., Fransen, T., & Romeiro, V. (2015). Bridging the gap between energy and climate policies in Brazil: Policy options to reduce energy-related GHG emissions. https://core.ac.uk/download/75782972.pdf (accessed on 30th September 2025).
39. Michaels, R. J. (2008). A federal renewable electricity requirement. https://core.ac.uk/download/71341672.pdf (accessed on 30th September 2025).
40. Mishra, S., Nayak, B. T. M., Pradhan, S., et al. (2023). National Conference on 'Renewable Energy, Smart Grid and Telecommunication-2023. https://core.ac.uk/download/588430838.pdf (accessed on 30th September 2025).
41. N/A. (2005). Renewables 2005: Global status report. https://core.ac.uk/download/71338835.pdf (accessed on 30th September 2025).
42. N/A. (2008). Charting our own course: Today's challenges, tomorrow's opportunities, December 2008. https://core.ac.uk/download/11350328.pdf (accessed on 30th September 2025).
43. N/A. (2008). Eighth National Garrett Morgan Symposium on Sustainable Transportation, Report MTI 08-03. https://core.ac.uk/download/70409892.pdf (accessed on 30th September 2025).
44. Ramamurthi, P. V. (2016). Political economy of renewable energy deployment in India: Case study of Karnataka. https://core.ac.uk/download/71338519.pdf (accessed on 30th September 2025).

45. Rydin, Y. (2010). Planning in a technological society: Discussing the London Plan. https://core.ac.uk/download/1760132.pdf (accessed on 30th September 2025).
46. Shaikh Salim, S. A. Z., Sharifmuddin, N. B., & Suwa, T. (2015). Concept study of microgrid dispatch strategy for solar thermal power plant with thermal storage. https://core.ac.uk/download/132281526.pdf (accessed on 30th September 2025).
47. Simao, A. (2006). WePWEP: Web-based participatory wind energy planning [1]. Background information on wind energy and wind farm siting. https://core.ac.uk/download/1671821.pdf (accessed on 30th September 2025).
48. Stein, A. (2009). Greening historic DC: Challenges and opportunities to incorporate historic preservation into the district's drive for sustainable development. https://core.ac.uk/download/70373622.pdf (accessed on 30th September 2025).
49. Wang, X. (2024). *Multi-Agent Systems for Smart Grids: Revolutionizing Energy Management.* https://core.ac.uk/download/630297906.pdf (accessed on 30th September 2025).
50. Werner, C., & Ames, J. (2001). The 2002 Farm Bill: Revitalizing the farm economy through renewable energy development. https://core.ac.uk/download/71341292.pdf (accessed on 30th September 2025).
51. Xiao, M. (2024). Application of power electronics technology in renewable energy systems. https://core.ac.uk/download/616937539.pdf (accessed on 30th September 2025).

CHAPTER 2

POWER SYSTEM FUNDAMENTALS

Contents

INTRODUCTION

Power system analysis is a natural component of electrical engineering that breeds the energy infrastructure upon which contemporary society depends. A power system comprises a number of components including generation, transmission, distribution, and utilization of electrical power. It is essential to be familiar with these components for understanding the process by which electricity serves the residential as well as commercial systems. The use of renewable energy technologies has also made the power system scenario more complicated with increasingly advanced control techniques and management of the grid. As conventional sources of energy dwindle, the move toward sustainable sources has fueled research and innovation in the sector. Improvement programs for the reliability and efficiency of grids are needed as society strives to mitigate the impacts of climate while striking a balance of rising energy demands, and investigation into power systems is vital in the solution of challenges in the future and in sustainability needs. Efficient operation of power systems relies on the interaction of generation and distribution networks. Power can be generated from various sources including fossil fuels, nuclear, and renewable sources such as solar and wind.

Figure 2.1. *Fundamentals of power system.*

Source: https://framerusercontent.com/images/f4cRcc8qnFkr5odnwBWb9cTQvrM.png.

All sources of energy have specific strengths and weaknesses, impacting the overall power grid efficiency and security. For instance, the integration of solar and wind energy presents new challenges to support energy storage and transmission with the uncertainty in supply. Additionally, advances in smart grid technologies are enabling enhanced communication and coordination among various parts of the power system. These advances not only optimize energy distribution but also optimize grid robustness to potential disruption. As power grids keep evolving, understanding the principles of power distribution, backed by technology advancements, is an essential readiness for next-generation power engineers to deal with future complexities of power systems. In power grids, digital signal processing and advanced control techniques are valuable additions to maximizing operation and reliability. Modern power networks are increasingly relying on data analysis and automation to manage the flow of electricity in the most efficient way. Through sophisticated algorithms, the system can be monitored real-time continuously and faults found and rectified in real-time. Moreover, technologies such as radar systems are also being studied for their application in ensuring safety and efficiency of grid operations.

The developments in these technologies show an ongoing trend towards smart systems capable of self-regulation. These developments are a shift towards the design and functioning of power systems towards making them more responsive and flexible in an increasingly dynamic energy regime. With this change, the learning and application of advanced technologies will become critical to the future engineering skill pool, leading to a maximized and resilient power infrastructure (Figure 2.1).

Definition and Importance of Power Systems

Power systems are the pillar of modern electricity supply, managing generation, transmission, and distribution of electrical power across long distances. In a broad sense, a power system consists of coordinated generation units, transmission lines, substations, and distribution networks to allow for the effective transfer of electrical power from generators to consumers. The significance of power systems lies not only in their capacity to supply the increasing demand for electricity but also in helping drive economic growth and industrialization. While the world population continues to rise, the need for a stable and secure power system is increasingly vital. Besides, renewable energy sources need new power system approaches to ensure stability as well as power efficiency, ultimately passing on to greener energy policies (Blass et al., 2007). The value of power systems extends beyond their working; it borders on the economic, environmental, and social values that they possess. An efficient power system improves productivity, advancing technological progress and quality of life. Economically, the non-stop availability of electricity supports industries from manufacturing to information technology, generating employment and investment opportunities. At the same time, effective power systems can contribute to environmental sustainability through the incorporation of renewable energy options and reducing the emission of green-house gases. As urban development advances and lifestyles evolve, power systems will have to develop more and more to fit shifting needs and fluctuating energy inputs, which underscores the value of strategic investment and innovation in grid operation and power supply (Diamant et al., 2008). In addition, power systems transformation is intertwined with technological progress and expanding world sensitivity to energy conservation. With the advent of smart grid technologies, power systems are becoming more and more efficient and responsive, allowing for real-time consumption monitoring and control of energy. The technologies not only enhance the reliability of the supply of electricity but also enlighten consumers through enhancing their knowledge of their energy use patterns. Moreover, since climate change is now an alarming issue globally, moving towards cleaner sources of energy in power systems is a key move to avoid its adverse effects. Therefore, understanding the definition and role of power systems is at the core of handling the complexities of energy needs with environmental and social demands, thereby ensuring a green future for the coming generations (Blass et al., 2007) (Figure 2.2).

Figure 2.2. *Importance of power systems.*

Source: https://www.pnnl.gov/sites/default/files/styles/hero_1600x1200/public/media/image/Realizing-a-secure-flexible-and-resilient-electric-power-system.jpg?h=96a96008&itok=L149iqhv.

Historical Development of Power Systems

The history of power systems goes back to the late 19th century when there was a remarkable technological development along with a growing need for electrification. With the advent of the electric generator, and later the establishment of electrical transmission systems, came the birth of a new age of energy supply. Initially, power systems were small and local and were driven by direct current (DC) technology, which had limited transmission distance and efficiency. The advent of alternating current (AC) systems, championed by names like Nikola Tesla and George Westinghouse, revolutionized the game of energy distribution. This progress enabled electricity to be passed along long distances, eventually to the construction of broad-scale power grids connecting regions and making universal availability of electrical energy possible and current power system infrastructure established. During the course of the 20th century, demand for electricity increased exponentially, necessitating enhancing power system design and operations. The interconnection of substations, transformers, and high-voltage transmission lines became imperative to support increasing consumption. Additionally, the mid-century saw the establishment of regulatory structures that influenced the ownership and operation of power stations and facilities. Utilities in most countries became central organizations tasked with generation, transmission, and distribution of electricity. This concentration created economies of scale and improved reliability but created issues of monopolistic behavior. Research on the financial stability and performance of these utilities began emerging since policymakers recognized the significance of energy supply to economic growth and stability, paralleled by research findings that institutions play a crucial role in financial system resilience (Michael, 2025). Over the past few decades, the historical development of power systems has been met with emerging challenges and opportunities triggered largely by technological advancements and environmental concerns for climate change. The evolution of renewable energy sources, including solar and wind energy, has profoundly transformed the landscape of power generation and delivery. Modern power systems are increasingly decentralized, with distributed energy resources being embedded in the conventional grid. With such advancements, the complexity of satisfying energy need in real-time calls for advanced grid management technologies and procedures. Sustainability and resilience are priority concerns in today's debate, as countries strive to satisfy energy needs along with environmental issues. Herein, the understanding of implications of flexible exchange rates and economic policy matters concerning energy markets will be central to future study of emerging economies (Miguel et al., 2025). The historical development of power systems thus accounts for a dynamic history of adjustment to technological, economic, and environmental pressures.

Overview of Power System Components

Power systems consist of intricate networks which facilitate generation, transmission, and distribution of electrical energy to meet the diversified demand of consumers. The generation part is the nucleus of such a network wherein it accommodates an array of sources of energy such as fossil fuels, nuclear, and renewable energy sources such as wind and solar. Each source of energy has inherent qualities that affect how electricity is produced and supplied into the power system. As power generation evolves further, more emphasis is laid on how to make the production of energy more efficient and integrate new technology. This shift towards modernizing power generation is important for optimizing the usage of resources and reducing environmental impacts, especially with pressure to convert to more sustainable energy sources building

(Fisher et al., 2011). Thus, understanding the working principles and technology behind power generation is key for stakeholders in the energy industry. Once it is generated, it must be sent over long distances to the consumers, hence the importance of transmission systems in electricity power networks. Transmission systems consist of high-voltage power transmission lines and transmission substations that facilitate the efficient transfer of electricity over extensive geographical areas. The systems must be designed and serviced in a manner to minimize energy loss and ensure reliability in electricity supply. Technological innovations in transmission technology, such as the use of smart grids, have entered into existence to enhance operational effectiveness and reliability. Moreover, integrating renewable energy resources into existing grids is an opportunity as well as a challenge because it requires enhancement in transmission infrastructure in order to handle variable outputs effectively (Hanafi et al., 2015). Therefore, the transmission aspect of the power system is a determining factor in the efficiency and longevity of electricity supply as a whole. Distribution networks form the final phase in energy supply, linking high-voltage transmission wires to end-users, ranging from residential to commercial and industrial consumers. The distribution systems transform high voltage electricity into lower voltage ranges feasible for consumer use through several substations and distribution lines. Distribution system reliability is important since power losses can disrupt everyday life and economic activity. Novel technologies, including distributed generation and smart metering systems, are being introduced into distribution systems in order to enhance performance and customer care. Furthermore, with growing electricity demand and the number of decentralized energy sources, distribution system optimization is required in order for them to be able to cope with the new dynamics. An understanding of the components and operations of distribution systems is essential to developing strategies by which the resilience and sustainability of entire power systems can be increased.

2.1. GENERATION OF ELECTRICAL POWER

Electrical power generation is a heterogeneous process that encompasses various technologies and methodologies, with a target to convert various energy sources into electricity. Traditional power-generating means such as fossil fuels, nuclear power, and hydroelectric power have existed for centuries. However, with growing awareness of climate change and the environmental context of the traditional sources, there has been an unprecedented focus on the utilization of renewable energy. Solar, wind, and biomass technologies are coming center stage, fueled by enhanced efficiency and reduced costs, and becoming increasingly feasible in today's energy economy. Moreover, the interconnection of distributed generation systems has introduced fresh dynamics, allowing power to be generated at the local level, reducing transmission loss, and enhancing the reliability of the grid. The shift towards renewables is reshaping the paradigmatic columns of power generation, and new strategies and policies are required to make this transition possible. One of the significant advancements in electrical power generation is the incorporation of energy storage systems, which address both unreliability and intermittency issues associated with renewable sources. Compressed Air Energy Storage systems, for example, can store excess energy produced during peak production hours and release it when demand exceeds supply. Such alignment of technologies significantly enhances the overall efficiency of power generation systems and allows for better grid management.

Figure 2.3. *Electric power generation.*

Source: https://www.cummins.com/sites/default/files/styles/hero_feature/public/2021-06/electric_powerlines_moving_snapshot_m_1.webp?h=d1cb525d&itok=cKNfnxlz.

A best-practice CAES model for application with photovoltaic systems demonstrates the technical and financial feasibility of such facilities. Through assessing various scenarios, such as the conditions of support based on incentives, evidence has been shown that the Levelized Cost of Energy (LCOE) can greatly be enhanced, and thus renewable investments become more viable in the energy competitive market (Bylykbashi et al., 2016). The achievements highlight the need for ongoing innovation in the field of energy storage in order to unlock the full potential of renewable energy production. Educational course development in energy engineering reflects the growing importance of power generation in a renewable environment. Incorporating courses such as sizing and design of renewable energy systems in university programs reflects the growth in complexity of the industry and the need for capable specialists who can effectively address such an issue. For instance, the Escola d'Enginyeria de Barcelona Est has introduced a comprehensive bachelor's degree in energy engineering focused on renewable energy challenges and their implementation in modern energy systems (Martínez et al., 2017). Not only does this educational change give students professional exposure in the new energy industry but also bridges the academic chasm among theoretical learning and actual application. Through the development of a new generation of engineers skilled in renewable technology and alternative energy solutions, academic institutions are helping to drive the power industry towards a more sustainable and efficient future (Figure 2.3).

2.1.1. Types of Power Generation (Thermal, Hydro, Wind, Solar)

The energy generation scenario consists of several fundamental types, which have their specific functions in the energy mix. Thermal energy generation, primarily fueled by fossil fuels such as coal, natural gas, and oil, uses combustion to convert thermal energy into electricity. Although it is a significant contributor as a steady supply of energy, thermal generation is a significant environmental issue with heavy emissions of greenhouse gases and air pollution. Mitigation of its effects involves adopting carbon capture technology and utilizing less carbon-intensive fuels. However, as renewable sources of energy gain ever-increasing prominence, the long-term viability of thermal generation is a key subject of debate, particularly energy security and climatic obligations. Hydropower is another significant source of power generation utilizing flowing water's energy to produce electricity. It is commonly acclaimed for its ability to provide a stable and controllable power supply, with the potential to meet peak demand well. The universal use of reservoirs allows energy to be stored and controlled during times of low output and thereby stabilized the grid. But the ecological impacts including dam construction and river ecosystem disturbance cannot be overlooked. New technologies, such as run-of-the-river systems, are being explored in order to maintain small ecological footprints. Moreover, hydroelectric facilities must cope with changing climate trends, which have the potential to impact water supplies and thus the capacity to produce electricity, making it crucial to invest in resilient infrastructure. Wind and solar power are now integral components of modern electrical systems, notably in terms of moving away from fossil fuel reliance. Wind energy utilizes turbines to convert wind kinetic energy into power and solar energy uses photovoltaic cells to harness solar light. Both technologies are characterized by low operational emissions and environmentally friendly resource use, which is appealing for addressing the urgent demand for climate action. However, their output needs efficient solutions for storing energy and grid management to ensure reliability.

There is more distributed generation on the grid that alters dynamics, as evidence suggests there can be resolution to transient stability problems with efficient energy storage and optimizing the addition of renewables to existing networks (Froome et al., 2025). Solar generation, additionally, has both positively and negatively affected energy storage system profitability and operation efficiency in specific markets, with the mixed interaction of generation types (Bertsch et al., 2018).

2.1.2. Role of Renewable Energy in Power Generation

The role of renewable energy in power production is emerging with increasing demand for clean and sustainable sources of energy. The renewable energy technologies such as solar, wind, and hydropower offer an efficient method of stopping the emission of greenhouse gases by providing safe energy. The discontinuation of fossil fuel use and switching to renewables not only stops the impacts of climate change but also enhances the energy independence of the majority of nations. As countries strive to meet global climate objectives, such as those made in the COP21 agreement, integration of renewables into the installed power systems becomes an emerging choice for green development. Indeed, this shift aligns with international efforts to limit a rise in environmental temperature to below 2 °C, indicating the need for stepped-up innovation and investment in renewables technologies (Fisher et al., 2011). In the last few decades, technological advancements in renewable energy technologies have increased their adoption at a faster rate, causing cost reduction and improvement in efficiency. Not only does the technology of energy-efficient systems reflect in generation but also in energy usage by industries. For instance, retrofit technologies that contribute to industrial energy efficiency innovations can enhance overall power utilization performance in factories (Fisher et al., 2011). Moreover, the ability to hybridize renewable energy and traditional power generation technologies offers unproblematic conversion across forms of energy, ultimately promising increased output and reliability. This integrated approach not only lowers operational costs but also contributes to environmental conservation, evidencing the varied advantages of using renewable energy as a source of power generation. The role of renewable energy in sustainable power generation is logically related with the notion of energy resilience. Through diversifying energy mixes, states can decouple themselves from finite fossil fuel reserves while building a flexible and adaptive power grid capable of addressing dynamically shifting energy demands. Additionally, with more focus on the energy-water-environment nexus, renewable energy sources can be integrated natively into existing frameworks that operate to maximize efficiency and minimize carbon emissions across industries. This combined view broadens the sustainability of energy systems along with community well-being based on these resources. As the world embarks on the journey of sustainable development, embracing alternative renewable energy will be vital in the achievement of overall environmental goals and safeguarding generations to come (Ang et al., 2017) (Figure 2.4).

Figure 2.4. *Role of renewable energy in power generation.*

Source: https://cdn.shopify.com/s/files/1/0493/9834/9974/files/renewa ble-energy.jpg?v=1689067201.

2.1.3. Challenges in Power Generation

The power generation map is constantly changing, yet it is also faced with overwhelming challenges that limit its efficiency as well as sustainability. Among the central of these challenges is the reliance on traditional energy sources, namely fossil fuels, which not only contaminate the environment but are also trapped in regulatory and market forces for a reduction. As the world moves in the direction of renewable energy, generators need to change their infrastructures in a way that they incorporate technologies that ensure energy efficiency and minimize emissions. For example, retrofitting industrial buildings with energy-efficient technology can improve performance while reducing the carbon footprint. This shift in strategy emphasizes the imperative necessity to harmonize the immediacy of energy demands with long-term ecological goals, creating a complicated challenge for power generation systems (Fisher et al., 2011). Another acute challenge to power generation is increasing power grid sophistication that has historically been optimized for one-way energy transfer. The introduction of Smart Grids (SGs) aims to solve this through bidirectional communication between the consumer and the utility, enhancing efficiency and reliability in energy delivery (Copeland et al., 2019). The introduction of SGs is accompanied by challenges like the requirement of large-scale upgrades to current infrastructures, interoperability between different technologies, and vulnerability reduction with increased connectivity. Further, the utilization of renewable energy sources necessitates sophisticated grid management to address variability in supply, prompting new approaches in energy distribution and storage. Such demands are ambitious in terms of investment and coordination among stakeholders, underlining the imperative of collaboration in overcoming technological challenges in power generation today. Additionally, the integration of IoT devices with Smart Grids is a revolutionary means of monitoring and managing systems of energy. While IoT-driven SGs are brought with the benefits of improved operational performance and real-time analysis, there exist cybersecurity, data privacy, and technological competency issues to be implemented in such high-tech systems (Copeland et al., 2019). The need to protect power generation plant facilities from cyber-attacks along with customer data confidentiality adds an additional layer of complexity. Also, in order to efficiently apply IoT technologies, there is a need to train personnel and provide high-level expertise in handling data. These intricate challenges must be overcome not only to enhance the reliability and sustainability of the generation of power but also to enable the smooth and secure transition to intelligent, more resilient grids (Figure 2.5).

Figure 2.5. *Challenges in power generation.*

Source:https://www.iiot-world.com/wp-content/uploads/2018/07/shutterstock_545497132-e1532937714142.jpg.

2.2. TRANSMISSION OF ELECTRICAL POWER

Electrical power delivery is a central aspect of modern power systems, characterized by its intricate network of high-voltage transmission lines and substations that transport electricity over vast distances. The underlying principles of electromagnetic induction serve as the basis of this system of transmission, facilitating the translation of mechanical energy into electrical energy. High-voltage transmission is used in order to minimize losses caused by conductor resistance; high voltage electricity minimizes current, thus keeping I^2R losses at a minimum. In addition, technological advances such as in the high-voltage direct current (HVDC) system have been found extremely effective in long-distance transmission, bypassing the limitations that come with traditional alternating current (AC) systems. Understanding these intrinsic principles ensures effective design and operation of power systems capable of supplying adequate electricity in a more globalized world. One example of significant innovation in the transmission of electrical power is the application of cell-free massive multiple-input multiple-output (MIMO) systems.

Figure 2.6. *500 kV three-phase electric power transmission lines at Grand Coulee Dam. Four circuits are visible, while two additional circuits are obscured by trees on the far right. The entire 6809 MW nameplate generation capacity of the dam is carried by these six circuits.*

Source: https://upload.wikimedia.org/wikipedia/commons/thumb/c/ca/500kV_3-Phase_Transmission_Lines.png/1920px-500kV_3-Phase_Transmission_Lines.png.

These systems employ an immense number of distributed antenna access points (APs) to serve numerous single-antenna users with minimal resource consumption. These systems not only enhance spectral efficiency but also enable more efficient power allocation mechanisms, which have a very real impact on energy efficiency. For instance, the closed-form expression of spectral efficiency based on these systems reveals their ability to consider channel estimation errors along with power control, which are most crucial to optimal energy efficiency in contemporary power systems. By focus on balance between power availability and user limitations, such novel transmission approaches encourage sustainable energy consumption and thus lead the way for a maximized electric grid able to address current and future energy demands (Duong et al., 2017). In education, the integration of power system fundamentals into engineering programs lays stress on skill in the art of electrical power transmission. Through programs such as the new bachelor's degree in energy engineering, universities are training the engineers of tomorrow with the knowledge required to navigate the complexities of renewable energy systems as well as traditional power transmission methods. Sizing, design, and optimization courses pertaining to these systems allow students to understand the challenges and innovations that will shape the transmission of energy in the future. Furthermore, the realistic applications of concepts enable graduates to make substantial contributions to the evolving practice, concentrating on measures that are sustainable in line with world energy demands. As institutions of learning embrace transformation from the revolution in this sector, they play a central role in creating professionals who will lead the transmission scene of electrical power (Martínez et al., 2017) (Figure 2.6).

2.2.1. Transmission Lines and Their Functionality

Transmission lines are important components of power system network responsible for the efficient transmission of electrical power from the generation to substations and consumers. Transmission lines are built to carry alternating current (AC) at various voltages over long distances. The impedance of transmission lines is one of the most important factors influencing their performance, which sets the power loss and voltage standing wave ratio. An understanding of the resistance, inductance, and capacitance properties of transmission lines is very important in providing effective and reliable operation. All these factors impact the overall performance, and therefore it is required that the designers design and optimize their devices for specific applications and situations in the power system infrastructure, ultimately contributing to the reliability of electricity supply in the current energy-thirsty world. The operation of transmission lines is also impacted by the emergence of technologies such as the Internet of Things (IoT) and fog computing. The operation of transmission lines can be enhanced by their operators through real-time monitoring of the condition of transmission lines to avert faults before they happen. IoT sensors can provide valuable data on the condition of transmission infrastructure that can inform maintenance decisions and enhance overall reliability. For example, fog computing provides the ability to analyze data collected through far-end sensors along transmission lines in real time, enabling quick responses to operational issues. The integration of new technology increases operational efficiency along transmission lines and supports the final goals of efficiency and sustainability in power systems to result in a smarter grid infrastructure highlighted in existing research (Akrivopoulos et al., 2017). Moreover, cooperation among different technological systems such as radar and communications may introduce optimizations to the transmission line processes. Wherein communications and radar share the same frequency channels, the two may cooperate rather than being sources of interference, as per the theory of multiuser detection radar (MUDR) (Bliss et al., 2014). By viewing these systems as a unified system rather than separate systems, it is possible to maximize the performance of these systems using synchronized optimization techniques. This new approach can be used for transmission lines because it has the ability to maximize the efficiency of signal transmission and the features of communication, which are essential for real-time monitoring and control of the power systems. Such synergy is symbolic of the possibility of cross-disciplinary solutions to achieve a better functionality in the intricate web of modern-day power distribution networks (Figure 2.7).

Figure 2.7. *Transmission lines.*

Source: https://circuitglobe.com/wp-content/uploads/2016/04/transmission-lines-1-compressor.jpg.

2.2.2. High Voltage Transmission and Its Advantages

The advent of high voltage transmission has revolutionized the effectiveness of power systems globally. Increasing the levels of voltage, electrical energy can be transmitted over great distances with minimal losses of energy. This primarily lies in the fact that greater voltage implies lesser current for equivalent power. Consequently, resistive losses in the transmission lines, which are directly proportional to the square of the current, suffer significant reduction. Effectively, this allows the utility companies to deliver electricity from faraway generation facilities—solar arrays and wind farms—to cities more efficiently. In addition, high voltage transmission lines have a more robust and stable grid with power outages and power fluctuations that are inherent in traditional voltage lines. On top of reduced energy losses, high voltage transmission offers substantial economic advantages. By minimizing transmission losses, utility operators cut down on operational costs, which can consist of increased generation capacity, smaller infrastructure investment, and reduced maintenance expenditure. Furthermore, high voltage systems enhance integration of renewable sources into the grid, providing a cleaner energy mix and enhanced sustainability. This transition is in line with global initiatives against climate change because utilities can harness geographically spread renewable resources without being limited by the requirement to be near the urban center.

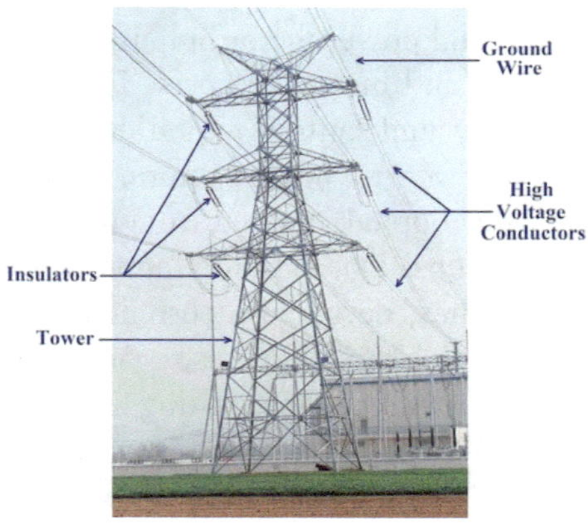

Figure 2.8. *Typical high voltage power transmission system.*

Source: https://eepower.com/uploads/articles/principles-of-high-voltage-transmission-part-1-fig1.jpg.

The financial advantages result in an energy mix diversified to ensure a secure energy future that caters to market requirements and environmental visions. Finally, the technologies employed to create high voltage transmission systems continue to prove to be more effective and efficient. Such technologies as superconducting cables and high-performance power electronics maximizing the passage of electricity demonstrate ongoing system evolution. System reliability is improved with better control of power flows, allowing more effective management of supply and demand patterns. A notable example of this trend is the technology of high voltage direct current (HVDC), capable of efficient power transmission over long distances and connecting isolated grids. These developments not only enhance the economic justification for high voltage transmission but also support infrastructure that is responsive and adaptive to changing patterns of energy consumption and generation, validating its pivotal role in future power systems management (Figure 2.8).

2.2.3. Losses in Transmission and Mitigation Strategies

The efficiency of power transmission systems is normally compromised by losses in electricity transmission from generation to consumers. These losses, typically categorized as resistive losses and reactive power losses, can greatly impact the operational efficiency of power systems. Resistive losses arise due to the inherent resistance in transmission lines, which dissipates electrical energy as heat. As the power transmission distance increases, these losses become even greater and need careful design considerations. Reactive power losses not only reduce the total power factor but also lead to inefficiencies that necessitate more generation and increased costs. They need to be minimized for optimum power delivery and cutting operational costs along the power grid, ending up in a cleaner energy infrastructure. To minimize transmission losses, various measures are adopted, ranging from infrastructure improvement to advanced technical solutions. One possible approach is the installation of high-voltage transmission lines, decreasing current flow for the same power level and consequently lowering resistive losses. Additionally, the application of smart control systems and advanced forecasting techniques can enhance grid management and overall efficiency. These systems facilitate real-time control according to various load demands and source capacity, thus optimizing the allocation of resources. Distributed generation resources and integration of renewable energies also become significant, as these systems can reduce the distance energy travels, directly opposing losses faced during long-distance transmission. Application of these advanced methodologies within the power industry can significantly reduce the losses for which traditional transmission systems are beset. Inverter-based technologies in modern power systems introduce challenges and opportunities in handling transmission losses. The quality of the energy conversion processes from direct current (DC) to alternating current (AC) and vice versa from AC to DC relies substantially on the control algorithms of inverters. According to research, the voltage-drive mode provides good quality AC output but induces large DC bus ripple to jeopardize power quality and system life (Burt et al., 2011). In contrast, sinusoidal current generation techniques demonstrate less ripple with satisfactory power quality. Researching new methods, such as cognitive interference alignment techniques in cognitive radio networks, offers a look into potential advancements toward reducing cross-tier interference and maximizing spectral efficiency, which is critical to future multi-tier energy networks (Debbah et al., 2013). Hence, drawing on these benefits can present a comprehensive methodology for reducing transmission loss while maximizing the reliability and sustainability of the power system.

2.3. DISTRIBUTION OF ELECTRICAL POWER

The transmission of electric power is part of the complete subject of power system fundamentals, emphasizing efficient and reliable power transmission of electricity from generating stations to consumers. This begins at substations where electricity in high voltage is reduced in voltage for consumption. The structure of distribution networks matters since these consist of various components like transformers, distribution lines, and protective devices. Good control of this system reduces losses and provides voltage levels that are favorable to safety and performance. Other technologies, such as smart grid systems, further complement traditional networks through real-time monitoring and adaptive control to achieve maximum power distribution for dynamic demand balancing (Mohamed et al., 2018). The incorporation of renewable energy sources into power distribution remains a pressing issue with evolving global energy paradigms towards sustainability. There are newer power grids increasingly utilizing alternative energy resources like solar and wind, so new thought patterns must be given to traditional distribution models. Innovation here is not theoretical but occurs in the form of educational programs, such as UPC's Bachelor's Degree in Energy Engineering with courses on the sizing and design of renewable energy systems (Martínez et al., 2017). This educational program prepares future engineers with the theory and practical skills necessary in developing and running such shifting energy landscapes. The incorporation of renewable technologies into power distribution plans underscores the need for an integrated strategy towards energy management, facilitating resilience and adaptability. The intricacies of electrical power distribution are further aggravated by the dynamic nature of consumer demand and regulation settings. As more devices are connected through the Internet of Things (IoT), demand for electricity becomes more and more unpredictable and fluctuates significantly throughout a day. Utilities wrestle with managing supply and demand while keeping their networks reliable. In addition, regulatory policies have a tendency to dictate how distribution networks operate, demanding higher efficiency, reliability, and transparency. Therefore, power distribution should not only cater for today's demands but also pre-plan for future evolution, providing a structure that sustains continued innovation and adaptability to the dynamics of the market. Such flexibility is essential as power networks absorb the intricacies of contemporary energy requirements and changing environmental imperatives (Mohamed et al., 2018) (Figure 2.9).

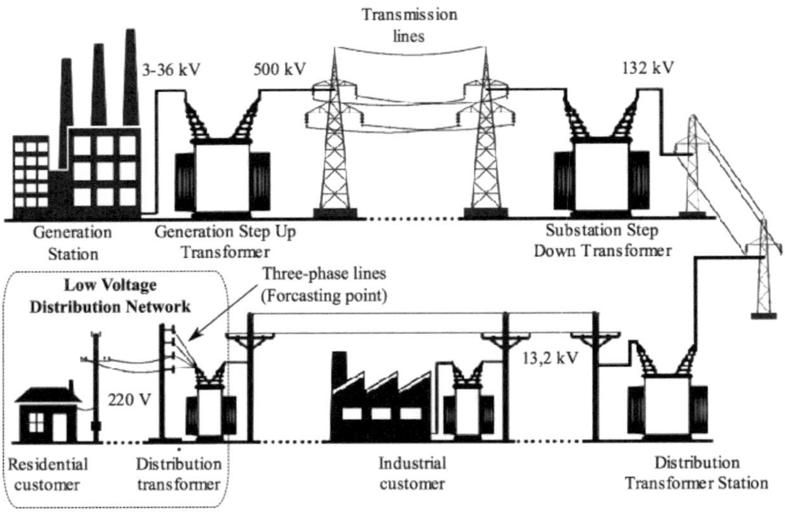

Figure 2.9. *The electric power distribution system.*

Source: https://www.researchgate.net/profile/Victor-Jimenez-5/publication/321025503/figure/fig2/AS:560093718876160@1510548156567/The-electric-power-distribution-system.png.

2.3.1. Distribution Networks and Their Configuration

The distribution network topology plays a pivotal role in the reliability and efficiency of power systems. The networks transmit electricity from customers to substations, forming the final stage of energy supply. Distribution network topology forms part of the fundamental constituents of these networks that are capable of significantly influencing parameters such as load balancing, loss minimization, and service reliability. There are a number of configurations, including radial, loop, and meshed systems, each with benefits and shortcomings. Radial configurations, for instance, are inexpensive and simple but result in all downstream customers suffering from a point failure. Loop and meshed networks provide greater reliability with several paths of electricity flow, hence mounting resistance against disturbances. This versatility is crucial, particularly as electricity use rises and becomes more complex in modern urban settings, demanding innovative approaches to distribution network planning. As distribution networks develop, embedding sophisticated technologies becomes a focal point in their design. Advanced smart grid technologies enhance monitoring, automation, and real-time analysis functionality, transforming conventional distribution networks into smart ones. Smart meters, auto distribution management systems, and communication interfaces allow resources to be allocated, and loads managed effectively, responding rapidly to fluctuating patterns of consumption. These types of adjustments permit operating efficiencies and enable utilities to improve grid stability management with accompanying support of the integration of renewable energy sources like solar and wind. Further, the implementation of concepts such as Massive MIMO, characterized by high-capacity antennas and precise signal processing, expands the scope for distributed communication in power systems, eventually leading to more interaction between distributed energy resources and the grid (Björnson et al., 2019).

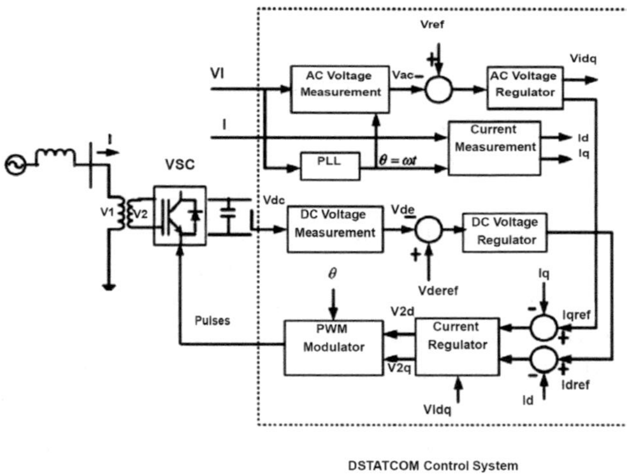

Figure 2.10. *System configuration of a distribution network.*

Source: https://www.researchgate.net/profile/Amin-Noshadi/publication/269219659/figure/fig2/AS:295208173490176@1447394521288/System-configuration-of-a-distribution-network.png.

Ongoing innovation in technologies is a progressive re-mapping of distribution networks in terms of prevailing energy requirements. Functional needs also play a significant role in determining the layout of distribution networks. As these networks expand to accommodate increasing energy demands, the threat of cyber-attack and industrial control system vulnerability should not be overlooked. Efficient inventory management of networked equipment is essential to support operating integrity and integrity in distribution networks. Passive network monitoring methods can prove to be effective in assessing vulnerabilities without generating interfering network traffic, thus sustaining performance in critical infrastructure (Auffret et al., 2019). Through techniques such as Media Access Control address-based identification, utilities can maintain extensive information about devices connected, providing for timely vulnerability scanning. Not only does this multi-faceted approach to configuration strengthen distribution networks but also meets evolving regulatory standards, required to ensure the sustaining of consumer confidence in an ever-more electrified world (Figure 2.10).

2.3.2. Role of Substations in Power Distribution

Substations are strategic locations in the power distribution network, facilitating conversion, regulation, and supply of electric power from the transmission lines to local networks. Performing primarily as step-down substations to step down high-voltage power generated at substations to lower voltages that are suitable for end-use by consumers, substations also provide other services such as voltage regulation and circuit protection. The integration of innovative technology within substation operation has significantly enhanced their efficiency and reliability. More sophisticated capabilities such as remote monitoring and automatic transfer are now common, facilitating real-time adjustments to power flow and reducing outage duration. Such an example is the evolution of substations into Smart Grids, where automation and communication advancements facilitate better energy distribution alongside feeding back up renewable energy sources. As technological advancements continue, substations will increasingly play a pivotal role in determining the nature of power systems today, ensuring efficient energy supply to consumers. The integration of renewable sources of energy into the power distribution system has positioned substations in a more dynamic role within power systems. With the increased prevalence of distributed generation of energy, substations are not only needed to execute heavy voltage transformation but are also needed to enable efficient integration and control of such resources. This necessitates sophisticated control schemes with the capability to balance the output of various renewable sources like solar and wind power, which are inherently variable. Substations, therefore, must serve a dual purpose: while they manage energy distribution from traditional sources, they must be able to manage stable interfaces for the intermittent energy supplies offered by renewables. Such a transition phase in managing energy underscores the requirement for smart technologies in substations that enable load management and optimize energy flow. These advances highlight the growing need for substations to continue evolving in order to support a more sustainable and secure energy future, thus maximizing their role in power distribution. The rural distribution grid infrastructure is another feature of how substations adapt to the power needs of the contemporary world. In rural areas, where energy supply is usually limited, sophisticated substations can leverage state-of-the-art power electronics to maximize energy storage and delivery. By means of the amalgamation of telecommunications and advanced management systems, such as decentralized SCADA systems, substations are capable of significantly enhancing the operational efficiency of rural grids. Besides making the energy distribution function smarter through automation, this smart infrastructure also allows for enhanced management of demand-supply dynamics in these geographically dispersed areas. The innovations realized in initiatives such as the European Smart Rural Grid project show a real effect of such new architectures. They highlight the key contributions substations can make in bringing rural energy systems into the modern age and integrating distributed energy resources effectively, demonstrating their multifaceted value in the overall context of power system basics (Figure 2.11).

Figure 2.11. *The role of transformer substations in power distribution.*

Source:https://utbtransformers.com/wp-content/uploads/2024/05/two-men-in-hard-hats-standing-in-front-of-an-electrical-substation-with-a-transformer-power-distribution-equipment-in-background.jpg.

2.3.3. Smart Grids and Modern Distribution Techniques

The move from traditional power grids to Smart Grids is a paradigm shift in electricity distribution and utilization. Unlike their predecessors, Smart Grids facilitate two-way communication between consumers and energy providers to allow for greater cooperation and efficiency in energy management. This transformation is significant because the challenges of energy demand fluctuations, outdated infrastructure, and ecological concerns cannot be resolved appropriately by traditional grids. The integration of high-tech technology into grid systems not only enhances the delivery of energy but also optimizes maintenance and decision-making activities. As noted in the discussion on modern energy system development, existing power structures are weighed down by severe limitations that Smart Grids are set to alleviate through improved informational communication as well as linking (N/A, 2010). Executing sophisticated distribution techniques into Smart Grids is dependent on tight integration of Internet of Things (IoT) technologies. These technologies, exemplified in smart meters, sensors, and automation equipment, enable real-time monitoring and control of electricity distribution throughout the whole grid network. Through the use of IoT, utilities can reap and analyze vast amounts of data, enabling dynamic responses based on consumption levels and grid operational metrics. Moreover, the connectivity enhances reliability and defense against potential outages, a growing concern in today's power systems. IoT integration enables not only smarter grid but also facilitates consumers in making intelligent energy choices, thus leading to energy efficiency and conservation (N/A, 2010). The ability to communicate and respond in real-time is a drastic change in the delivery of energy. In considering the future of Smart Grids, it is essential to tackle existing innovations in distribution techniques and challenges that are likely to emanate. While the technological foundations of Smart Grids keep advancing at a fastened pace, issues such as interoperability, data management, and cybersecurity remain major concerns for stakeholders. In countering these challenges, an inclusive strategy that confronts regulatory frameworks, industry standards, and consumer engagement is paramount. Recent research still examines potential innovations in load balancing, distributed power supplies, and demand response strategies. According to broad reviews of IoT-aided Smart Grid systems, current trends will also significantly influence future innovations in this field, creating wiser and greener ways to energy (Copeland et al., 2019). Ultimately, the success of Smart Grids will depend on how well technological innovations match policy and social imperatives to create a robust and efficient energy network (Figure 2.12).

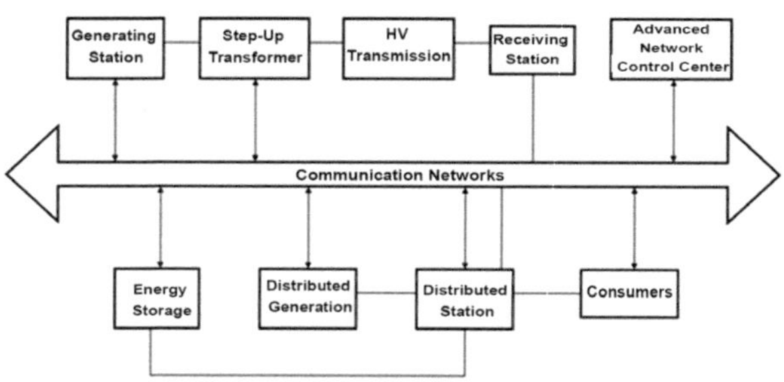

Figure 2.12. *Defining a smart grid.*

Source: https://www.tutorialspoint.com/smart-grid-technology/images/what-is-smart-grid.jpg.

2.4. POWER SYSTEM PROTECTION

Dependable operation of electrical grids relies greatly on efficient power system protection. The primary purpose of protection systems is to rapidly detect faults and disconnect faulty parts, thereby lessening equipment damage and sustaining stability across the system. Smart sensors, as well as monitoring systems, are increasingly employed to enhance protection strategies. By virtue of utilizing real-time data, utilities can make informed decisions regarding intervention needed, which has an immediate impact on operational efficiency. An integral component of this process is understanding the phenomenon of partial discharge, which is often indicative of aged or degraded high-voltage equipment. Research that takes the nature of partial discharge under realistic conditions, as noted in research conducted at Southampton, demonstrates how having a complete understanding of the nature of such events can guide improved maintenance programs and maximize asset life, to support a sound energy infrastructure (Hao et al., 2011).

Figure 2.13. *Power system protection.*

Source: https://egycet.org/wp-content/uploads/2022/02/power-system-protection.jpg.

Moreover, the use of wireless sensor networks (WSN) is transforming the area of power system protection. WSNs enable remote monitoring and data collection, providing operators with the information required without risking operations. Having the capability to significantly reduce response time to significant events, this technology can enable speedy response and minimize potential damages. Implementing WSNs brings with it a host of security risks, particularly with regard to unauthorized access and data integrity. These vulnerabilities must be mitigated to ensure that the worth of enhanced connectivity is realized without sacrificing system security. With the use of multilevel security measures, utilities can defend against potential intrusions, hence safeguarding the integrity of their infrastructure for power protection as public and private sectors further invest in smart grid (Delic et al., 2006). Lastly, power system protection is an ever-developing field that balances traditional methods with existing technology improvements in order to better deliver reliability and security. The study of partial discharge events provides extremely valuable data that can be used to guide maintenance practices so that aged equipment is adequately monitored and addressed. Additionally, the incorporation of wireless sensor networks is an excellent progress in power system capability to dynamically respond to real-time data. Though security issues are intrinsic with such technologies, the global comprehension of security can offer security provisions to ensure such innovations are advantageous to the global performance and resilience of power systems. Such twin focus on preventive maintenance and resilient protective means will be the backbone of future innovation in the realms of power system fundamentals to more sustainable energy solutions (Figure 2.13).

2.4.1. Importance of Protection Systems in Power Networks

The protective systems in power networks are key mechanisms that are designed to minimize the risks of electrical faults and failures. They are pivotal elements that ensure the reliability and stability of the electric supply, hence safeguarding equipment, and personnel against potentially catastrophic events. Effective protection devices not only sense faults such as overloads or short circuits but also enable immediate disconnection of engaged elements in the network to avoid large-scale outages. When well-placed, such systems improve overall power system resilience that is increasingly critical with increasing demands for uninterrupted power in most industries. In addition, integration of sensor networks contributes to the above features by allowing immediate data harvesting and therefore better decision-making during emergency situations (Cleary et al., 2013). Moreover, protection systems become increasingly important as the complexity in modern power grids increases, where they include numerous varying technologies, including renewable energy sources. With the evolving energy landscape, the systems must be adjusted to take in the unique challenges of distributed generation, storage, and intermittent energy supplies. The wind and solar sources case, for example, is providing the sort of fluctuation that could only be accommodated by high-end protection schemes to maintain operation stability. Correct application of protection schemes cancels out the threat of the fluctuations and prevents equipment failure or damage, all for the safety of the electrical grid. This adaptability not only protects physical infrastructures but also enforces the overall trust and belief of customers in the ability of power grids to offer sustainable energy solutions. As a secondary role to safeguarding physical infrastructure, protection systems are significant in maintaining the economic efficiency of power grids. By reducing the frequency of blackouts as well as their durations, these systems decrease operational disruptions and the associated costs. For instance, effective fault detection and isolation schemes can significantly minimize the time taken to return supply by maintenance teams, hence improving customers' satisfaction and minimizing economic loss for electricity firms. Moreover, since power systems are increasingly becoming interconnected on the international platform, the need for robust protection schemes is more critical. Ignoring such protection requirements could result in serious economic implications both locally and globally, making it imperative to implement comprehensive protective interventions that are in line with global standards (Cameron et al., 2001) (Figure 2.14).

Figure 2.14. *The importance of protection systems in power networks.*

Source:https://scontent-del1-2.xx.fbcdn.net/v/t39.30808-6/481164982_649567420797400_3585929081282030734_n.jpg?_nc_cat=100&ccb=1-7&_nc_sid=833d8c&_nc_ohc=kRGiGgOZ9G IQ7kNvwHl-jaj1&_nc_oc=Adn0-APvgqJ_6aUxjuhi6hw8Fs1A0IbluHC4F0pYbxN2F4XSbgQDJQEivhQzZBANdiw&_nc_zt=23&_nc_ht=scontent-del1-2.xx&_nc_gid=Rfae9rbNVihwybyjujOYfw&oh=00_AfWs4J4GYOBBhJ2p_oppNMhkENFUZEQoDfd-voYXF4cTp-A&oe=68B879A6.

2.4.2. Types of Protective Devices (Relays, Circuit Breakers)

The function of protective devices in power systems cannot be exaggerated as they have the responsibility of ensuring the stability and security of electrical networks. Protective relays are employed to detect abnormal conditions such as overloads, short circuits, or equipment failure as the primary protection measure. Protective devices function by constantly monitoring electrical parameters and cross-referencing them with pre-calibrated values. When a fault occurs, relays drive a response action, typically the tripping of circuit breakers, which disconnects affected parts from the power grid. This synchronized effort avoids the danger of equipment damage and reduces the potential for prolonged outages. Recent advances in technology, such as those in microprocessor-based relay development, have enhanced the performance and functionality of such protective devices, for instance, through initiatives at Cal Poly San Luis Obispo to enhance the preparedness of engineering students for industry challenges (Kurth et al., 2019). Alongside relays, circuit breakers are fundamental devices that provide the physical means of interrupting electrical flow under faulty conditions. Unlike conventional fuses that must be replaced after use, circuit breakers can be re-energized and reclosed once a fault has been cleared. This makes them more efficient and versatile at handling transient faults that may occur in power systems. Various types of circuit breakers including air-insulated, oil-insulated, and vacuum types exist, each suited for specific voltage levels and conditions. Their technology provides quick loss of power to prevent extensive damage. For instance, in DC distribution systems, advanced circuit breaker technology can effectively handle faults and ensure maximum operating reliability of vital services (Khanlar et al., 2018). Power systems render themselves stronger against disturbances resulting from external and internal sources through the integration of these protection devices. The evolving power system environment demands ongoing emphasis on learning and implementing capable protective measures. The incorporation of equipment such as circuit breakers and relays is not a matter of technical difficulty but one that impacts the overall safety and efficiency of the infrastructures. Training programs complemented by practical lab-based exercises, including the electric power protection lab at Cal Poly, provide an environment where students can engage directly with such technology. Training to work and test microprocessor-based relays in combination with typical circuit breakers places future engineers in the best position to solve practical challenges in the power sector. While learning about working principles and debugging methods, they are in a better position to assist in creating advanced protection systems that safeguard electrical equipment while pushing the art of power delivery and management (Kurth et al., 2019).

2.4.3. Fault Analysis and Protection Coordination

Effective protection coordination and fault analysis are major contributors to the successful operation of power systems. Fault analysis refers to the systematic observation and testing of abnormal electrical conditions that can lead to power system failure. Engineers must understand the nature and causes of faults—short circuits, ground faults, and open circuits—so they could design and implement appropriate protective measures. By careful fault analysis, power utility providers can respond to potential dangers and enhance the reliability of the system against failure. The basis for designing protection relaying schemes that ensure the reliability of electrical grids, and consequently a stable and reliable supply of power, resides in this forward-thinking strategy. Protection coordination takes this analysis to the next stage with the guarantee that protective devices operate in a particular and harmonized way when the system is experiencing unusual conditions. The primary objective is to keep the failed area of the network separate from the rest of the network while not causing any interruption in service to the healthy areas. Most protection devices, such as circuit breakers and relays, must work together, often subject to time-current characteristics that dictate when a device will trigger under specific fault conditions. Such coordination is achieved with great attention to many variables, including system configuration, fault currents, and time delay settings for protective devices. Proper protection coordination ultimately reduces the likelihood of prolonged outages and equipment damage, hence public safety, and power system integrity. The importance of fault analysis and protection coordination extends beyond efficiency in operation; it also involves matters of regulatory compliance and the evolving nature of energy systems. As pressures of the 21st century redefine the power business, e.g., through the integration of renewable energy systems, the complexity of managing faults and system protection increases. Designers must respond to such dynamic conditions using advanced analytical tools and techniques capable of anticipating and responding to systemic change. It takes new strategies to address crises driven by new configurations and increasing interdependencies across power systems, as traditional practices would not be sufficient in the wake of current challenges. This development highlights the need for continuous enhancement and innovation in protection coordination and fault analysis to evolve with the demands of modern power systems as well as promote sustainability and resilience (Bezpartochnyi et al., 2019; Lagadec, 2025).

2.5. POWER SYSTEM CONTROL

Effective control of power systems is essential to attaining the stability and reliability of electricity networks. Power system control is primarily aimed at being balanced according to supply and demand for electricity while minimizing service interruptions. AGC and LFC are some of the methods applied in this regard and are geared to work together with the aim of modulating the output from power plants according to real demand. Furthermore, developments in communication technologies have made it possible to design more advanced control systems that can handle massive amounts of data to improve decision-making. These together constitute the foundation of contemporary power system operations, providing efficient transition under peak demand or sudden outages, thereby emphasizing the need for strong control mechanisms towards ensuring power system integrity. The integration of renewable energy sources in conventional power systems poses challenges as well as opportunities for power system control. As the proportion of renewable generation, e.g., wind and solar, rises, conventional control techniques must accommodate the intrinsic variability and volatility of these sources. This necessitates better forecasting techniques and developing advanced grid management systems capable of responding dynamically to fluctuating input from renewable sources. Additionally, the emergence of smart grid technologies brings new regimes of real-time supervision and control, enhancing the efficiency of energy distribution and consumption patterns. Through these advancements, power system control can become more adaptable and robust, maintaining the system in good safety while accommodating cleaner forms of energy. Beyond technical efficiency, implications of power system control also touch upon economic and environmental considerations. Ideal control systems allow for the optimization of energy resources, saving utility and consumer costs. Further, through greater grid reliability, these systems also foster environmental sustainability by allowing a greater percentage of renewable energy sources into the system. The use of data-based understanding presented by new control systems allows for resource optimization, conserving waste and emissions involved in energy production. The integration of sophisticated technologies and strategic control methods, as presented in industry records and academic case studies, highlights the central role that power system control plays not only in the operational realm but also in establishing a more sustainable energy future (N/A, 2016; Rummel et al., 2019).

2.5.1. Overview of Control Systems in Power Networks

Power control systems are at the heart of electrical grid operation and management to ensure stability, reliability, and efficiency. These systems are essential for coordinating every aspect of the power system, including generation, transmission, and distribution networks. Through advanced feedback mechanisms, control systems monitor parameters such as frequency and voltage to allow operators to respond rapidly to disturbances. The interconnectivity among system components necessitates the use of proper control policies, which can include centralized and decentralized systems. As there is a growing need for power and complexity in power systems, the design and optimization of these control systems are more important than ever before, and they form a fundamental part of power system fundamentals. Furthermore, modern advanced control systems leverage new technologies to render power grids more resilient. For instance, sensor network deployment enables real-time gathering of data, which facilitates more informed decision-making and proactive management against prospective disturbances. Autonomous integration of the sensor network not only aids in system performance monitoring but also forms input into environmental sensing applications that are of immense relevance in modern society. The amount of data generated by such sensors is so large that it enables operators to identify trends and anomalies more effectively, securing the grids operational health. Control system development is therefore intrinsically interconnected with technology advancements, demonstrating an unwavering commitment to improve the reliability of power grids and support the shift towards sustainable energy supplies. Beyond stability and dependability, power network control systems are essential for achieving ultimate economic efficiency. By the use of sophisticated algorithms for demand prediction and load balancing, these systems allow utilities to optimize resource utilization, reduce operating costs, and improve efficiency. Other sophisticated control techniques such as model predictive control and adaptive control respond dynamically to fluctuations in system conditions to make sure that energy generation is coordinated with consumption rates. Secondly, with the increasing application of renewable energy resources in power networks, the control systems must respond to the natural variability of the resources. This is not just an issue of controlling generation levels in real time but also requires robust forecasting techniques. Lastly, new control approaches become essential to enable contemporary power grids to keep up with increased consumer demands while remaining environmentally friendly as well as economically sustainable.

2.5.2. Load Frequency Control and Voltage Regulation

Load frequency control (LFC) is a critical power system management tool, which ensures a stable and firm power supply in case of fluctuating demand. LFC systems, through continuous monitoring and control of the generation, maintain system frequency within controlled limits. Synchronous generation and load demand have to be equalized to prevent frequency fluctuations that lead to unstable operation and potential blackouts. Effective LFC policies maximize both the power system reliability and the efficiency of energy utilization. Renewable energy sources pose an additional layer of complexity in terms of variations in LFC because they are intermittent. Therefore, new control policies are necessary to mitigate such variations without compromising the integrity and performance of the system in accordance with the overall goals of modern power management systems as outlined by (Martínez et al., 2015). Voltage regulation complements load frequency control to manage the voltage levels throughout the power distribution system. Stable voltage ensures the functioning of electrical equipment and prevents damage from changes in voltage levels. Voltage regulation techniques often involve transformers, capacitors, and voltage regulators, which work in synergy to compensate for changes due to varying loads and system dynamics. Stringent voltage regulation systems can enhance the overall power quality, enabling the integration of sensitive electronic equipment that demands strict voltage tolerances. Like LFC, the deployment of renewable sources of energy necessitates new voltage regulation strategies, placing emphasis on real-time monitoring and adaptive controls that can respond quickly to dynamically changing conditions. The coordination between load frequency control and voltage control is thus critical in the maintenance of the reliability of modern power systems, signifying interdependency of the system as introduced by Martínez et al. (2014). Coordination between load frequency control and voltage regulation is crucial to the success of power systems, particularly during an era where renewable energy integration remains the trend. Both of these functions, though distinct, complement one another in ensuring frequency stability and voltage regularity, which are the cornerstones of successful power supply. As power systems evolve with growing electrical needs and new technology, research and development into integrated schemes has become a focus point among researchers and engineers. Such integrated schemes use sophisticated algorithms and real-time data analysis, with dynamic adjustment of frequency and voltage values. These systems can substantially alleviate the issues brought about by the intermittency of renewable sources, ensure efficient energy delivery, and enhance grid stability. The integration of new control mechanisms is a reflection of the significance of sound management strategies in realizing the goals of modern power systems, as noted by Martínez et al. (2014, 2015).

2.5.3. Role of SCADA Systems in Power System Management

SCADA systems play a crucial role in managing today's power system as they enable real-time monitoring and control of the electrical grids. SCADA systems enhance the reliability and efficiency of power supply through the provision of detailed information about the grid's operations to the operators. With rising complexity due to renewable sources of energy, SCADA systems enable the merging of the resources into the existing grid system. By virtue of data transmission control from elements of the power system, SCADA systems facilitate an improved response to supply and demand energy oscillations. This is crucial in avoiding the risk associated with outages and equipment malfunction, thus ensuring a stable and secure energy supply. The success of microgrid implementation, particularly in addressing the challenges posed by traditional power distribution networks, underscores the significance of SCADA systems. In situations where there is increased utilization of renewable resources, complexities in management and control reach the forefront. SCADA systems lead to the type of system that can be employed in the development of centralized energy management systems (CEMS) which facilitate coordination and optimization of power flow within microgrids.

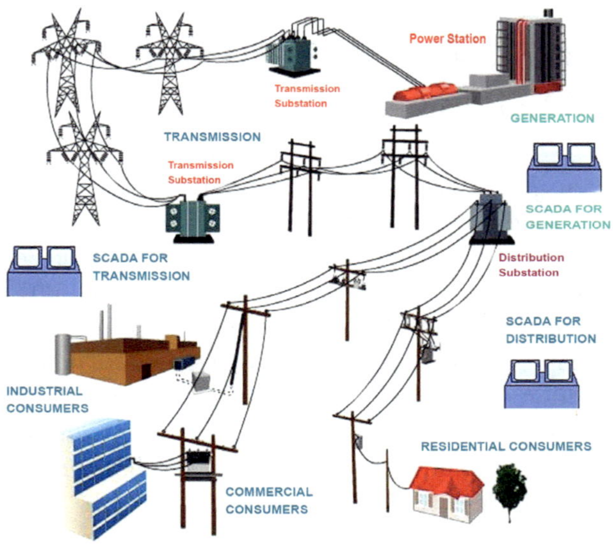

Figure 2.15. *SCADA for electrical power industry.*

Source: https://www.researchgate.net/profile/Khairy-Sayed-2/publication/312000929/figure/fig4/AS:777948504272897@1562488785562/SCADA-for-Electrical-Power-Industry.jpg.

For instance, Sapienza University of Rome's LAMBDA MG testbed laboratory demonstrates a case study of the actual world success facilitated by SCADA technology in demonstrating how a CEMS can deliver balanced energy distributions with decreased reliance on external grid connections (Carni et al., 2020). Not only does such SCADA integration enable microgrids to operate smoothly but also dynamically adapt based on varying network conditions. Furthermore, SCADA implementation is based on the implementation of Smart Grid technologies, which rely on advanced methods of communication for enhanced decision-making in power systems. Powerline Communication (PLC) is particularly notable since it enables real-time data transfer for precise load management and power distribution. PLC enables control devices and intelligent meters to communicate effectively, thus leading to automatic responses to changes in demand or supply (Baig et al., 2012). Through incorporation of intellect into grid operations, SCADA systems render traditional power management proactive and responsive. This evolution not only increases operation efficiency but also enables the transition to a more sustainable energy system, supporting the crucial importance of SCADA in contemporary power system management (Figure 2.15).

2.6. ECONOMIC AND ENVIRONMENTAL CONSIDERATIONS

The alignment of economic and environmental factors is the determinant aspect of modern power systems. The integration of renewable energy, and notably offshore wind power, has been proven as one of the major strategies towards enhancing sustainability while fostering economic growth. The EU set ambitious targets for consumption from renewables in 2007 within energy production, and offshore wind has the potential to play a significant role in reaching these targets. These schemes emphasize the potential of greater electricity generation and greater return on investment (ROI) that ensues from the prolongation of the life of wind turbines. By prolonging their life beyond the typical 25 years, capital expenditures can be distributed over a longer output, minimizing the levelized cost of electricity (LCOE). This multi-faceted strategy not only advances financial viability but also intersects with major environmental objectives through decreased use of fossil fuels and greenhouse gas emissions (Kolios et al., 2015). Secondly, the environmental consequences of energy systems further imply that insights into consumer behavior and its connection with sustainability are more critical than ever. A sustainable energy system not only intends to minimize its environmental impact but also aims to actively engage consumers in energy-saving conduct. Addressing the intricate relationship between consumers' conduct and massive district heating networks reveals a series of prime challenges. These vary from addressing the intricacies of the environmental factors that govern conduct changes to the ethical considerations of the influence of designers on consumer choice. Through the use of imaginative visualization concepts, it is possible to establish an instant connection between consumers and energy providers for making smart decisions regarding heating and water use. This mutual connection not only strengthens consumers but also mobilizes consumption practices towards sustainability goals, finally leading to a more mindful and green society (Balksjö et al., 2013). The economic and environmental dimensions of power systems are thus interconnected, with policy, technology, and consumer engagement implications. When nations and regions struggle to transition towards cleaner energy mixes, the role of innovative design strategies is more crucial than ever. Ensuring consumers are enlightened enough to understand their energy consumptions environmental impact can instigate valuable changes in behavior. In addressing issues of sustainable design in energy use, stakeholders can maximize the performance of district heat systems while creating a culture of conservation. Such considerations extend even beyond financial assessment, necessitating thorough analysis of how economic incentives may coexist with environmental sustainability. Finally, the simultaneous emphasis on economic sustainability and environmental responsibility is essential to creating robust power systems that respond to today's energy needs without compromising ecological integrity.

2.6.1. Cost Analysis of Power Generation and Distribution

Cost evaluation of power production and transmission is key to the determination of the economic viability of various sources of energy as well as their integration into an existing infrastructure. Critical to this analysis is the evaluation of variable renewable energy sources (vRES), which are increasingly becoming relevant to the power system. Integrating vRES, such as solar and wind, impacts the cost of running conventional power generation units, primarily coal and gas. Notably, this growth in vRES capacity has been associated with a 3–22% reduction in hedgeable profit margins for dispatchable power generators, significantly affecting their financial stability and operational planning. Stakeholders who are interested in reconciling cost-effectiveness with ecological responsibility, in accordance with overall climate objectives and commitments, must appreciate these dynamics. Moreover, uncertainty in energy markets like fossil fuel price uncertainty and carbon emissions complicates the finance environment for power generation. VRES relationship with traditional sources of energy leads market players to weigh their investment plans carefully, especially with changing government policies and consumer demands for clean power. As mentioned, the decline in hedgeable profit margins is attributed not solely to the rise of vRES but rather is fueled by the intricate interplay of fuel prices, policy regimes, and electricity and CO_2 allowance market demand. Therefore, conducting thorough cost analyses requires multi-dimensional consideration of these diverse and integrated factors to better outline the cost of energy in the future and possible investment. Lastly, the implications of energy efficient technologies in the context of power generation and supply cannot be overlooked. The use of energy-efficient systems on industrial structures is an example of a militant move towards reducing overall costs for energy consumption and wastage. Reducing the cost aside, retrofitting also meshes with sustainability objectives that are increasingly becoming necessary in the context of modern energy discourse. Evaluation of the cost-effectiveness of various energy-saving technologies requires an understanding of both upfront capital expenditure and eventual operating costs. By focusing on improved efficiency, stakeholders are in a position to generate maximum profitability while, concurrently, assisting in the resolution of a declining carbon footprint, illustrating the manner in which prudent investment in technology can yield economic and environmental benefits in the evolving pattern of power systems.

2.6.2. Environmental Impact of Power Systems

The power system's environmental impact is multifold, including land use effects, air and water pollution, and biodiversity. Modern approaches to energy generation, particularly fossil fuel-fired power plants, are significant contributors to greenhouse gas emissions, leading to global warming. These emissions, primarily carbon dioxide, result from the burning process, which not only reduces air quality but also exerts enormous pressure on ecosystems. Moreover, land use conversions for power generation plants interfere with natural ecosystems, endangering fauna and lowering biodiversity. Thus, a serious analysis of the environmental impact related to various power systems is needed to support the shift towards cleaner energy practices in the context of power system foundations and environmental management (Fisher et al., 2011). Renewable energy sources have gained increasing focus in recent decades because of their capability to offset the harmful impacts of conventional power generation systems on the environment. Solar, wind, and hydropower offer alternatives that have the potential to significantly reduce carbon emissions and decrease the dependence on fossil fuel. But this transition to these clean processes is not without its environmental limitations too. For example, the construction of wind farm projects is able to disrupt indigenous birdlife and alter land usage patterns. Further, huge solar panels occupy a great amount of area, potentially resulting in habitat loss. So, even if renewables are a critical response to climate change, their installation must be undertaken in a manner so as not to unleash unintended ecological effects (Ang et al., 2017). Balancing these against one another is the biggest challenge for policymakers and energy planners. The application of emerging technologies in enhancing power system efficiency is also included in environmental impact reduction. Energy-efficient retrofitting of industrial facilities is one of the methods emphasizing the necessity of existing infrastructure optimization (Fisher et al., 2011). Furthermore, advances in energy production and management, such as developing high-efficiency thermodynamic cycles and biomimetic materials, can play a significant role in hindering undesirable resource utilization. For instance, improved water desalination technologies, powered by renewable technology, can help decrease the overall energy needed for purification to reduce related environmental pressures in freshwater supply and energy production. Therefore, technological innovations' adoption will be central to leading power systems towards more sustainable behaviors that not only satisfy energy demands but also guarantee environmentally friendly outcomes (Ang et al., 2017).

2.6.3. Policies and Regulations Affecting Power Systems

The regulation framework of power systems is largely driven by national and international policies that promote efficiency, security, and environmental conservation. These policies are legislated by governments to guarantee the complex interplay between energy supply, demand, and customer safety. The regulatory bodies often provide standards that dictate how power systems can operate, influencing elements such as grid reliability, energy price, and renewable energy integration. Specific areas of emphasis include the encouragement of competition within energy markets and the creation of incentives for clean technology. Consequently, the power system landscape continually undergoes change, responding to policy and consumer fluctuations. The interplay between innovation encouragement and grid stability demonstrates regulators' challenges in adjusting to an evolving energy landscape. The energy sector experience of the trend of liberalization and deregulation points towards the importance of well-designed policies. While the take-up of capital account liberalization by various countries has been a general trend, evidence has emerged that they have destabilizing consequences unless well-regulated control is applied to them. These findings suggest that an integrated approach is needed, whereby capital account liberalization is succeeded by enhancing macroeconomic stability and sound internal financial systems (Daianu et al., 2002). Policymakers are therefore forced to accord priority to phased implementation of reform measures, with needed structural reforms being carried out in order to bring about transitions to more open markets. This specific sequence of reforms is important not merely to enhance market efficiency, but to ensure consumers and promote sustainable energy practices globally. Regulation responses to commodity price volatility therefore highlight the vital nexus between power grids and overall economic policy. Recent crises have revealed that commodity price volatility, particularly in energy markets, has far-reaching spillovers on economic stability as well as food security (Chadwick et al., 2017). Governments have begun to impose regulation to curb excessive speculation in the derivative markets in response, an indicator of growing realization of the need for policies to stabilize energy prices. Certain regulatory measures may, however, not capture the roots of price volatility, and thus the call for wider reforms emerges. In addressing the improvement of a greater understanding of the influences of market dynamics on base prices, stakeholders would be in good position to fashion regulations for safeguarding both the energy sector and the general economic landscape. The evolving dynamics between policy, regulation, and market behavior continue to present significant challenges in effectively managing power systems.

SUMMARY

In capturing the complexity and inherent nature of power systems, it is evident that understanding the underlying principles is crucial to practitioners as well as policymakers. The intricate interaction between various elements, e.g., generation, transmission, and distribution networks, plays a central role in sustaining system reliability and efficiency. As energy needs rise in direct proportion with technological growth, stakeholders have to remain proactive in addressing the challenges posed by the aging infrastructure and marrying renewable sources of energy. Identification of these forces serves not only to guide strategic choice but also aligns with the greater aims of sustainability and innovation in the power sector. By placing these fundamental principles first, the future power system can be addressed with improved vision and preparedness. In addition, research on specific emerging trends of power systems, such as smart grid technologies and distributed power resources, identifies opportunities and issues that are on the horizon. These emerging technologies hold the potential for enhancing system resilience and operating efficiency but also necessitate a rethinking of regulatory frameworks and market design. The underlying importance of cybersecurity and data management cannot be overstated, with increased interconnectivity of systems exposing the market to more cyber-related risks. It is imperative for industry leaders to aggressively seek strategies that include technological advancements as well as risk mitigation to drive a better future. The overall strategy recognizes the necessity for continuous evolution of power system fundamentals as determined by market conditions as well as advancements in technology. Lastly, this exploration of the fundamentals of power systems in summary highlights the necessity of collaboration between multiple types of stakeholders, including government, private enterprise, and consumers. Achieving a balanced drive toward a more sustainable energy future will require work with regards to mismatches in access to energy and education/workforce development. Through the strengthening of the connections between practice and theory, the power system discussion can facilitate an enhanced and more robust energy environment. Thus, as the industry keeps developing, ongoing study and debate are paramount in solving the complexity that accompanies power systems as well as fostering an atmosphere encouraging innovation and growth. To this end, considerations of economic fundamentals also further reinforce the call for flexibility to offset macroeconomic forces and systemic factors, as noted in currency crisis and market volatility research studies (Hunter et al., 2013; Moore et al., 2009). One beginning point for power systems is electricity generation, transmission, and distribution, focusing on the fundamental operational phases in power. Generation typically involves the conversion of disparate energy sources—such as fossil fuels, nuclear energy, and renewables—into electricity. Because the sources differ in terms of their characteristics, the balance between them is vital to meet demand and ensure sustainability. Moreover, the electricity must be transmitted over long distances by a network of high-voltage power transmission lines for it to be efficient and minimize energy losses. Such interconnection is symptomatic of the basic necessity of the availability and resilience of infrastructure supporting firm power supply across wide geographic regions. According to Blass et al. (2007), effective management of such processes is critical to maximize energy consumption and achieve strategic effects in power system planning. The intricacies of power systems do not stop at generation and transmission but extend to the distribution phase, where electricity is delivered to end-users. Distribution phase requires a strong distribution network able to absorb fluctuations in load and provide voltage stability. Distribution systems are mostly segregated into primary and secondary circuits with varying functions in the supply of power to domestic, business, and industrial consumers. Effective distribution management not only increases consumer satisfaction but also contributes to the general efficiency of the power system. As noted in other related literature, the increasing penetration of renewable energy sources necessitates advanced distribution methods, such as smart grids, which facilitate real-time monitoring of information and quick response to outages and shifts in demand (Lazhentsev et al., 2015). These advances are a significant combination of technology and traditional systems for advancement of overall performance. Stability and reliability of power systems are of utmost significance in achieving operational superiority. Stability is an indicator of the ability of the system to remain in balance in the event of disturbances such as sudden changes in load or generation. Reliability is an indicator of the ability of the systems to supply electricity reliably and uninterruptedly. Various techniques, including contingency analysis and fault detection, are employed to analyze and improve the stability and reliability of electricity grids. Moreover, regulatory structures typically propel these endeavors through guaranteeing availability of resources and robustness against potential breakdowns. With the integration of new management principles and technology, the electricity sector can enhance its systems against adversity and enhance sustainability performance. This combined view emphasizes the essence of interdisciplinarity in the underlying foundations of power systems, triggering the pursuit of future research in energy transition activities due to the evolving global energy industry.

REFERENCES

1. Akrivopoulos, O., Antoniou, A., Chatzigiannakis, I., Tselios, N., et al. (2017). On the deployment of healthcare applications over fog computing infrastructure. https://core.ac.uk/download/98343526.pdf (accessed on 30th September 2025).
2. Ang, L., Burhan, M., Ng, K. C., Shahzad, M. W., et al. (2017). Energy-water-environment nexus underpinning future desalination sustainability. https://core.ac.uk/download/287613577.pdf (accessed on 30th September 2025).
3. Auffret, P., Bilodeau, O., LaPorte, D., Kollman, E., Caselli, M., Hadžiomerović, D., Zambon, E., et al. (2019). Efficient passive ICS device discovery and identification by MAC address correlation. http://arxiv.org/abs/1904.04271 (accessed on 30th September 2025).
4. Baig, S., & Yazdani, J. (2012). Performance analysis of discrete wavelet multitone transceiver for narrowband PLC in smart grid. https://core.ac.uk/download/9637097.pdf (accessed on 30th September 2025).
5. Balksjö, T., de Jong, A., & Katzeff, C. (2013). Challenges in energy awareness: A Swedish case of heating consumption in households. https://core.ac.uk/download/11435470.pdf (accessed on 30th September 2025).
6. Bertsch, V., & Spodniak, P. (2017). Determinants of power spreads in electricity futures markets: A multinational analysis. *ESRI WP580*, December 2017. https://core.ac.uk/download/286054622.pdf (accessed on 30th September 2025).
7. Bertsch, V., Devine, M., & Spodniak, P. (2018). The profitability of energy storage in European electricity markets. *ESRI WP605*, December 2018. https://core.ac.uk/download/269020276.pdf (accessed on 30th September 2025).
8. Bezpartochnyi, M., Britchenko, I., & Tkachenko, S. (2019). Leading methods for promoting finished product quality. https://core.ac.uk/download/305120435.pdf (accessed on 30th September 2025).
9. Björnson, E., Hoydis, J., Marzetta, T. L., Sanguinetti, L., et al. (2019). Massive MIMO is a reality -- What is next? Five promising research directions for antenna arrays. http://arxiv.org/abs/1902.07678 (accessed on 30th September 2025).
10. Blass, E. (2007). Professional learning and work-based learning: Divergence in rhetoric, convergence in reality. https://core.ac.uk/download/1638646.pdf (accessed on 30th September 2025).

11. Bliss, D. W. (2014). Cooperative radar and communications signalling: The estimation and information theory odd couple. http://arxiv.org/abs/1403.1476 (accessed on 30th September 2025).
12. Bordo, M. D. (2025). Growing up to financial stability. https://core.ac.uk/download/pdf/6586430.pdf (accessed on 30th September 2025).
13. Burt, G. M., Finney, S. J., & Roscoe, A. J. (2011). Trade-offs between AC power quality and DC bus ripple for 3-phase 3-wire inverter-connected devices within microgrids. https://core.ac.uk/download/9033319.pdf (accessed on 30th September 2025).
14. Bylykbashi, K., Capata, R., & Testa, F. (2016). A model proposal for the electric energy valorization in a PV power plant equipped with CAES system. https://core.ac.uk/download/74318484.pdf (accessed on 30th September 2025).
15. Cameron, G., Cornfield, D., Freymond, N. C., Palmer, S., et al. (2001). Positive possibilities for child and family welfare: Options for expanding the Anglo-American child protection paradigm. https://core.ac.uk/download/157834832.pdf (accessed on 30th September 2025).
16. Carni, D. L., Kermani, M., Manzo, F., Martirano, L., et al. (2020). A nearly zero-energy microgrid testbed laboratory: Centralized control strategy based on SCADA system. https://core.ac.uk/download/323192286.pdf (accessed on 30th September 2025).
17. Chadwick, A. (2017). Regulating excessive speculation: Commodity derivatives and the global food crisis. https://core.ac.uk/download/111558324.pdf (accessed on 30th September 2025).
18. Cleary, J., Collins, F., Diamond, D., Fay, D., et al. (2013). Distributed environmental monitoring. https://core.ac.uk/download/11311498.pdf (accessed on 30th September 2025).
19. Copeland, R., Crespi, N., Rehmani, M. H., Saleem, Y., et al. (2019). Internet of Things-aided smart grid: Technologies, architectures, applications, prototypes, and future research directions. http://arxiv.org/abs/1704.08977 (accessed on 30th September 2025).
20. Daianu, D., & Vranceanu, R. (2002). Opening the capital account of transition economies: How much and how fast. http://www.wdi.umich.edu/files/Publications/WorkingPapers/wp511.pdf (accessed on 30th September 2025).
21. Debbah, M., Maso, M., & Vangelista, L. (2013). A distributed approach to interference alignment in OFDM-based two-tiered networks. http://arxiv.org/abs/1302.4785 (accessed on 30th September 2025).

22. Delic, H., Ersoy, C., Onel, T., Onur, E., et al. (2006). Security in wireless sensor networks: Issues and challenges. https://core.ac.uk/download/300359201.pdf (accessed on 30th September 2025).

23. Diamant, E. (2008). I'm sorry to say, but your understanding of image processing fundamentals is absolutely wrong. http://arxiv.org/abs/0808.0056 (accessed on 30th September 2025).

24. Díaz González, F., Gallart, R., Girbau Llistuella, F., Heredero Peris, J., et al. (2018). Smart grid architecture for rural distribution networks: Application to a Spanish pilot network. https://core.ac.uk/download/157811722.pdf (accessed on 30th September 2025).

25. Duong, T. Q., Larsson, E. G., Matthaiou, M., Ngo, H. Q., et al. (2017). On the total energy efficiency of cell-free massive MIMO. https://core.ac.uk/download/144581336.pdf (accessed on 30th September 2025).

26. Fisher, D., Hand, J., Liesen, R., Wilson, E., et al. (2011). Screening of energy efficient technologies for industrial buildings' retrofit. https://core.ac.uk/download/9033399.pdf (accessed on 30th September 2025).

27. Fletcher, S., Longstaff, A. P., Myers, A., Pislaru, C., et al. (2012). Five-axis machine tool condition monitoring using dSPACE real-time system. https://core.ac.uk/download/5224559.pdf (accessed on 30th September 2025).

28. Froome, C., Foster, J., Zhao, J., Wagner, L., Skoofa, L., & Wild, P. (2025). Market and economic modelling of the intelligent grid: End of year report 2009. https://core.ac.uk/download/pdf/6633330.pdf (accessed on 30th September 2025).

29. Hanafi, S. (2015). Kesediaan pelajar dari aspek kemahiran teknikal terhadap pembentukan kebolehkerjaan di Kolej Vokasional Wilayah Selatan. https://core.ac.uk/download/12006836.pdf (accessed on 30th September 2025).

30. Hao, L., Hunter, J. A., Lewin, P. L., Michel, M., et al. (2011). Partial discharge testing of defective three-phase PILC cable under rated conditions.

31. Hunter, J., & Menla Ali, F. (2013). The monetary model of the US Dollar–Japanese Yen exchange rate: An empirical investigation. https://core.ac.uk/download/13641667.pdf (accessed on 30th September 2025).

32. Khanlar, A. (2018). Resistive solid state protective device. https://core.ac.uk/download/217194710.pdf (accessed on 30th September 2025).

33. Kolios, A., & Martinez-Luengo, M. (2015). Failure mode identification and end of life scenarios of offshore wind turbines: A review. https://core.

ac.uk/download/42144154.pdf (accessed on 30th September 2025).

34. Kurth, A., Terre, R., & Tham, V. (2019). Cal Poly EE 518 Electric Power Protection Laboratory. https://core.ac.uk/download/219382079.pdf (accessed on 30th September 2025).

35. Lagadec, P. (2025). A new cosmology of risks and crises: Time for a radical shift in paradigm and practice. https://core.ac.uk/download/pdf/6801653.pdf (accessed on 30th September 2025).

36. Lavrinenko, A. V., Babicheva, M., Babicheva, R., Born, P., Brongersma, M., Cai, W., et al. (2012). Plasmonic modulator optimized by patterning of active layer and tuning permittivity. https://core.ac.uk/download/13797189.pdf (accessed on 30th September 2025).

37. Lazhentsev, V. N. (2015). Theoretical results of research on spatial and territorial development (with examples on the European North of Russia). https://elar.urfu.ru/bitstream/10995/47432/1/r-economy_2015_v1_4_02.pdf (accessed on 30th September 2025).

38. Martínez García, H. (2014). Linear-assisted DC/DC converters with modified current-mode control applied to photovoltaic solar systems. https://core.ac.uk/download/41777805.pdf (accessed on 30th September 2025).

39. Martínez García, H. (2015). Capacitorless DC–DC regulator as a candidate topology for photovoltaic solar facilities. https://core.ac.uk/download/81581474.pdf (accessed on 30th September 2025).

40. Martínez García, H. (2017). Renewable energies course for Energy Engineering students at the Technical University of Catalonia (UPC). https://core.ac.uk/download/157812010.pdf (accessed on 30th September 2025).

41. Mohamed, A. M. (2018). Facial expressions emotional recognition with NAO robot. https://core.ac.uk/download/12005717.pdf (accessed on 30th September 2025).

42. Moore, T., & Wang, P. (2009). Can the persistence of a currency crisis be explained by fundamentals? Markov switching models for exchange market pressure. https://core.ac.uk/download/337511.pdf (accessed on 30th September 2025).

43. N/A. (2010). TechNews digests: Jan–Mar 2010. https://core.ac.uk/download/4151979.pdf (accessed on 30th September 2025).

44. N/A. (2016). *SciTech News*, *70*(4). https://core.ac.uk/download/83049836.pdf (accessed on 30th September 2025).

45. Rummel, K. (2019). The fundamentals of radar with applications to autonomous vehicles. https://core.ac.uk/download/213463684.pdf (accessed on 30th September 2025).
46. Savastano, M. A., & Edwards, S. (2025). Exchange rates in emerging economies: What do we know? What do we need to know? https://core.ac.uk/download/pdf/6822565.pdf (accessed on 30th September 2025).

CHAPTER 3

CHARACTERISTICS OF RENEWABLE ENERGY SOURCES

Contents

INTRODUCTION

The development of alternative energy resources is becoming increasingly necessary in the midst of acute environmental conditions and the urgent need to transition from fossil fuels. Renewable energy is characterized by its natural recharging and comprises solar, wind, hydro, geothermal, and biomass. As the energy demand around the world is growing and global warming accelerates, these sources present a plausible alternative that not only produces power but also reduces greenhouse gas emissions. The intrinsic characteristics of renewable resources are their variability, opportunity for decentralization, and less environmental footprint compared to conventional sources like coal and natural gas. This shift away from fossil fuels serves not only to reduce global warming but also to improve energy security and economic stability through diversification of energy supplies and reducing reliance on imported energy resources. The inherent variability of renewable energy resources presents opportunities and challenges to them integrating into the power system. Variability has been defined as the differences in power production on account of environmental conditions, varying solar radiation to solar panels and intermittent wind speed for wind turbines.

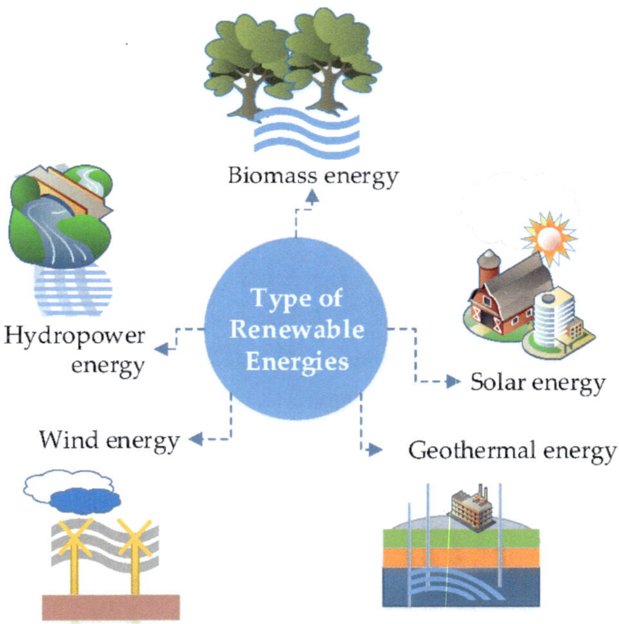

Figure 3.1. *Types of renewable energy resources: Biomass, solar, geothermal, wind, and hydropower.*

Source: https://www.mdpi.com/resources/resources-08-00149/article_deploy/html/images/resources-08-00149-g001.png.

Definition of Renewable Energy Sources

Renewable energy resources are defined by their natural ability to be replenished over a human timescale, and they are primarily derived from continuously renewing processes such as sunlight, wind, water, biomass, and geothermal heat. This definition not only identifies the sustainability of the energy resources but also distinguishes them from finite fossil fuels, which are exhausted upon use and extraction. Renewable energy contributes to a portfolio of energy diversification that minimizes environmental degradation most commonly associated with traditional energy production. The classification of renewable energy resources extends beyond their immediacy of replenishment to include their capacities, technologies, and applications that are influential in shifting towards a low-carbon economy. Information regarding the types of renewable energy sources involves understanding their diverse nature and functionality, including the inherent variability inherent in weather and climate. For instance, solar and wind are variable power resources with problems relating to reliability and efficiency.

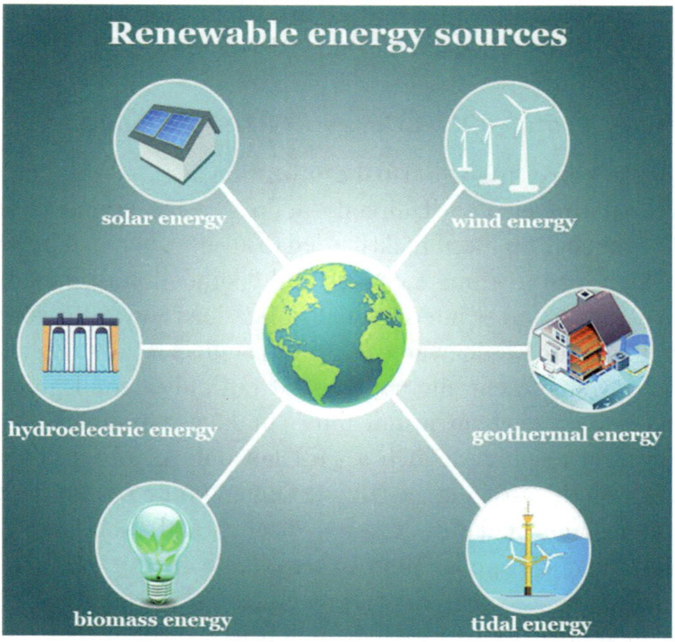

Figure 3.2. *Renewable energy sources.*

Source: https://www.priyamstudycentre.com/wp-content/uploads/2020/11/renewable-energy-sources.jpg.

Their combination is capable of producing more robust energy systems. Hybrid systems combining several renewable energy technologies have the potential to mitigate these problems related to variability because each source is complementary. This new approach not only promotes energy security but also maximizes resource efficiency and utilization, thereby calling for integrated research in energy systems that entail renewable resources that enable such natural limitations and maximize their advantages (Beluco et al., 2019). Additionally, the successful use and integration of renewable energy demand sophisticated management systems, particularly when it comes to small-scale energy systems. As the adoption of renewable technology increases, the ability to support their integration with the existing grid infrastructure becomes increasingly significant. Middleware architectures may be used to address interoperability among various devices, offering a solution to the complexity introduced by heterogeneous energy systems. Middleware architectures facilitate data sharing and resource management necessary for realizing the full capability of diverse renewable sources. Enhanced information and infrastructure development are, therefore, critical in supporting the transition to renewable energy systems to help them operate efficiently and contribute to energy requirements and sustainability aspirations (Kammen et al., 2018) (Figure 3.2).

Importance of Renewable Energy in Modern Society

In contemporary society, transition to the utilization of renewable energy sources is of gigantic significance as it directly addresses the crying environmental problems. The reliance on fossil fuels has led to all-time highs of greenhouse gas emissions, which lead to climate change and degradation of the environment. As nations attempt to meet international obligations toward carbon footprint reductions, stimulation of renewable energies is indispensable. Solar, wind, and hydroelectric power present alternatives that not only curtail greenhouse gas emissions but also bring about sustainability in the use of readily available and naturally renewing resources. Such a shift not only combats climate change but also deals with the growing worldwide demand for cleaner energy sources, consequently pushing societies towards cleaner and more resilient energy methods. Moreover, the applicability of renewable energy extends into socio-economic arenas, namely gender empowerment and equality. As has been documented in more current research, renewable technologies can alleviate burdens disproportionately borne by women, especially in developing regions. For instance, greater access to new renewable energy technologies can reduce women's time and effort requirements for domestic energy needs, thus allowing them to engage more readily in economic and educational programs. Gender analysis incorporated in renewable energy initiatives enhances the effectiveness of such measures by guaranteeing equality and consideration for women's needs. Therefore, as society develops with the implementation of renewable energy, it also progresses gender equity and social empowerment to foster an equal economic base. Apart from environmental and social concerns, renewable energy sources are a major force in the encouragement of local economic development and energy autonomy. The UK experience in Kirklees is a prime illustration of the interconnection of small-scale renewable projects and their socioeconomic effectiveness. Small-scale renewable energy initiatives were discovered to be more financially, socially, and environmentally sustainable than the large-scale ones, despite the latter being presented in numerous instances as being more financially favorable. The call for stakeholder involvement in the formulation of standards for evaluating renewable energy projects comes with this, because local contexts and needs are believed to be given proper consideration. Finally, the promotion of renewable energy not only helps in attaining energy goals but also promotes local economies, hence increasing the overall sustainability and resilience of contemporary society (Figure 3.3).

Figure 3.3. *Importance of renewable energy in modern society.*

Source: https://www.hulkapps.com/cdn/shop/articles/scaled_IMAGE-FILE-UPLOADS-the-integral-role-of-renewable-energy-in-modern-society-212351_990x.jpg?v=1722238405.

3.1. TYPES OF RENEWABLE ENERGY SOURCES

The renewable energy landscape is diverse, comprising many different types that harness natural phenomena to generate energy in a sustainable way. One of the most visible segments is solar power, which utilizes photovoltaic cells to convert sunlight into electricity. The technology has come a long way, becoming more efficient and affordable over the past decades, and is being used more and more for both home and commercial use. In addition, solar thermal systems can also capture and concentrate sunlight to provide heating solutions. This versatility underlines the great potential of solar energy to limit greenhouse gas emissions and fulfill energy independence, in accordance with broader environmental agendas. As long as society's reliance on fossil fuels remains a top issue, the spread of solar technology provides the foundation for a crucial shift toward cleaner energy programs that fulfill growing energy demands without compromising the integrity of the planet. Wind power is also a critical source of renewable energies, utilizing the kinetic energy from winds flowing to generate electricity. Installed onshore or offshore wind turbines transform winds force into mechanical power to drive generators. The energy is highly efficient with the advancement in the design of the turbines leading to greater capacity and reliability.

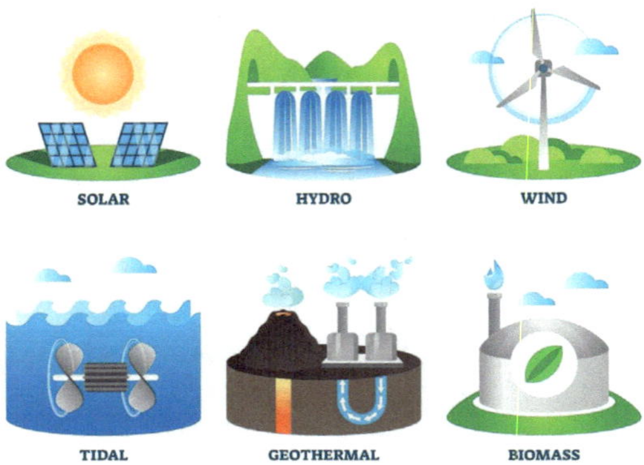

Figure 3.4. *Types of renewable energy sources.*

Source: https://www.inspirecleanenergy.com/_next/image?url=https%3A%2F%2Fimages.prismic.io%2Finspirecleanenergy%2Ftypes-of-renewable-energy-sources.jpeg%3Fauto%3Dcompress%252Cfo rm at%26rect%3D0% 252C0%252C1 840%252C 1036%26w%3D1840%26h%3D1036&w=1920&q=50.

Various research has demonstrated the scalability of wind powers and how it can provide a significant percentage of energy in various regions of the world. However, integrating wind power into the existing energy infrastructure is not straightforward, such as the need to use efficient energy storage systems and grid connection regulations. See the factors, as introduced in studies intended to connect distant energy sources to distribution networks, which highlight the potential and the limitation of wind power as a pillar of the renewable energy mix (Abdullah, 2018). Biomass energy, from organic matter, is an increasingly vital component of the renewable energy mix. It is possible to generate this fuel from agricultural waste, wood by-products, or other biodegradable materials, cutting down on waste and generating energy that will be usable. Being flexible, the biomass can be made to convert to a variety of sources of energy, including heat, electricity, and biofuels, and thus provides a comprehensive solution to fossil fuel use. Further, the application of phase change materials to energy systems, as seen in recent studies, points to new ways of enhancing the energy efficiency of buildings using biomass-based energy (Aguiar et al., 2015). This reflects the wider trend under renewable energies: the drive for sustainability, efficiency, and lowest available environmental impact, which will become paramount to future energy policy (Figure 3.4).

3.1.1. Solar Energy

Solar energy is a revolutionary energy source in renewable energy, by virtue of its abundance and potential for sustainability. The immense amount of energy produced by the sun makes provision for extensive conversion technology, by means of solar cells that convert solar radiation into usable electric power. The technology has phenomenal environmental advantages by virtue of its lack of greenhouse gas emissions when compared to fossil fuels. As solar energy systems become better in efficiency and cost, their installation is gaining momentum globally. The transition shift towards this source of power is a notable step towards efforts to mitigate global climate change, as governments and businesses commit themselves to the installation of solar facilities in order to promote cleaner alternatives. Thus, the growing focus on solar energy indicates an important characteristic of renewable sources: their potential to provide long-term solutions to contemporary energy concerns. Notwithstanding the clear advantages of solar energy, a number of challenges make it less adopted and effective. Significantly, the variability of solar radiation presents a major obstacle to ensuring a continuous supply of energy. Such fluctuations require the incorporation of stable energy storage mechanisms in order to ensure an uninterrupted energy supply, especially in times of low solar exposure. More sophisticated systems that combine battery storage, and other complementary renewable technologies such as wind power, have been proposed to create more resilient energy networks. Hybrid power systems that incorporate solar, wind, and storage systems, for instance, are more efficient, effectively balancing energy consumption and production. Strategic control of energy flow through advanced control systems enhances the dependability of such hybrid schemes, echoing the dynamism of solar energy in the entire renewable environment (Izadian et al., 2014). Further study of solar technology is increasingly enhancing its efficiency and applicability. Recent advances in the design of solar cells, particularly through the use of nanostructures, are making significant steps towards maximizing energy conversion efficiencies. These advances build upon the power of quantum confinement to enhance charge carrier behavior in photovoltaic materials for more effective energy harvesting and utilization. Especially, silicon nanowires are being prepared as promising candidates for next-generation solar cells due to their excellent optoelectronic characteristics, as demonstrated through a series of studies (McCartor et al., 1997). As future technological innovations bring greater efficiency, solar energy will have an even greater role in the renewable energy portfolio. Emphasizing the ongoing expenditure needed in research and development, the advancement of technologies for solar energy illustrates the evolving nature of renewable energy supplies (Figure 3.5).

Figure 3.5. *Solar energy.*

Source: https://greenhomesystems.com/wp-content/uploads/2023/09/blog-cover-photo-98.jpg.

3.1.2. Wind Energy

Wind power has emerged as one of the most promising sources of renewable energy, and it has significantly contributed to the transition towards sustainable energy systems. It is everywhere and cost-effective, making it a favorite among efforts to obtain clean energy alternatives. Wind power harnesses kinetic energy in wind through turbines, converting it into electricity. This process is not only economical but also lessens greenhouse gases as opposed to fossil fuels. Its use, however, is not free from problems. One of the characteristics is the natural intermittence of wind, which has the potential to undermine power generation reliability, and hence timely planning and coordination with other power systems must be ensured in order to give a reliable supply (Izadian et al., 2014). The advancement of advanced technologies in wind power has enabled dramatic advancements in harnessing this source of energy. For instance, modern wind farms have sophisticated control systems that enable optimization of energy production from varying wind speeds. Technologies such as these can optimize wind turbines to operate under optimal conditions even under unpredictable environmental conditions. Wind power can also effectively mitigate the issue of intermittency when combined with solar power systems. This complementary synergy ensures a more reliable power supply, taking advantage of the complementary nature of the solar and wind resources, wherein low wind duration may coincide with high solar radiation, and vice versa (Izadian et al., 2015). Moreover, incorporation of energy storage systems complements the variability of wind energy in modern power systems. Battery storage systems are employed to serve as a buffer for the wind turbine output, which varies, and supply the demand with a stable source. Excess electricity during high wind generation can be stored and utilized during low generation. This system is particularly necessary to supply electricity in micro-grids with over one source of clean energy. With wind, solar, and storage-based architectures, energy providers can enhance grid stability and reduce reliance on traditional sources of energy, ultimately contributing towards a greener future (Izadian et al., 2014) (Figure 3.6).

Figure 3.6. *Wind energy.*

Source: https://renown-electric.com/wp-content/uploads/wind-turbine-500x699.jpg.

3.1.3. Hydropower

Hydropower, a significant component of renewable power resources, exploits the power generated from flowing water, primarily in dams and river currents. It is highly efficient in generating energy, with the rate of conversion potentially hitting 90% when it is channeled into electricity. Its application of flowing water provides the hydropower facility with the potential to provide stable energy supply, especially where there is an abundance of water sources. Apart from that, the systems can also be readily hybridized with other renewable sources because existing studies determine the advantage of hybridization between solar, wind, and hydropower systems. With this, energy generation can better adapt to changing patterns of demand, addressing the intermittency issue of most renewable technologies (Beluco et al., 2019).

Figure 3.7. *Hydropower.*

Source: https://www.innovationnewsnetwork.com/wp-content/uploads/2020/08/Hydropower.jpg.

Hybridization enhances energy security because it also helps provide a smoother transition from fossil fuels to cleaner energy sources. Hydropower plant construction and expansion are central elements of the worldwide shift towards renewable energy. Despite environmental concerns regarding hydropower impacts, particularly on water ecosystems, advances in new technology and engineering approaches have made more environmentally friendly hydropower possible. For instance, incorporation of fish ladders and other mitigation methods is supposed to limit disruption to migratory fish and ensure biodiversity. Additionally, other competing technologies, such as run-of-river hydroelectric power plants, minimize the need for major dams and are still able to provide high energy outputs. Although hydropower currently contributes less than 10% to the global energy mix, as outlined in current literature, its room for growth is still significant due to added investment, technological innovation, and growing knowledge of its contribution towards secure energy supply (Aheicheva et al., 2018). Moreover, hydropower's ability to function as a baseload power source cannot be overlooked. Although the majority of renewables that are weather-dependent, such as solar or wind, cannot offer the steady amount of electricity for as long as they receive enough water flow, hydropower is able to do so as long as there is water flowing. That reliability makes it a cornerstone of a diversified energy mix, particularly for regions that have variable renewable energy production. With energy needs continuing to increase globally, integration of hydropower into larger energy systems may be vital in maintaining economic activity and ensuring adequate energy supply. Optimization of economic efficiency of hydropower systems remains a research focus that has increasingly formed the basis for improved project development emphasizing sustainability while addressing the imperative energy requirements of an evolving world (Aheicheva et al., 2018) (Figure 3.7).

3.2. ENVIRONMENTAL BENEFITS

The environmental benefits of renewable energy technologies are significant in addressing the widespread global concerns about climate change. The utilization of wind, solar, and hydroelectricity serves to reduce greenhouse gas emissions—certainly critical in the war against the rising temperatures and devastating weather patterns characterized by increased flooding, storms, and droughts. The European Union's proactive measures, as embodied in the 6th Environment Action Program, emphasize the need to combat climate change and promote greener energy sources. By reducing carbon dioxide emissions, renewable energy can detoxify emerging economies from fossil fuels for a greener tomorrow. Despite all these technological improvements, challenges still exist regarding penetration of green goods into the market such that current consumer involvement is very low despite general support for environmental goods, showing that intervention by the public would be required in order to enhance adoption rates (Barrett et al., 2013). Switching to renewable energy is based on an acknowledgment not only of environmental need but also of locational benefits. Renewable energy projects can add to air and landscape quality, stimulate biodiversity, and even enable long-term employment opportunities. Studies of public attitudes towards different renewable technologies show differential tastes based on regional location. For instance, in rural society, the residents may be open to moderate environmental impacts from wind farms or biomass power stations as being beneficial to the local economy. Conversely, urban residents are more concerned with maintaining ecological integrity and aesthetic qualities of surroundings, an indication of a difference in environmental trade-off values (Bergmann et al., 2004). Such values indicate the necessity of fine-tuning renewable energy projects to the needs of communities to facilitate increased acceptance and backing. The environmental benefits of renewable resources extend beyond mere emissions reductions to broader ecological sustainability and resilience. The application of renewable technology can lead to significant improvement in local air quality, enhancing public health and reducing healthcare costs stemming from pollution-related diseases. Furthermore, as economies transition towards these cleaner sources of energy, they promote more sustainable land use policies that preserve ecosystems and wildlife habitats. By investing in renewable energy, society undertakes an interdisciplinary approach to environmental stewardship that harmonizes economic development with natural conservation. The cumulative effect of these projects not only addresses near-term climate requirements but provides the foundation for a resilient and environmentally sustainable future, stressing the pivotal role that renewable energy has in shaping an environmentally sustainable future (Barrett et al., 2013; Bergmann et al., 2004) (Figure 3.8).

Figure 3.8. *Environmental benefits of renewable energy.*

Source: https://www.solarunion.com/cms/thumbnails/00/540x303/images/blog/shutterstock_535718161.1).2303160936550.jpg.

3.2.1. Reduction of Greenhouse Gas Emissions

Greenhouse gas emission reduction is one of the top priorities in ending the global climate disaster, as evident in many international and regional environmental policies. One of the top strategies is the transition out of fossil fuels into renewable energy, such as wind, solar, and biomass. This procedure is not only required for climate change prevention but also to ensure sustainable use of energy that can support a better environment. The European Union, having realized the immense role of carbon dioxide emissions in causing climate change, has set tough targets to reduce emissions by 20% by 2020 through promoting sustainable energy infrastructures and green consumer economies. These notwithstanding, the proportion of green products remains low, indicating the need for more public intervention to engender consumer engagement and patronage of renewable energy products, which are critical to greenhouse gas abatement (Barrett et al., 2013). Biomass energy is an example of one such renewable product that has shown promise as a viable alternative to conventional fossil fuels. Used primarily in rural areas with an excess of agricultural waste, biomass can be converted into electricity and heat, contributing significantly to the reduction of greenhouse gases.

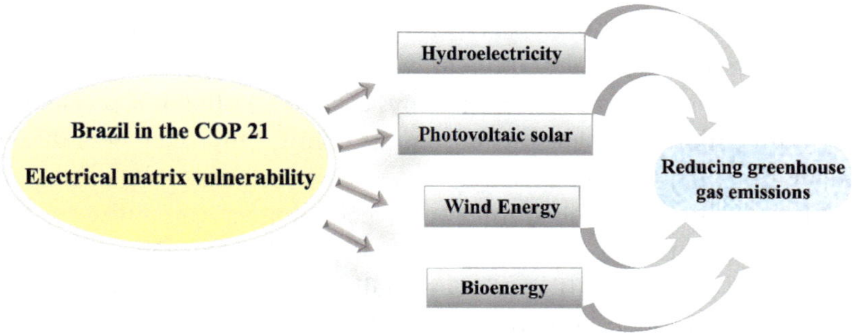

Figure 3.9. *Renewable energy in reducing greenhouse gas emissions in Brazil.*

Source: https://ars.els-cdn.com/content/image/1-s2.0-S2211464520300191-fx1_lrg.jpg.

In utilizing biomass energy systems, such as centralized co-generation facilities, emissions associated with traditional energy production can be significantly minimized. This dual-purpose generation of electricity along with heat not only maximizes the efficiency but also upholds the image of biomass as a green energy source. The comparative evaluation of various biomass supply chains puts forward the economic, environmental, and social dimensions of sustainability in which optimization of such parameters can render biomass energy more efficient in reducing the emission of greenhouse gases (Rentizelas et al., 2014). Policy backgrounds and the overall public attitude also play a vital role in playing their parts in promoting renewable energy technologies. Green consumption goods and services must be promoted to drive market growth in renewable energy industries. However, as can be seen from the low market penetration level, there is normally a gap that exists between consumer intentions and actual market practices. Such a gap suggests that additional work must be done to inform and encourage consumers into actively seeking out renewable energy options. By the bridging of this gap, governments can enhance the awareness and involvement of the public, thereby supporting a solid foundation for the application of renewable energy sources globally, further moving the world towards the achievement of greenhouse gas emission reduction (Barrett et al., 2013) (Figure 3.9).

3.2.2. Decreased Air and Water Pollution

The transition to clean power sources represents a pivotal step toward eliminating air and water pollution, fundamentally transforming the environmental landscape. Whereas traditional fossil fuels release heavy amounts of emissions, including sulfur dioxide and nitrogen oxides, renewable technology such as wind, sun, and water generates energy with negligible emissions. This reduction in poisonous emissions is not only essential to human health but also to ecological stability. By the application of clean energy sources, countries can significantly minimize the frequency of toxic pollutants causing smog and respiratory illness. As an illustration, air quality improvements due to the implementation of renewable energy have been witnessed in most parts of the world as a way of demonstrating such technologies' capability to produce cleaner environments and enhancing the quality of life. Furthermore, the shift to renewable energy also affects water quality very significantly. Traditional forms of energy production, particularly oil and coal, are prone to leading to runoff that contaminates water bodies around them with heavy metals and other toxic chemicals. In contrast, the vast majority of renewable energy systems consume little water when in operation and have fewer possibilities for the discharge of contaminants. For example, hydropower, if well regulated, has the potential to maintain regional ecosystems through maintaining river flow and protecting aquatic habitats. These schemes like the National Emission Ceilings Directive (NECD), which demands rigorous reduction quotas of air pollutants in emissions, go hand in hand with the utilization of renewables perfectly, emphasizing the need for green activities contributing towards air as well as water quality. With these initiatives, the role of renewable energy in playing a significant role beyond energy supply towards safeguarding the environment gains primary significance. The reduction in water and air pollutants as seen is evidence for the capability of renewable energy to respond to environmental challenges. Evidence shows a high reduction of pollutants like $PM_{2.5}$ and nitrogen oxides in areas that adopt renewable energy, with fewer related health issues. For instance, in Italy, long-term monitoring of ecosystems indicates remarkable recovery from acidification due to reduced sulfur dioxide emissions, indicating major environmental benefits associated with the application of clean energy technologies. Further, programs set up for monitoring the health and ecosystem effects of pollution, as mandated under such directives like the NECD, play a key role in providing understanding and means to counteract the effects of air pollution. So, the big picture is then straightforward: the application of renewable energy sources not only diminishes pollution but also serves as a catalyst for broader environmental rejuvenation and public health gains, ultimately yielding a sustainable future for all.

3.2.3. Conservation of Natural Resources

Natural resource conservation is critical towards realizing a future that is sustainable in terms of energy, especially as societies increasingly depend on renewable sources of energy. As the economic paradigms evolve, energy demands rise, and this, therefore, amplifies the worlds dependence on fossil fuels whose levels are finite. Renewable energy, such as solar, wind, and hydroelectric power, is critical in battling this unsustainable alliance. Not only do they provide an alternative to conventional energy systems, but they can also reverse environmental degradation. Attempts at achieving maximum economic effectiveness in alternative systems, as given in the example of solar and wind energy production, constitute a significant drive toward a green and sustainable energy future (Aheicheva et al., 2018).

Figure 3.10. *Renewable energy poses challenges for wildlife conservation.*

Source: https://cnr.ncsu.edu/news/wp-content/uploads/sites/10/2019/11/11132021-wind-turbines-unsplash-featured.jpg.

Examining the value of renewable energy programs, researchers have discovered an interlinking of environmental factors and rural quality of life. Different forms of renewable energy technologies, such as onshore and offshore wind farms, are assessed for their impacts on local systems, air quality, and fauna. Evidence indicates that rural communities will balance economic gains versus environmental charges in determining new energy programs, recognizing the long-term employment generation these provide. Conversely, cities are more prone to carry the sensitivity of environmental disturbances, and they give top priority to the preservation of landscapes and wildlife (Bergmann et al., 2004). Such dualism illustrates the necessity of policymakers to balance economic growth and environmental protection to make the effective implementation of conservation policies achievable. Conservation of natural resources is not strictly controlled by energy production but also has a strong relationship with broader social values and interests. As energy systems from renewable sources continue to expand, sustainability concepts need to become institutionalized in local and national decision-making. Education and sensitization campaigns are important in cultivating a sense of appreciation for preserving natural resources, to ensure that communities realize long-term implications of their energy choices. The shift to renewable sources can only be achieved if accompanied by a dedication to sustainable resource management, moving forward practices that preserve ecosystems while meeting energy needs. Ultimately, natural resource preservation is a key component in the successful integration of renewable energy sources, moving ahead towards a more sustainable tomorrow that benefits humans as well as the planet (Figure 3.10).

3.3. ECONOMIC ADVANTAGES

The economic advantages of renewable energy sources are not just mere cost savings, such as job creation, energy independence, and improved energy security. With renewable technologies such as solar and wind technology still evolving, the cost of such technologies is already on par with conventional fossil fuels. Price competitiveness is particularly valuable in emerging economies where energy demands are growing rapidly. In particular, SMRs are a revolutionary solution that benefits from the strengths of blending renewable energy. They have the miniaturized and flexible nature that allows them to be deployed in areas that do not accommodate large grids, thereby increasing access to clean energy in remote locations. Moreover, SMRs' lower capital and shorter construction period can accelerate the establishment of energy infrastructure, which is vital for countries that aim to alleviate energy poverty and enhance economic stability (Black et al., 2015). Apart from immediate economic impacts, renewable energy projects have the capacity to enhance the local economy by opening up employment and further investment. Utilization of renewable energy sources as a component of the energy mix promotes development of technology, which, in turn, creates new avenues of opportunities for industries of manufacturing, installation, and maintenance. For instance, utilization of landfill gas to produce energy, as in Heraklion, Greece, illustrates this quite well. The potential of the generation of electricity with the help of biogas not only provides a profitable return on investment with payback periods ranging from 3.5 to 6 years but also enhances employment opportunities at the local level and minimizes waste (Skoulou et al., 2016). These initiatives not only contribute to the proportion of renewable energy in the energy sector but also facilitate socio-economic development by revitalizing localities and providing sustainable employment opportunities. Furthermore, the green footprints of renewables result in great economic benefits for countries and communities. By saving carbon emissions and preventing climate change, renewables are able to ensure public health and local environments that otherwise go unnoticed in the immediate monetary benefits of traditional energy sources. This long-term avoidance strategy maintains public support for renewable schemes, which in turn promotes more sustainable policies. As seen from SMR and landfill gas use considerations, the long-term yields and financial viability of renewable enterprises can encourage administrations to prioritize clean energy investment. It is essential to consider the long-term economic benefits—ranging from infrastructure enhancement to public well-being—while countries seek sustainable energy futures (Black et al., 2015; Skoulou et al., 2016) (Figure 3.11).

Figure 3.11. *Economic benefits of renewable energy.*

Source: https://www.certrec.com/wp-content/uploads/2024/09/Exploring-the-Economic-Benefits-of-Renewable-Energy-and-Its-Impact-on-Global-Markets-Certrec-1024x636.jpg.

3.3.1. Job Creation in the Renewable Energy Sector

The renewable energy sector is now the leading force of employment generation, transforming local and international economies at their very center. The shift to renewable energy, particularly wind and solar energy, is central to achieving energy self-sufficiency, as well as climate change reduction. With policies in various countries favoring the production of renewable energy, tremendous jobs are generated in various sectors. These include research and development positions, manufacturing, installation, and maintenance work. Rural regions, which tend to be sites for renewable energy projects, can benefit economically as employment numbers grow in these categories, resulting in local community advantages like improved quality of life and enhanced local investment. The research evidence supported by organizations like the Scottish Economic Policy Network indicates that economies are prone to weigh economic gains of such projects in mind, so they sacrifice some aspects of the environment in order to gain employment opportunities (Bergmann et al., 2004). Moreover, the government's reliance on job creation in the sector of renewable energy cannot be overstated. Effective long-term support policies by governments are essential to foster an ecosystem where local value chains can grow and develop, giving rise to local labor markets.

Figure 3.12. *Job creation in the renewable energy sector.*

Source: https://www.canadianbiomassmagazine.ca/wp-content/uploads/2022/09/Renewable-energy-jobs-2022.jpg.

The wind sector, for example, offers proof that investing in clean technologies not only increases economic resilience but also stimulates cross-border trade and investment flows. Furthermore, as elaborated by research on the wind industry, the evolution of sound carbon pricing mechanisms renders renewable energy competitive, further increasing demand for support jobs (Kirkegaard et al., 2025). The intricacy of job creation in this sector points to the need for sustained policies that reduce entry barriers and drive innovation, and which maximize employment opportunities. With increasing opportunities for employment in renewable energy, there is an increasing demand for skilled human resources that can respond to the industry's evolving demands. The industry is increasingly requiring professionals with expertise in engineering, environmental science, and technology, among other fields. Such institutions of learning and vocational training programs are therefore essential to formulating a workforce that is proficient in renewable energy technology. By aligning curricula with the needs of industries, such programs can produce graduates fit for work. Moreover, as the global energy industry is heading toward sustainability, opportunities for international collaboration in workforce development increase. As countries share best practices and create training programs, opportunities for a skilled global workforce in renewable energy will expand, eventually leading to more employment on a global scale while, concurrently, addressing the urgent concerns of climate change and energy security (Figure 3.12).

3.3.2. Decreased Reliance on Fossil Fuels

The transition to less fossil fuel dependency is inherently coupled with greater utilization of renewable energy sources, hence the necessity to counteract climate change. Fossil fuels have dominated the global energy economy for centuries, culminating in soaring greenhouse gas emissions and environmental pollution. The diversification of the energy base through the use of renewable energy types such as solar, wind, and water can be a solution to these effects. In regions like the United States' Interior West, policy frameworks are evolving to spur cleaner energy options that challenge the traditional reliance on fossil fuels. By developing policies to spur investment in renewable energy, stakeholders can facilitate the transition of the economy to a sustainable energy economy while reducing emissions and increasing energy independence simultaneously (Driver et al., 2004). Also, nations around the world are making historic decisions about their energy future, most importantly in regard to investment in renewable infrastructure. Brazil is one of the outstanding examples of a country at a crossroads, weighing the potential for a low-carbon economy against the risk of continued fossil fuel dependence. The coming decade is a crucial time for Brazil, in which investment choices worth about 0.5 trillion U.S. dollars can either solidify carbon-based practices or stimulate development in clean energy industries. Moving away from fossil fuels and towards renewable energy will demand strategic planning and the enactment of strong policies to encourage energy efficiency, innovation, and sustainability. The implications of these decisions extend beyond economic competitiveness and into the broader climate change mitigation paradigm, and they demonstrate the inherent interconnection between energy policy and environmental responsibility (Lucon et al., 2015). Ultimately, reduced reliance upon fossil fuels is not merely an issue of energy supply but is instead an integration of environment, economics, and social factors. The move toward renewables is a crucial shift in energy practice that will make the future more secure and sustainable. Countries choosing to switch to alternative forms of energy will need to overcome many challenges, including putting money into technological innovation, providing regulation assistance, and gaining public approval. Public awareness campaigns and neighborhood communications will play central roles in recalibrating attitudes toward renewable energy as viable alternatives. By addressing these obstacles wisely, societies can make the shift to renewable energy more seamlessly, fueling local job creation and strengthening energy security while making serious progress toward lowering overall emissions and safeguarding the planet for coming generations (Figure 3.13).

Figure 3.13. *Reducing reliance on fossil fuels.*

Source: https://green.org/wp-content/uploads/2024/01/Reducing-Reliance-on-Fossil-Fu.jpg.

3.3.3. Long-Term Cost Savings for Consumers

The long-term cost savings from renewable sources of energy are a significant advantage to consumers making the switch from traditional sources of energy. One of the principal drivers of cost savings is the reduction of recurring fuel costs, particularly evident in solar and wind energy. Renewable sources of energy are free from price volatility that drains consumer wallets compared to fossil fuels. With inputs such as sunlight and wind, which are natural and extremely abundant for free, consumers can stabilize the price of energy in the long term. This predictability eventually leads to stable energy bills, thereby eliminating the economic burden created due to variable pricing in traditional energy markets. Financial innovations such as solar energy service firms are gaining traction to make it easy to switch to renewable energy and optimize consumer savings. For example, as the Baja California Sur proposals identified, such companies can provide installation and maintenance of solar photovoltaic systems at no upfront expense for consumers. The expense is rather offset against saved payments on electricity, with homeowners getting credits on their energy bills ahead of time. Such programs not only generate consumer trust in renewable technology but also create long-term sustainable financial systems that promote mass adoption of renewable energy products, ultimately generating long-term household and community saving. Moreover, good policies such as feed-in tariffs (FiTs) can further make the financial profitability of renewable energy programs while concurrently generating substantial consumers' savings. With the use of a FiT model, governments can inspire investments in clean technologies by making participants sell surplus energy to the grid at fixed rates. Such programs, based on foregone generation costs, can potentially reduce the cost of energy to consumers by a considerable amount. Secondly, such schemes' earnings can be invested back into more renewable schemes, enabling a cycle of cost reduction and maturity of advanced energy technology to ensue, so that the long-term economic benefits are spread sufficiently to the consumer base (Andrea et al., 2011; Bermudez-Contreras et al., 2016).

3.4. TECHNOLOGICAL INNOVATIONS

The realm of renewable energy resources is rapidly evolving and greatly influenced by technological innovations that enable increased efficiency and sustainability. These have played a central role in pushing the energy sector towards sustainable practices. An important consideration in this regard is the use of strategic niche management, which facilitates the emergence of grassroots innovations driven by civil society. Such grassroots innovations are likely to stem from local initiatives in the field of community-owned renewable energy production and energy efficiency measures. Through aggregation of learning from such grassroots efforts, stakeholders have been able to cross-leverage best practices and implement winning models in other contexts. Over time, this process not only makes it easier to start innovative projects but also involves communities in and encourages engagement with the broad sustainability transformation, hence advocating for localized technological interventions as a core characteristic of renewable energy transformation (Hargreaves et al., 2013). Also, the patterns of technological innovation for the instance of renewable energy are very much connected with policy regimes and market forces. Innovative processes such as the development of quality-adjusted patents in renewable technology are influenced considerably by environmental policy and market regulation. The empirical evidence is such that countries with liberalized energy markets and lower barriers to entry have higher levels of green patent production. Policy inducement in this case is an innovation driver, and it suggests that governments are key to determining the incentives for technological innovation. The intersection with policy-conducive environments on top of determinants such as income inequality and current knowledge bases further highlights the complexity of innovation dynamics. These are findings that demonstrate that a favorable policy environment is essential in facilitating an innovation environment, which in turn stimulates the development and dissemination of renewable energy technologies (Nicolli et al., 2025). In addition to that, the integration of technological innovations into renewable energy not only addresses environmental problems but also strengthens social cohesion. Societies involved in renewable energy programs mostly feel empowered and included in the game, mobilizing people's action towards sustainability. Tacit knowledge acquired through such local schemes cannot be appreciated enough; however, it is still hard to transport it to other contexts. Replication of such programs is based on trust, belief, and appropriate networking between actors. Policymakers must ensure that they are proactive in developing conducive environments to foster these elements, which are indispensable for guaranteeing the long-term success of grassroots innovations. In a global society that is increasingly focused on sustainable solutions, technological innovation of renewable energy sources is not only a solution to ecological issues but also a tool for creating resilient active communities in an effort to attain collective sustainability goals (Hargreaves et al., 2013).

3.4.1. Advances in Solar Panel Efficiency

Recent technological advancements in the efficiency of solar panels have transformed the landscape of renewable energy. New technologies such as bifacial solar panels and tandem cell architecture have emerged as key game-changers to explore more energy conversion from inputs. Bifacial panels utilize sunlight from both sides, enabling higher energy production from reflected light. Tandem solar cells, which stack multiple photovoltaic materials, can be more effective than standard silicon-based cells, however. These are not incremental improvements but represent a strategic breakthrough towards the best possible output of solar energy systems. As levels of efficiency rise, the practicability of integrating solar technology into many applications, from residential homes to utility-scale installations, becomes more possible, opening doors to greater use of solar power as a base load source of electricity. Coupled with advances in material science and design, it is critical to integrate intelligent energy management systems for optimizing the performance of solar panels. Smart grid technology enhances the efficiency of monitoring and controlling energy consumption in real-time, ensuring that solar panel efficiency is utilized to the fullest. Such systems can adjust energy usage based on forecasting algorithms, reducing the use of fossil fuels and overall building energy consumption. As described by Suplee et al. (2020), the use of energy-efficient strategies with renewable sources virtually eliminates greenhouse gas emissions and relieves grid stress. With evolving solar technology, these smart systems become an essential addition to progress in the efficiency of solar panels to advance the argument for a system approach in order to achieve maximum benefit from renewable energy measures. Furthermore, experimental research on hybrid energy systems illustrates the on-the-ground implications of advanced solar technology in diverse environments. In a reviewed study via TRNSYS software, the performance of various hybrid systems, such as solar thermal collectors and photovoltaic panels, showed higher efficiency and lower emissions under varied climatic conditions in Spain. The study hints at the applicability of localized energy solutions as well as the potential for locally adapted hybrid systems to respond suitably to respective energy needs. Attention to quantitative records of primary consumption of energy, emissions, and life cycle cost adds a very important dimension to the renewable energy debate. As noted by Rodríguez et al. (2016), the comprehensive approach ensures policymakers, including regulators, and energy service engineers have adequate information at hand in making investment and policy decisions, ultimately driving future efficiency increases in solar panels within the broader sustainable energy practice (Figure 3.14).

Figure 3.14. *Advancing solar panel efficiency.*

Source: https://media.licdn.com/dms/image/v2/D4D12AQFSU1FsIt1uBA/article-cover_image-shrink_720_1280/article-cover_image-shrink_720_1280/0/1714545983947?e=2147483647&v=beta&t=zVIjXP3EV9Mkw5ZFa_G2G-q9qglGp0xqW-GHKmB1958c.

3.4.2. Development of Wind Turbine Technology

The technology development of wind turbines shows significant advancements in the areas of engineering and design, focusing on minimizing damage to the environment while maximizing efficiency and energy harvest. Stringently designed primitive wind turbines that operated at low efficiencies were mostly of limited application to energy needs. However, research and development innovations have allowed the development of advanced models that utilize aerodynamic blades specifically engineered to optimize wind power harvesting. Variable-speed turbines have made it easier to harness more energy from varied wind speeds, and hence they are a departure from the fixed speed. Second, improvements in materials, such as lightweight composites, have enabled the construction of larger capacity turbines with increased capacity, thereby making them more suitable as a primary renewable energy source compared to fossil fuels. Offshore wind technology represents a paradigm change in wind power use, particularly with its ability to harness more consistent and stronger winds that take place offshore. The United Kingdom of late has witnessed a rise in offshore wind farms installation, a revolution attributed to policy intervention as well as economic analysis that defined greater competitiveness in energy markets. Such rapid advancement is explained through the creation of a protected area that engaged core political and economic interests, facilitating long-term budgetary assistance despite rising initial costs. The intricacies of these actor networks require policy advocacy and coalition building in taking up renewable technologies. As offshore wind infrastructure grows, the economic and political arenas become more apparent in shaping the destiny of energy generation. Should there be future development, it will be necessary to understand the socio-environmental aspects of wind turbine technology. Some research has been done into the potential health impact of wind turbines, where noise pollution and its psychological impact are highlighted. Literature reviews reveal correlations between reported disturbance or sleeping disturbance and the presence of wind farms, yet evidence on direct causal relationships remains incomplete. Also, problems with low-frequency noise and infrasound due to wind turbines need strict investigation as well as means for preventing such adverse effects through appropriate design of wind farms. It is evident that the integration of public health concerns into wind turbine technology development is crucial in gaining community acceptance and sustainable long-term growth in renewable energy utilization (Baguley et al., 2013).

3.4.3. Innovations in Energy Storage Solutions

The evolution of energy storage solutions is an integral component of enhancing the efficiency and reliability of renewable energy. Battery technology, including lithium-ion and solid-state batteries, has seen spectacular improvements in efficiency and capacity. Excess energy generated during peak production, particularly in solar and wind power, can be stored by such developments, minimizing issues of intermittency. Apart from this, integrating advanced energy management systems makes it easier to optimize the release of stored energy in line with anticipated demand and user consumption (Rodríguez et al., 2016). According to research, these systems will yield significant savings on total energy expenses as well as carbon emissions, particularly for residential use, owing to consumer inclination toward sustainable solutions and reduced reliance on fossil fuels. Beyond batteries, other advanced storage technologies are pumped hydro and compressed air energy storage (CAES), which utilize surplus power to pump water up a hill or compress air in geologic caverns, respectively, for return of energy during periods of high demand. With their ability to provide high levels of energy and respond rapidly to fluctuations in energy supply, these technologies are most suited for the integration of renewable sources into the grid.

Figure 3.15. *Energy storage trends.*

Source: https://knowhow.distrelec.com/wp-content/uploads/2024/11/GettyImages-2155976732-1024x652.jpg.

However, there are environmental and geographical dimensions of such systems that are worth close examination; placement of pumped hydro systems, for instance, can be environmentally demanding. Nonetheless, the processes continue to evolve towards efficiency and sustainability, in accordance with goals for increasing the reliance on renewable resources while mitigating associated environmental impacts (Caird et al., 2007). The future of energy storage is also being shaped by research on hybrid energy storage systems, which entail a combination of technologies. Through integration of different methods, such as batteries and supercapacitors or hydrogen storage devices, hybrid solutions provide optimized performance and lifetime while overcoming the respective weaknesses of individual technologies. Such systems provide immediate energy pulses for short-duration applications while also providing longer-duration energy storage. Additionally, with growing demand for decentralization of power generation, particularly through the use of smart grid technologies, hybrid storage systems can lead to improving energy distribution networks into more stronger and efficient ones. These advancements also contribute towards greater integration of renewable energy sources and towards the attempts to become carbon neutral in the energy sector, thereby confirming the necessity of diversified energy storage solutions in the quest towards sustainable energy futures (Rodríguez et al., 2016) (Figure 3.15).

3.5. CHALLENGES AND LIMITATIONS

The transition to renewable energy resources, while promising so much in terms of environmental returns, is beset by numerous challenges and limitations. Perhaps the most pressing among these is that of intermittency characteristic of resources such as solar and wind, whose electric power is only available when weather conditions are favorable. Such uncertainty would complicate energy supply management, necessitating the formulation and deployment of highly advanced technologies in energy storage and grid management. As per the above, "Smart grids are set to be at the heart of any shift to a low-carbon energy future," indicating the importance of an advanced system capable of handling variable energy supplies. Also, integrating renewable energy means modifications in existing power grids, which could mean huge investments and efforts to automate such systems for greater efficiency and dependability (Copeland et al., 2019). Another glaring limitation of renewable energy sources is the technological and infrastructural constraints that accompany their adoption. To illustrate, the current grid of energy in most of the world relies primarily on uni-directional flow of electricity, resulting in inefficiency and wastage of energy. There must be a transition towards Smart Grids, which offer the possibility of bi-directional flow of energy, to avail of renewable energy technologies while meeting growing needs for energy. This transformation involves the installation of varied monitoring and control devices and requires connectivity and automation through Internet of Things (IoT) solutions. The technology complexity of these systems poses integration, standardization, and data management concerns, thereby necessitating collaborative effort among diverse stakeholders to facilitate successful implementation (Copeland et al., 2019). Aside from the technological hurdles, there are also socio-political hindrances to the mass adoption of renewable energy sources. The successful application of Smart Grids, for example, not only relies on technological innovations but also on the amount of acceptance from the people and governmental regulations. Concerns over data security, cybersecurity, and disruption to existing energy markets can deter businesses and consumers from embracing these technologies. Furthermore, the understanding of achieving a balance between regulation and competition in the energy sector provides another layer of intricacy to the transition towards renewable energy. Such intricate challenges translate that while the transition to renewable energy is necessary, it is equally critical to address these challenges comprehensively in order to facilitate a sustainable energy future (Axon et al., 2014).

3.5.1. Intermittency and Reliability Issues

One of the notable aspects of renewable energy sources is their inherent intermittency, particularly evident in the case of solar and wind energy. These sources depend significantly on weather conditions, i.e., their availability can fluctuate significantly in a day and in seasons. This fluctuation is problematic from the aspect of providing reliability of power supply, which is crucial for providing grid stability. Unlike fossil-fuel facilities that can provide a constant output, renewables require accompanying systems—such as energy storage equipment or reserve generation capacity—to meet demand when the renewables are not producing. As deployment of renewables expands in power portfolios, it is important to understand and compensate for their intermittent nature in planning future energy security. Reliability issues of renewable energy production are not merely a matter of supplying adequate supply but also issues of grid resilience and consumer confidence. For instance, a diversified fuel mix will enhance the resilience of the system to failure by interruption of any individual source. The UK Electricity System evokes this perspective, placing emphasis on the demand for fuel diversity to balance risks in relation to the intermittence of renewable generation. Policymakers must realize that as low carbon goals trend towards increasing penetration of renewables, long-term diversity of energy sources will prove critical in guaranteeing a stable electricity supply. Such dynamic relationships necessitate strategic frameworks that permit the unique problems presented by renewable intermittency and offer end-user dependability. Furthermore, the economic aspects of intermittency and reliability issues need to be explored as renewable energy technology improves. Investments in infrastructure, such as enhanced storage capacity and grid alteration, can be front-loaded. In exchange, however, the long-term cost benefit of an intermittency-free renewable energy system may be justified by the initial investment. Evidence shows that investment strategies that prioritize diverse energy sources also enhance overall grid stability and reliability in addition to supporting low carbon policies (Butler et al., 2025). Therefore, employing a methodical strategy in planning energy that takes diversity and reliability into consideration is likely to lead to stronger systems that are capable of meeting both current needs and future trends in energy (Butler et al., 2006). Embracing these factors in policy approaches is critical to the attainment of a sustainable and reliable energy environment.

3.5.2. High Initial Investment Costs

The transition to renewable energy has been largely blocked by the investment costs, which are a tall order for their mass scale deployment. The majority of renewable technologies, such as solar photovoltaics (PV) and offshore wind, call for huge capital investment for installation and infrastructure building upfront. The costs involved are not only the materials and technology needed but also the labor and regulatory aspects of putting in place renewable energy systems. Though the long-term benefits are minimized operating costs and environmental viability, the practicality of massive front-end costs may deter the private sector as well as the government from investing in such projects. This issue needs innovative financing methods and policy interventions, which can act as a bridge for overcoming the gap created by the high capital charges and hence ensure quicker shifts to cleaner fuels. Secondly, the difference in investment behaviors between nations further aggravates the issue of high initial costs.

For example, the German photovoltaic industry has experienced huge market expansion due to encouraging government policies as well as research and development expenditure, with the result being a thriving renewable sector. But this model is clearly differentiated from China's, whose development has been export-led, which implies dependence on foreign markets for growth and innovation. And so, these divergent paths highlight the significance of national strategies and investment climates. Nations with strong structures to offset high upfront costs are better able to scale up renewable energy use more effectively, whereas nations that have been facing financing issues may see their progress considerably weakened, eventually influencing world-wide renewable energy uptake rates (Arrighi et al., 2012).

In overcoming these high upfront costs, stakeholders will have to work together to create strategic alliances that make it easier to invest in renewable technology. For instance, initiatives like the Special Initiative on Offshore Wind at the University of Delaware aim to advance the installation of offshore wind power through leveraging knowledge and knowledge transfer among industry, government, and nongovernmental stakeholders (Ozkan et al., 2016). On a mutual platform, these initiatives can divide the cost and pool resources together to reverse costs. In addition, innovative financing tools such as green bonds, tax credits, and public-private partnerships can be a key to lowering the costs of financing and enabling more stakeholders to invest in renewable energy projects. By synergistic efforts to leap over the financial barriers, it is feasible to speed up the transition to sustainable energy systems while ensuring economic viability for all involved parties.

3.5.3. Land Use and Environmental Impact Concerns

The application of renewable energy resources raises serious land use and environmental effects concerns that need to be given every consideration. With the increased demand for green energy options, the deployment of technologies such as solar, wind, and biomass necessarily requires vast tracts of land to set up. This land conversion raises fundamental questions regarding habitat disruption, loss of biodiversity, and competition with food production requirements. For instance, the construction of huge solar farms in deserts can lead to soil erosion and disruption of the ecosystem. Therefore, the planning and site selection of renewable energy plants must be based on conservation and ecological balance to ensure proper utilization of land to prevent harming the environment and nearby communities (Blengini et al., 2015). Without such considerations, the benefits of renewable energy are far outweighed by environmental disturbances they cause. Reception of renewable energy projects by society is also context-dependent and varies depending on land use and environmental attitude. Public perception has been found to be a primary driving force in the successful implementation of utility-scale solar plants. Specifically, the majority of stakeholders are concerned with visual impacts, efficiency in the use of land, and potential degradation of natural environments through such ventures. In Australia, quantitative survey and qualitative interview-based research showed that community acceptance is based on transparency over the environmental impact of renewable projects (Ipsos, 2025). When stakeholders become informed of the ecological and economic benefits of renewable energy, i.e., reduced carbon emission and employment, they may become more supportive of these projects. It is thus critical to address land use and environmental concerns in being awarded social license to operate for renewable energy facilities. Second, using an integrated approach to environmental analysis of renewable energy projects will contribute to sustainability. This strategy involves the consideration of broader ecological factors within the land use context, including ecosystem services, resource criticality, and social impacts. The development of an integrated sustainability assessment framework (ISAF) enables decision-makers to quantify and address various sustainability challenges concerning energy generation (Blengini et al., 2015). Through the categorization and examination of the environmental, technical, economic, and societal characteristics of renewable energy resources, stakeholders are better positioned to make land use choices that account for more effectively the trade-offs. An integrated strategy not only decreases the negative ecological impacts, but also promotes more sustainable land management methods, by ensuring that development of renewable energy contributes positively to both environmental sustainability and people's well-being.

3.6. POLICY AND REGULATION

The relationship between policy and regulation is crucial in the typology and use of renewable energy. As countries increase their emphasis on low-carbon alternatives, legislation to stimulate investment in renewables has become necessary. Energy policy development in Scotland has led to stakeholders needing to consider environmental and technical challenges around hydropower, for example. The country's prioritization of high-profile renewable energy aims calls for a cohesive response that balances energy generation with environmental sustainability. However, such a call for policy coherence often emerges in the context of a challenge, particularly in situations where energy objectives compete with environmental protection needs. This evolving regulatory environment accentuates the necessity of establishing frameworks that can develop in sync with sustenance challenges of hydropower and cascading impacts on other renewable technologies (Nelson et al., 2013). Moreover, new regulatory approaches are required to spur innovation in renewable energy technologies. Research indicates that national energy market characteristics play a significant part in shaping innovation in the sector, particularly through policy levers best adapted to local contexts. Deregulated energy markets combined with robust policy systems define nations with higher rates of green innovation, including patent production for renewable technologies. The said research points out how a strong policy index creates favorable conditions for technology progress, especially when it is complemented by low market entry barriers. This dynamic captures the necessity for policymakers to methodically integrate market structure and demand mechanisms to foster innovation, and in the process, transform the renewable energy development landscape as a whole (Nicolli et al., 2025). Further, the manner in which renewable energy policies interact with demographical facets such as income inequality cannot be excluded when considering regulatory efficiency. Societies that have high income disparities may be limited from achieving unity of support for renewable schemes, deterring innovation and technology adoption. This means that successful policies on renewable energy not only address the technological factors but also take into account social cohesion to achieve broad-based support. Examination of interactions between market forces and social contexts is therefore likely to enable regulators to make more accurate anticipations of the impact of their policies on renewable energy performance. This type of multidisciplinary collaboration towards policy making is crucial in advancing sustainable practices among different renewable energy sources in order to effectively understand their character and trajectories in evolving energy markets (Nicolli et al., 2025).

3.6.1. Government Incentives for Renewable Energy

Policies by the government to promote the use of renewable energy help in transforming the energy landscape worldwide. Government policies usually take the guise of tax credits, subsidies, and grants that mitigate the costs of adopting renewable technology. By the reduction of upfront costs, governments facilitate increased investment in solar, wind, and other renewable power sources, stimulating growth in the industries. As an illustration, regions like Western Australia have distinct electricity supply infrastructure that needs differentiated government incentives with the ability to drive technology innovation in remote areas. Policies that make it easy to integrate new renewable technologies into advanced networks have the potential to greatly promote energy security while meeting environmental sustainability targets, thereby demonstrating a holistic approach to energy policy with room for greater climate ambitions (McHenry et al., 2010). Moreover, strategic use of government incentives develops a culture of responsiveness and innovation in the renewable energy sector. As the European Union's biofuels policy stipulates, government measures not only initiate the creation of new markets but also stimulate research and development that roots technology in environmental sustainability. The expectation that market forces will drive attractive innovation brings about synergies between technoscience and natural resources, enhancing the drivers to renewable energy adoption. The relationship is typical of the necessity to incentivize not just energy projects, but the overall ecosystem conducive to them, resulting in lower carbon footprints and enhancing energy resilience in a constantly changing global landscape (Birch et al., 2012). Lastly, the long-term effectiveness of government incentives rests on their ability to accommodate technological advancements and market influences. As renewable technologies advance, incentives need to be reassessed constantly to mimic market positions and public policy objectives. It is essential that this adapting approach sustain momentum in sectors that are susceptible to changes in economic viability due to technology innovation or changes in the availability of resources. By establishing a flexible regulatory system, governments are in a good position to support sustainable energy practice and ensure the continued growth of renewable sources as part of a complete energy plan. Knowledge regarding the character of renewable energy sources plays a central role here because it highlights the requirement for policy measures supplemented with sound scientific research and public engagement.

3.6.2. International Agreements and Commitments

Multilateral commitments and agreements are central to deciding the destiny of renewable sources of energy. The agreements, typically entered into through collective international action, are designed to address the burning question of climate change while promoting sustainable development. Through offering a system of shared endeavor, nations can conclude binding volumes of greenhouse gas emissions reduction and promote the take-up of renewable energy technologies. The Kyoto Protocol, for instance, introduced legally binding goals to developed nations, paving the way for phasing out the use of fossil fuels globally.

Similarly, the Paris Agreement has already pushed states to make voluntary commitments to make robust climate pledges, inspiring a cooperative and responsible mindset. Hence, global agreements are indispensable instruments that not only drive policy initiatives but also stimulate investment in renewable infrastructure. Implementation of these international commitments can be enhanced by various policy instruments designed to internalize the externalities associated with power production, i.e., non-renewables. Economists typically emphasize embracing instruments such as feed-in tariffs, green certificates that are tradable, and competitive bidding programs, which support the transition to renewable energy.

By aligning economic instruments with environmental objectives, policymakers are able to mitigate the undesirable effects of energy production on society and the environment. In addition, effective internalization of externalities comprises an arduous trade-off across a range of criteria such as efficiency, equity, and technological progress. The empirical evidence obtained from studies such as the ExternE project is significant in that it uncovers the socio-economic impacts of these policy measures, hence keeping policymakers informed of the most appropriate ways to foster efficient renewable energy programs (Longo et al., 2025). While countries navigate the intricacies of climate agreements and commitments, an action-oriented approach has been the primary paradigm to facilitate the creation and dissemination of clean energy technologies. This approach centers on the needs of tangible actions that can be taken at the national and global levels.

The reports of efforts such as the G8+5 Task Force underscore the necessity to link targets with unshakeable and commensurate policies so that developing and developed nations are able to join together to address pressing energy needs. By stimulating innovation and supporting the installation of renewable energy technology, these commitments not only contribute to emissions reduction but also facilitate structural transformation in existing energy systems. Finally, international community initiatives to seek out such action-oriented programs reflect growing recognition of economic development's correlation with environmental responsibility and the imperatives for international conventions in achieving sustainable energy futures (Egenhofer et al., 2008).

3.6.3. Local and State-Level Initiatives

Local and state-level initiatives play a pivotal role in driving the use of renewable energy, supporting experimental projects that are usually driven from grassroots sources. All these can be achieved in several formats, including community-owned power projects, education programs that support energy efficiency, and local policy frameworks aimed at increasing the use of renewable energy. The diversity in these individuals' movements not only leverages solutions to specific community needs but also engenders high-level participation and investment from local residents. This participatory mechanism generates trust and collective ownership within the community, key determinants of the success of renewable energy projects. In addition, as empirical research in strategic niche management has demonstrated, local initiatives create productive knowledge exchange networks, enabling the best practices to be replicated and evolved in other settings and thereby enhance their sustainability effects (Hargreaves et al., 2013). Challenges of local and state initiatives are often intertwined with worldwide societal forces in energy policy and system transition. Community Action Initiatives (CAIs) mirror the manner in which grassroots movements operate towards successful change both in the energy sector as well as citizens' consciousness toward sustainability. While CAIs are credited with contributing to energy transition, they are successful due to appropriate mobilization processes entailing not just technology-based solutions, but also intrinsic social forces such as communal bonding and activism. This dual orientation ensures that projects are both innovative and locally specific, thus increasing their sustainability and scalability. These bottom-up projects bring us back to how bottom-up initiatives supplement top-down policies for an inclusive and efficient shift toward renewable energy (Arrobbio et al., 2020). An analysis of the success of such neighborhood projects is a complex interaction between innovation, policy support, and local involvement. The evidence shows that intermediary institutions can be helpful in the formulation of such projects by providing access to needed resources, technical support, and knowledge transfer bridging among stakeholders. But the success of renewable energy projects also rests as much on tacit knowledge and trust that are formed in communities, which are similarly harder to institutionalize. It is these intangible social relations that can ultimately determine the sustainability of renewable energy projects with regards to funding uncertainties and political regime shifts. Through further research in local project experience, researchers and policymakers are capable of creating a more profound sense of the conditions that are necessary to facilitate renewable energy innovations to flourish among diverse groups of people (Hargreaves et al., 2013).

SUMMARY

In general, the character of renewable energy sources reveals a complex interplay of ecological, economic, and social issues that must be handled with care. Based on the study, the public will balance perceived environmental limitations of renewable energy initiatives with alleged economic benefits, particularly in rural communities where long-term job opportunities and the development of infrastructure can yield impressive returns in local well-being. The findings based on the Scottish Economic Policy Network-sponsored research identify that rural societies may be willing to accept limited environmental sacrifices in exchange for economic growth, demonstrating a nuanced attitude contradicting the traditional viewpoint on rejection by local communities of renewable projects (Bergmann et al., 2004). As such, the identification of such dynamics is crucial for policymakers who aim to promote sustainable energy alternatives. Furthermore, the variability of renewable sources necessitates innovative solutions to energy generation and delivery. The hybridization of multiple renewable sources is increasingly becoming a viable solution to tackle temporal and spatial incompatibilities between supply and demand of energy. According to recent literature, the integration of solar, wind, and hydroelectric systems can best leverage the complementary strengths of each power source to yield more stable and reliable energy outputs. This is an approach that not only optimizes resource utilization but also enhances the resilience of energy systems to variability caused by weather fluctuations and demand variability, paving the way for a more sustainable energy future (Beluco et al., 2019). The completeness of hybrid systems alleviates the constraints on each renewable resource individually. Lastly, the transition to renewable energy sources has tremendous implications for human development and environmental sustainability. Encouraging greater acceptance of renewable energy plans, particularly in cities, entails addressing public concerns over landscape and wildlife disruptions, as urban respondents will tend to appreciate these more than perceived monetary benefits. To this end, effective communication strategies and community involvement are important in transcending such barriers as a way of facilitating the economic aspirations of rural communities and the environmental interests of urbanites. With the advancement of renewable energy technology, an effort requiring cooperation among stakeholders at all levels will be essential to promoting acceptance and making feasible the integration of renewable solutions in the international energy picture.

Renewable energy technologies can renew naturally after some time and contrast with fossil fuels, which are limited and get depleted upon usage. Solar energy, wind energy, hydro energy, geothermal energy, and biomass energy are some renewable energy sources. One of the most important features of renewable energy is sustainability; they can be utilized without causing permanent damage to the environment. For instance, solar panels capture sunlight and produce electricity without creating harmful emissions, a prime advantage in the fight against climate change. Secondly, renewable energy sources offer improved energy security in terms of diversifying the energy supply and reducing dependence on imported fuel, thus enhancing the resilience of local economies (Adam et al., 2007). In addition to sustainability, renewable energy sources are also flexible and scalable. Renewable technologies can be installed on various sizes and configurations, therefore allowing for small-scale plants as well as large utility-scale ones. Wind turbines and solar panels, for instance, can be installed on single homes or integrated into large solar farms or wind farms. The capability to scale encourages local energy generation, whereby communities can participate in their own energy generation and ownership. As societies move towards renewable energy, the potential of decentralized energy systems arises, allowing local communities to be empowered and foster self-sufficiency without sacrificing sustainable development goals (Dechezleprêtre et al., 2013). The cost feasibility of renewable energy sources is another desirable characteristic as technological developments continue to lower costs, becoming increasingly more feasible than traditional fossil fuels. The cost reduction in solar panels and wind turbines has led to greater utilization of these technologies, and this has led to the expansion of the renewable energy sector. Additionally, government incentives and policies favoring clean energy technologies contribute to their economic viability. Renewable energy investment not only creates employment opportunities but also local economic spurring through innovation and infrastructure creation. Thus, the economic dimension of renewable energy helps achieve long-term sustainability objectives while tackling pressing environmental challenges and, in the end, towards a more stable and resilient energy system (Adam et al., 2007).

REFERENCES

1. Abdullah, E. (2018). *Integration of Renewable Energy Sources in the Distribution Network.* https://core.ac.uk/download/186449299.pdf (accessed on 30th September 2025).
2. Adam, D., Giddings, B., BRE, Bunce, M., Underwood, C., Cloke, P., et al. (2007). Renewable energy in remote communities. https://core.ac.uk/download/4149082.pdf (accessed on 30th September 2025).
3. Agnolucci, P., Arnstein, L., Bäckstrand, K., Bana e Costa, C. A., Brans, J. P., Cavallaro, F., et al. (2007). Is small beautiful? A multicriteria assessment of small-scale energy technology applications in local governments. https://core.ac.uk/download/51100.pdf (accessed on 30th September 2025).
4. Aguiar, J. L. B. de, Cunha, S. R. L., & Kheradmand, M. (2015). Mortars with phase change materials: Contribute to sustainable construction. https://core.ac.uk/download/55636410.pdf (accessed on 30th September 2025).
5. Aheicheva, O., Bukhkalo, S. I., Byalskiy, Y., Derkunska, O., et al. (2018). Distance education possibilities analysis for integrated innovative projects. https://core.ac.uk/download/187298594.pdf (accessed on 30th September 2025).
6. Allen, A., Corubolo, E., & da Silva, N. (1999). Environmental problems and opportunities of the peri-urban interface and their impact upon the poor. https://core.ac.uk/download/1668830.pdf (accessed on 30th September 2025).
7. Anav, A., Ciancarella, L., D'Elia, I., De Marco, A., et al. (2019). Impacts of air pollution on human and ecosystem health, and implications for the National Emission Ceilings Directive: Insights from Italy. https://core.ac.uk/download/188832483.pdf (accessed on 30th September 2025).
8. Arrighi, G., Boyle Godfrey, B., Bradsher, K., Diao, Y., Du, J., Yeung, G., et al. (2012). Geographical interdependence, international trade and economic dynamics: The Chinese and German solar energy industries. https://core.ac.uk/download/9551789.pdf (accessed on 30th September 2025).
9. Arrobbio, O., Candelise, C., Delvaux, S., Gilcrease, W., et al. (2020). Collective action and social innovation in the energy sector: A mobilization model perspective. https://core.ac.uk/download/289098471.pdf (accessed on 30th September 2025).
10. Axon, C., Balta-Ozkan, N., Cipcigan, L., Connor, P., et al. (2014). Scenarios for the development of smart grids in the UK: Literature review.

https://core.ac.uk/download/155776124.pdf (accessed on 30th September 2025).

11. Baguley, D., Fiumicelli, D., Moorhouse, A. T., von Hünerbein, S., et al. (2013). Health impacts of wind turbines. https://core.ac.uk/download/12799959.pdf (accessed on 30th September 2025).
12. Barrett, S., Chichilnisky, G., Krahmann, E., Ellerman, A. D., Graham, V., Helm, D., Kent, J., et al. (2013). Green consumer markets in the fight against climate change. https://core.ac.uk/download/29140000.pdf (accessed on 30th September 2025).
13. Beluco, A., Canales, F. A., Guezgouz, M., Jurasz, J., et al. (2019). A review on the complementarity of renewable energy sources: Concept, metrics, application and future research directions. https://core.ac.uk/download/286642706.pdf (accessed on 30th September 2025).
14. Bergmann, A., Hanley, N., & Wright, R. E. (2004). Valuing the attributes of renewable energy investments in Scotland. https://core.ac.uk/download/9019727.pdf (accessed on 30th September 2025).
15. Bermudez-Contreras, A., & Ivanova-Boncheva, A. (2016). Financing schemes for residential, grid-connected PV solar systems in BCS, Mexico. *International Journal of Energy Production and Management*, *1*(2), 1–16. https://elar.urfu.ru/bitstream/10995/57408/1/ijepm_2016_v1_2_01.pdf (accessed on 30th September 2025).
16. Birch, K., Levidow, L., & Papaioannou, T. (2012). Neoliberalising technoscience and environment: EU policy for competitive, sustainable biofuels. *Science as Culture*, *21*(3), 307–326. https://core.ac.uk/download/2832274.pdf (accessed on 30th September 2025).
17. Black, G., Shropshire, D., Solan, D., et al. (2015). Carbon free energy development and the role of small modular reactors: A review and decision framework for deployment in developing countries. *Energy Policy*, *85*, 266–278. https://core.ac.uk/download/61747430.pdf (accessed on 30th September 2025).
18. Blengini, G. A., Dewulf, J., Latunussa, C., et al. (2015). Toward an overall analytical framework for the integrated sustainability assessment of the production and supply of raw materials and primary energy carriers. *Resources Policy*, *46*(Part 1), 12–22. https://core.ac.uk/download/55743314.pdf (accessed on 30th September 2025).
19. Butler, L., Grubb, M., & Sinden, G. (2006). Diversity and security in UK electricity generation: The influence of low carbon objectives. *Energy Policy*, *34*(18), 4050–4062. https://core.ac.uk/download/42280807.pdf (accessed on 30th September 2025).

20. Butler, L., Grubb, M., & Sinden, G. (2025). Diversity and security in UK electricity generation: The influence of low carbon objectives. *Energy Policy*, *83*, 221–234. https://core.ac.uk/download/pdf/7093211.pdf (accessed on 30th September 2025).

21. Caird, S., & Roy, R. (2007). Designing low and zero carbon products and systems: Improvements based on consumers' experience of adoption and use. *Energy and Buildings*, *39*(5), 618–626. https://core.ac.uk/download/82905287.pdf (accessed on 30th September 2025).

22. Clancy, J., Oparaocha, S., & Roehr, U. (2004). *Gender Equity and Renewable Energies: Thematic Background Paper*. ENERGIA. https://core.ac.uk/download/pdf/11462744.pdf (accessed on 30th September 2025).

23. Copeland, R., Crespi, N., Rehmani, M. H., et al. (2019). Internet of Things-aided smart grid: Technologies, architectures, applications, prototypes, and future research directions. *IEEE Access*, *7*, 62962–63003. http://arxiv.org/abs/1704.08977 (accessed on 30th September 2025).

24. Dechezleprêtre, A., Michaelowa, A., Hoekman, B. M., Ai, C., Goenner, C. F., Hendrix, C. S., Popp, D., et al. (2013). External sources of clean technology: Evidence from the clean development mechanism. https://core.ac.uk/download/42368335.pdf (accessed on 30th September 2025).

25. Driver, B., Berry, D., Nielsen, J., Lehr, R., & Innis, S. (2004). *A Balanced Energy Plan for the Interior West*. Land and Water Fund of the Rockies. https://core.ac.uk/download/71349464.pdf (accessed on 30th September 2025).

26. Egenhofer, C., & Fujiwara, N. (2008). *Making the Most of the G8+5 Climate Change Process: Accelerating Structural Change and Technology Diffusion on a Global Scale* (CEPS Task Force Reports, 5 June 2008). https://core.ac.uk/download/5082065.pdf (accessed on 30th September 2025).

27. Fay, G., Villalobos Meléndez, A., & Schwörer, T. (2012). *Power Cost Equalization Funding Formula Review*. https://core.ac.uk/download/162576082.pdf (accessed on 30th September 2025).

28. Gong, J., Niu, Z., Zhou, S., et al. (2014). Energy-efficient antenna selection and power allocation for large-scale multiple antenna systems with hybrid energy supply. *IEEE Transactions on Wireless Communications*, *13*(11), 6316–6329. http://arxiv.org/abs/1408.5583 (accessed on 30th September 2025).

29. Hargreaves, T., Hielscher, S., Martiskainen, M., et al. (2013). A grassroots sustainable energy niche? Reflections on community energy case studies.

Energy Policy, *61*, 610–620. https://core.ac.uk/download/17209211.pdf (accessed on 30th September 2025).

30. Ipsos. (2025). *Establishing the Social Licence to Operate Large Scale Solar Facilities in Australia: Insights from Social Research for Industry*. https://core.ac.uk/download/pdf/30672015.pdf (accessed on 30th September 2025).

31. Izadian, A., & Shen, D. (2014). Combined generation and optimization of a wind-solar-battery power system. *Renewable Energy*, *69*, 67–76. https://core.ac.uk/download/46959232.pdf (accessed on 30th September 2025).

32. Izadian, A., & Shen, D. (2015). Extremum power seeking control of a hybrid wind-solar-storage DC power system. *Renewable Energy*, *76*, 194–201. https://core.ac.uk/download/46963007.pdf (accessed on 30th September 2025).

33. Kammen, D. M., & Rodriguez-Molina, J. (2018). Middleware architectures for the smart grid: A survey on the state-of-the-art, taxonomy and main open issues. *Renewable and Sustainable Energy Reviews*, *78*, 998–1010. https://core.ac.uk/download/323310963.pdf (accessed on 30th September 2025).

34. Kirkegaard, J. F., Weischer, L., & Hanemann, T. (2025). *It Should Be a Breeze: Harnessing the Potential of Open Trade and Investment Flows in The Wind Energy Industry*. https://core.ac.uk/download/pdf/6486668.pdf (accessed on 30th September 2025).

35. Longo, A., & Markandya, A. (2025). *Identification of Options and Policy Instruments for The Internalisation of External Costs of Electricity Generation. Dissemination of External Costs of Electricity Supply: Making Electricity External Costs Known to Policy-Makers (MAXIMA)*. https://core.ac.uk/download/pdf/6264371.pdf (accessed on 30th September 2025).

36. Lucon, O., Fransen, T., & Romeiro, V. (2015). *Bridging the Gap Between Energy and Climate Policies in Brazil: Policy Options to Reduce Energy-Related GHG Emissions*. https://core.ac.uk/download/75782972.pdf (accessed on 30th September 2025).

37. McCartor, G., Leutwyler, H., Pauli, H. C., Wilson, K. G., Nakanishi, N., Steinhardt, P., Dirac, P. A. M., et al. (1997). Optical absorption characteristics of silicon nanowires for photovoltaic applications. *Journal of Applied Physics*, *82*(2), 1234–1241. https://core.ac.uk/download/25215464.pdf (accessed on 30th September 2025).

38. McHenry, M. P. (2010). Diverse, remote and innovative–Prospects for a globally unique electricity network and market in Western Australia.

Energy Policy, *38*(3), 1221–1230. https://core.ac.uk/download/11237237.pdf (accessed on 30th September 2025).

39. Nelson, E. (2013). Hydropower in Scotland: Linking changing energy and environmental agendas with sustainability outcomes. *Renewable and Sustainable Energy Reviews*, *28*, 132–139. https://core.ac.uk/download/20443681.pdf (accessed on 30th September 2025).
40. Nicolli, F., Vona, F., & Nesta, L. (2025). Determinants of renewable energy innovation: Environmental policies vs. market regulation. *Ecological Economics*, *110*, 57–67. https://core.ac.uk/download/pdf/6441011.pdf (accessed on 30th September 2025).
41. Okie, A. M., Hibbard, P. J., Darling, P. G., & Tierney, S. F. (2011). The economic impacts of the Regional Greenhouse Gas Initiative on ten Northeast and Mid-Atlantic states. https://core.ac.uk/download/71358044.pdf (accessed on 30th September 2025).
42. Ozkan, D., McClellan, S., & Kempton, W. (2016). *Massachusetts Offshore Wind Future Cost Study*. https://core.ac.uk/download/75785617.pdf (accessed on 30th September 2025).
43. Rentizelas, A. (2014). Assessing the sustainability of biomass supply chains for energy exploitation. *Journal of Cleaner Production*, *76*, 22–29. https://core.ac.uk/download/19909242.pdf (accessed on 30th September 2025).
44. Rodríguez Jara, E. Á., Romero Rodríguez, L., Salmerón Lissén, J. M., et al. (2016). Analysis of the economic feasibility and reduction of a building's energy consumption and emissions when integrating hybrid solar thermal/PV/micro-CHP systems. *Applied Energy*, *178*, 672–687. https://core.ac.uk/download/288003155.pdf (accessed on 30th September 2025).
45. Skoulou, V. K., & Vourdoubas, J. (2016). Case study of viability of bioenergy production from landfill gas (LFG). *Renewable Energy*, *92*, 355–361. https://core.ac.uk/download/151162915.pdf (accessed on 30th September 2025).
46. Smith, A., Bilgili, M., Verhees, B., Breton, S., Shaw, C., et al. (2013). From laggard to leader: Explaining offshore wind developments in the UK. https://core.ac.uk/download/20267315.pdf (accessed on 30th September 2025).
47. Suplee, D. (2020). Net energy index: A new way to measure energy efficient buildings. *Energy and Buildings*, *211*, 109818. https://core.ac.uk/download/322850910.pdf (accessed on 30th September 2025).

CHAPTER 4

SOLAR POWER INTEGRATION

Contents

INTRODUCTION

The integration of solar power into the energy grid is a paradigm change in the generation and utilization of electricity. While the world becomes increasingly conscious of climate change, attempts to transition towards clean energy sources like solar power have never been more ferocious. In addition to addressing environmental concerns, this approach fortifies energy security and spurs sustainability. Solar power, with its ubiquity and purity, can be a significant contributor to energy independence and a reduced dependence on fossil fuels. But ramping up of solar panel installations alone will not do; it has to be a master plan that includes efficient incorporation of available electricity infrastructure so that the variability of solar generation does not threaten grid reliability. Therefore, the study of novel solutions and technologies for rigorous integration of solar power has become an essential field of research and policy-making interest. For easy integration of solar energy to be enabled, certain challenges have to be addressed, and particularly so in the case of distribution networks and grid management. For instance, as California has a goal of 33% renewables by 2020, renewable sources such as solar make traditional voltage management procedures more difficult.

Figure 4.1. *Solar power integration in the Seychelles Islands.*

Source: *https://journals.openedition.org/factsreports/docannexe/image/4148/img-1.jpg.*

These systems will need to adapt to provide appropriate voltage profiles with increased solar and electric vehicle technology. It becomes clear that it is necessary for there to be better Volt/VAR Optimization and Control (VVOC) technologies since these will play a crucial role in ensuring the system's reliability and efficiency in light of growing solar use. Understanding and filling in the gaps in research concerning these emerging technologies is essential in enabling the maximum potential for solar energy integration into the existing energy mix. This underscores the need for innovative approaches in the maximization of the efficiency and contribution of solar energy to the power grid (Arghandeh et al., 2013). Furthermore, as solar power becomes more and more a permanent component of the energy mix, both end-use electrification and promoting energy efficiency also come to the forefront. Californias experience in reducing greenhouse gases demonstrates the interconnection of solar power with broader attempts at sustainability. Demand responsive (DR) strategies, including the use of controllable building technologies, are some of the available options to address critical resource adequacy, power quality, and grid reliability needs. For instance, model predictive controller-based dynamic facades can effectively cut peak demand while offering optimal energy consumption. Such technologies facilitate enhanced solar power integration as well as enhanced overall grid performance in load shaping. Through combining solar energy with innovative building technologies, recipients can take full advantage of renewable energy resources while ensuring the resilience and stability of the power electrical grid (Blum et al., 2020) (Figure 4.1).

Definition of Solar Power Integration

Solar power integration refers to integrating solar energy systems into existing energy infrastructure to create a more efficient and more sustainable energy model. The process does not necessarily mean utilizing photovoltaic panels in the direct sense but also involves advanced technologies like energy storage systems, which are crucial to eliminating the intermittency of solar energy generation. Successful deployment of solar energy requires alignment of technological innovation with policy frameworks such that energy grids are able to absorb the variable nature of solar generation without affecting their reliability and efficiency. Coordination achieves an overall enhancement in the resilience of energy systems to facilitate greater penetration of renewable sources as an acceptable alternative for traditional fossil fuel-based technologies. Another critical field of solar power integration includes the reliance on energy storage facilities in a bid to counter the challenges posed by the non-dispatchability of solar power. Emerging technologies in energy storage have become crucial in facilitating easier integration of large amounts of solar power into national grids. Such systems, for example, store excess power produced during daylight hours and release it when demand is high. As indicated in previous studies, energy storage methods—such as the use of transcritical carbon dioxide cycles with thermal energy storage—offer promising opportunities for efficiency improvements in solar power applications (Villanueva et al., 2019). In addition to enhancing energy management, these advancements also contribute to reducing the levelized cost of electricity, making solar power more feasible in terms of commerce within the energy sector. Besides the advances in energy storage, embedding solar energy in energy systems needs robust infrastructure and regulatory frameworks. Adding solar energy to the energy system must be done in consideration of areas such as grid compatibility, interconnection standards, and market incentives. Additionally, research and development are constantly in pursuit of solving technical problems related to deploying solar technologies. As demonstrated in various studies, solar integration strategy implementation of novel concepts, such as solar power satellite systems and modular design techniques, can contribute to a radical shift in the impact of solar integration strategies (N/A, 2025). Ultimately, however, the transition towards an improved interconnectivity of solar power is a technologically, socially, and economically imperative change that demands the concerted efforts of stakeholders toward an energy future that is sustainable (Figure 4.2).

Figure 4.2. *Solar power integration.*

Source: https://theenterpriseworld.com/wp-content/uploads/2024/01/5.3-Maximizing-solar-power-generation-through-optimal-system-design.jpg.

Importance of Renewable Energy Sources

The role of renewable energy sources cannot be overstated in the current debate on climatic stability and energy security. With dwindling and increasingly detrimental conventional fossil resources, nations must transition to sustainable resources. Renewable energy systems, particularly solar power, harness naturally replenished resources to generate electricity without depleting finite resources or contributing to greenhouse gas emissions. The integration of solar energy into the existing energy systems enables easier transitions towards cleaner technologies that not only reduce ecological imprints but also increase energy self-sufficiency. From the long-term trajectory of energy consumption and environmental delivery, the role of renewable sources in balancing economic growth with sustainable strategies will be central to delivering a more secure foundation of infrastructure for future generations.

One of the most important challenges for mass adoption of renewable energy is that solar energy is intermittent, making complex energy storage technologies necessary. There is a need for technological advancements in terms of energy storage so that the non-dispatchable nature of solar energy can be mitigated so that electricity is available during peak sunshine hours. Methods such as maintaining excess energy in reserve when there is low demand for it to use subsequently will balance supply and demand more effectively. Some of the new technologies, such as for transcritical carbon dioxide charge and discharge cycles incorporating thermal energy storage, present promising avenues toward achieving this purpose. Through research into economically viable energy storage systems, the integration of solar power can significantly enhance grid resilience while ensuring an assured supply of energy, as indicated by research citing efficiency gains where solar input into energy storage systems is integrated (Villanueva et al., 2019). Additionally, determining the spatial extent and architectural configurations pertaining to renewable energy systems improves their integration into urban growth. Comparative studies of fossil-fuel systems with regenerative systems provide crucial insights into the planning and organization of energy-efficient landscapes. The unique attributes of regenerative energy sources provide adaptive solutions specific to particular geographic and socio-economic conditions, maximizing the locality's energy autonomy. Regional case study from South Limburg illustrates this potential through verification of the need for close harmonization of energy systems with local infrastructures and environments (Koh et al., 2009). By prioritizing renewable sources, particularly solar energy, municipalities can not only meet present energy needs but also make sustainable energy practices community matters, paving the way for environmental stewardship globally (Figure 4.3).

Figure 4.3. *The importance of renewable energy sources.*

Source: https://green.org/wp-content/uploads/2024/01/OIP-3-17.jpg.

4.1. HISTORICAL CONTEXT OF SOLAR POWER

The historical background to solar energy is important to comprehend its incorporation into modern energy systems. Solar technology dates back to the 19th century when the photovoltaic cell was invented in 1839 by Alexandre Edmond Becquerel. His find formed the basis of harnessing solar energy, even if it was mostly experimental for more than a century. The oil crises of the 1970s were a stimulus, fundamentally changing the public and government focus towards non-traditional sources of energy. The demand for energy independence and ecologically aware consciousness caused investment to increase in solar R&D, setting the stage for current solar technology.

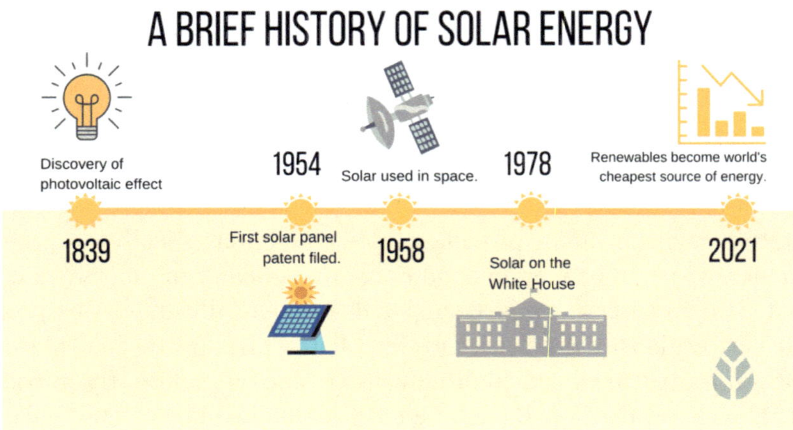

Figure 4.4. *The evolution of solar cells.*

Source: https://indosolar.co.in/wp-content/uploads/2024/06/History-of-Solar-3-1024x576-1.png.

These early breakthroughs initiated the innovations that would make solar power a potential form of power in today's energy mix. The evolution of solar energy has kept step with technological advancements and changing regulatory regimes. In the latter part of the 20th century, that is, since the 1970s, countries were forced to diversify their energy mix in response to oil price shocks, which heavily influenced energy policy and market forces. The transition of the energy system to sustainable forms relied on the uptake of new technologies as well as policy support from governments. This shift showcased the interaction between technological progression and market forces, so that more competition and investment fueled innovation in solar energy technology. As noted from historical accounts, policy changes have been instrumental in optimizing solar integration, displaying how certain policies can foster an environment conducive to the growth of renewable energy markets such as solar power (Bhuyan et al., 2025). Investment in research and development has played a key role in the growth of solar power technology. One would refer to federal investment in wind power as an excellent case study, where substantial economic payback was achieved from focused public spending and research and development (Millstein et al., 2020). In a similar vein, financial support and strategic investment in solar power have brought tremendous development, eventually contributing to better photovoltaic cells and lower production costs. This track record of technology development not only provided backup for the viability of solar powers but also contributed to drastically cutting the cost of the sector, making its application more universal. The compound influence of these policy trends and past innovations positioned solar power as an essential component of the energy solutions of the world. Reading this history is vital in being able to understand fully the nuances of incorporating solar power into the current energy blend (Figure 4.4).

4.1.1. Early Developments in Solar Technology

The history of solar technology traces its roots to the 19th century when science first started to deconstruct the possibilities of solar energy. One of the most significant milestones was the development of the photovoltaic cell by Edmond Becquerel in 1839, which showed the way that sunlight could be translated into electricity using the photovoltaic effect. This innovation laid the groundwork for future innovations in solar technology, proving the feasibility of harvesting a common natural resource. However, it was not until the middle of the 20th century that advances in materials and manufacturing techniques made solar cells accessible to practical applications. The introduction of silicon solar cells during the 1950s was the initial significant breakthrough, enabling greater efficiency and scalability for solar systems, and the future direction towards integrating solar energy into overall energy policy. Throughout the latter half of the 20th century, the integration of solar technology became increasingly viable due to technological advancements and rising energy demands. As the globe grappled with oil deficits during the 1970s, the demand for alternative sources of energy, particularly solar energy, picked up. Researchers continued to improve photovoltaic materials further, and their conversion efficiencies improved from the initial 6% to over 15% over the next decades. This was an era when government policies supporting the establishment of renewable energy sources were taking shape. Significantly, offering incentives for solar system installation promoted bulk uptake. Also, the evolution of solar thermal systems brought forth another way of using solar power, taking solar technology further and offering an alternative to photovoltaic. For the future, solar technology's advancements predict revolutionary possibilities for future energy sources, particularly when new materials are included. Organic thin-film electronic devices and bio-responsive hydrogels research indicate a promising track towards enhancing solar utilization. Those materials can lead to the production of shape-memory solar cloths and surfaces that are both functional and visually integrated into daily objects, which can redefine urban solar power usage. Such adaptive materials' advanced functionalities can make possible more personalized energy solutions and support energy efficiency in residential and commercial sectors. As the demand for renewable sources of energy rises, their integration into solar power systems will be critical for achieving energy independence and managing the challenges of climate change (Blain et al., 2005; Oliver et al., 2009) (Figure 4.5).

Figure 4.5. *Edmond Becquerel.*

Source: https://gienergy.com.au/wp-content/uploads/2021/05/becquerel.jpg.

4.1.2. Milestones in Solar Energy Adoption

The process of adoption of solar energy has been characterized by key milestones that mark technological advancement and changing social attitudes toward non-conventional energy. Among the key milestones was the period in the mid-1970s when the U.S. government began investing significantly into photovoltaic (PV) technology in the form of contract with institutions like the Jet Propulsion Lab. It was under this program that the technologies necessary for solar energy to be a viable ground source were developed. As outlined in the Flat Plate Solar Array Project, goals included a cost of $1.07 per peak watt on the commercial market in 1985 and a minimum module conversion efficiency of 10% with a 20-year lifetime. This joint effort not only set the foundation for future innovation but also put solar technology on the market as a competitive force within the overall energy industry (Callaghan et al., 2025). Yet another turning point in the milestones of solar energy adoption was felt when solar technology was integrated into architecture and urban planning. Building Information Modeling (BIM) witnessed the introduction of more integrated, efficiency-oriented energy design, making low-energy building commissioning and operating less complex. This paradigm shift was compared to the microelectronics industry, were standardization and systematic quality control enhanced product efficiency and reliability immensely. Using controls and operating codes from simulations, the construction industry came to establish means that would ensure more efficient incorporation of solar systems into new constructions. These advances not only encouraged a greener city lifestyle but also established solar energy as an essential aspect of modern building design (Tuohy et al., 2009). As public interest in global warming intensified over the past few decades, cumulative progress in embracing solar power mobilized the global campaign for renewable energy. Legislative measures like the Solar Investment Tax Credit (ITC) spurred the solar industry forward with economic incentives toward residential and commercial installations. With legislative support of that kind coupled with technological advancements that lowered costs, the application of solar panels grew exponentially across the United States and globally. Consequently, a cultural revolution is taking place whereby solar energy is not just viewed as an option but also as an essential piece of a sustainable energy future. With industrial-government collaboration setting milestones, solar power integration into daily life is expanding and has reached a path that signifies a promising path toward world energy sustainability (Figure 4.6).

Figure 4.6. *Milestones in solar energy adoption.*

Source: https://www.newsband.in/uploads/blog_main_img/b5e4987b90a75f-71c17925e1d136efc6_1.jpg.

4.1.3. Evolution of solar power policies

The formulation of solar power policy is mainly characterized by differentiated national policies formulated according to precise economic, political, and environmental contexts. For example, Germany has been at the forefront of solar energy policy since the early 2000s. The German government implemented strict feed-in tariffs, which encouraged extensive utilization of photovoltaic systems and triggered rapid market expansion. This was followed by massive investments and technological advancement, primarily in the eastern regions of the country. These policies not only encouraged domestic energy generation but also developed a competitive market, although it subsequently led to trade disputes with nations such as China, an example of the complex interdependencies in global solar power dynamics.

Thus, Germanys pushy strategy was critical in shaping solar energy tales around the world, setting a model other nations began to employ as an effort to market their renewable energy schemes, although not with equal success (Arrighi et al., 2012). Chinas response to the global demand for solar energy, however, provided a rather different course. Having been behind in-home capacities for production at the outset, the country shifted to become the world's leading exporter of solar technology.

The force behind that shift was driven by state-sponsored investment and subsidies which encouraged rapid industrialization and technological advances in photovoltaics. Policy in China focused on scale and efficiency, driving manufacturing costs which effectively re-sized global solar competitiveness. But this aggressive growth was vilified by world markets, triggering huge trade tensions and retaliations by veteran solar powers like Germany. This evolving landscape illustrates the double-edged nature of solar power policymaking—as a backboard to creativity and a war zone of economic interests, ultimately determining how various countries integrate solar energy into their overall energy systems (Arrighi et al., 2012).

In highly renewable resource countries, such as Algeria, solar power policy development captures the intricate steps of effective energy transition. Even though fossil fuel is still the dominant power generator, the government of Algeria has recognized vulnerability to climate change and acted accordingly. With aggressive goals for renewable capacity by 2030, Algeria's policy instruments involve encouraging investment by means such as feed-in tariffs and tenders aimed at triggering private sector engagement. The government intends to deploy around 22,000 MW of renewable electricity, or 27% of national output. This development signals a broader shift towards the integration of solar power into a variety of energy mixes, demonstrating how local policy is guided by local needs as well as globally applied pressure to solve the climate crisis.

The development is evidence of the intersection of environmental imperatives and economic forces, signaling a turning point towards sustainability in power generation.

4.2. TYPES OF SOLAR POWER SYSTEMS

The application of solar power in energy systems entails a full understanding of the various solar power systems used. Among them, photovoltaic (PV) systems are the most common technology for solar energy conversion. They directly convert the energy of the sun into electricity using semiconductor materials, which induce an electric current of electrons when subjected to solar irradiance. PV systems are either off-grid or grid-connected. Grid-connected systems are fed directly into the electrical grid so that unused power is sold to consumers or the grid is accessed when production is low. Off-grid systems, instead, store energy in batteries for personal use. It is essential to realize the nuances between the configurations for a good incorporation of solar energy into existing power infrastructure and well-informed decisions for adoption and expansion. Another important category of solar energy systems is concentrated solar power, or CSP, that operates on a different principle than PV systems. CSP concentrates sunlight from a large area by means of mirrors or lenses to generate heat that drives a turbine connected to a generator that generates electricity. CSP technology is particularly suitable for large-scale power production and may be more efficient than traditional PV systems, particularly in sun-rich environments. CSP technology tends to require extensive land use and works best on a utility scale. Integration of CSP technology into regional energy plans can enhance the renewable energy mix as a whole, particularly when integrated with a hybrid power system comprising other renewables such as wind or hydroelectric power. The collective operation of such disparate energy sources can balance the inherent intermittent nature of solar production of power, resulting in further exploration of CSPs potential in future energy plans (Beluco et al., 2019). Lastly, solar thermal systems represent a vital technology in the solar power spectrum. Solar thermal systems employ solar energy for thermal applications, i.e., heating water for household use or industrial processes. Solar thermal systems are based on the technology of harvesting solar radiation and transferring the heat to a fluid, which can be utilized directly or be converted into electricity. The performance and efficiency of solar thermal systems depend on the local climate and are therefore highly site-specific. While comparatively lower rates of adoption compared to PV, solar thermal systems provide a practical answer to meeting energy demands, especially in urban areas where domestic service demands are diversified and extensive. Their integration into urban energy systems can optimize the overall sustainability of urban energy systems by comprehensively considering all sectors, e.g., domestic, and commercial demands, as seen in pioneering research on urban energy systems (Brazier et al., 2017) (Figure 4.7).

Figure 4.7. *Types of solar power systems.*

Source: https://www.solar-sky.com/wp-content/webp-express/webp-images/uploads/2024/01/What-are-the-different-types-of-solar-power-systems-3.jpg.webp.

4.2.1. Photovoltaic (PV) Systems

The application of photovoltaic (PV) systems in cities is a new approach to energy efficiency and facilitation of sustainable development. By applying these systems, buildings can become self-sufficient in terms of power supply, having a substantial contribution towards reduced fossil fuel dependency. Photovoltaic panels not only serve as a renewable energy source but also play a central function in passive building architecture. For instance, the utilization of photovoltaic shading systems would be able to minimize heat gain, which in turn minimizes air conditioning and other cooling requirements. This double functionality serves to underscore the importance of integrating PV technology into architectural buildings, as it not only maximizes energy production but also enhances the optimization of building performance, making traditional buildings greener and economically feasible (Bloem et al., 2005). Also, the growth of the PV industry, particularly in different global environments, is indicative of different adaptation measures and market circumstances. As an illustration, in Germany, strong local market demand has spearheaded the rapid growth of PV technology and infrastructure as well as active policy intervention toward renewable energy integration. This growth has been underpinned by strong research and development investment, bolstering the industry's innovation. On the other hand, the Chinese PV industry has experienced a fast growth, where the primary driver is the export market, emphasizing country differences in growth and sustainability drivers. The interrelationship between German and Chinese markets shows the broad patterns of solar power integration globally, depicting how the policies and economic strategies of different nations determine the competitiveness and technological evolution paths of photovoltaic technologies (Arrighi et al., 2012). The benefits of installing PV systems extend beyond conservation of energy; they also yield broader socio-economic advantages. Providing clean, renewable energy encourages green economy employment that is favorable to local economies and facilitates the shift to more sustainable labor patterns. In addition, with growing awareness of climate change, demand for renewable energy technologies will rise, and hence a growing market for PV technologies will follow. With increased awareness and education on solar power, individuals and enterprises can enjoy tax credits and subsidies, further driving the application of photovoltaic systems. Overall, the use of PV technologies maximizes energy autonomy while positioning society to adopt a more sustainable way of life through innovative and economically viable practices (Figure 4.8).

Figure 4.8. *Photovoltaic (PV) systems.*

Source: https://solargaininc.com/wp-content/uploads/2019/05/image.png.

4.2.2. Concentrated Solar Power (CSP) Systems

Concentrated Solar Power systems are a significant innovation in solar technology because it focuses sunbeam energy using mirrors or lenses onto a point to generate heat. The generated heat is usually harnessed to drive steam that turns a turbine to generate electricity, creating a renewable energy source that can reduce the use of fossil fuels. One of the advantages of CSP systems is that they can incorporate thermal energy storage (TES) within them, so they can generate electricity even in times when the sun is not available. This feature also makes them more reliable, particularly in regions with variable availability of the sun. However, as proposed by Nahdi et al. (2021), there are disadvantages against CSP systems with respect to investment cost and maintenance for optimal performance. These financial and technical problems need to be addressed for broader integration of CSP into the renewable energy portfolio. Integration of CSP systems with other energy solutions has appeared as a viable path to enhance their efficiency and effectiveness.

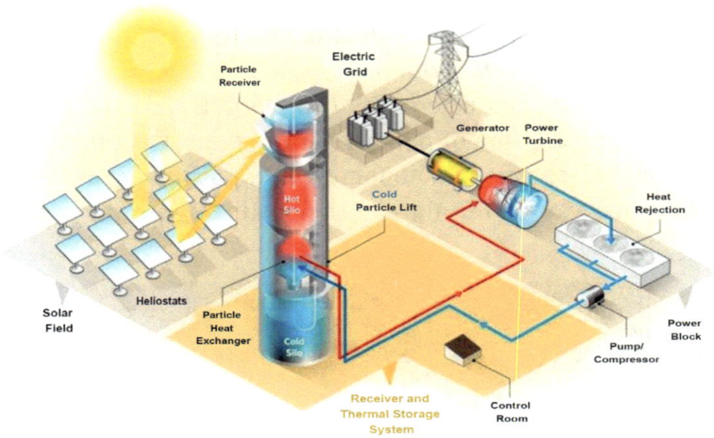

Figure 4.9. *The basic principle of concentrated solar power (CSP) systems.*

Source: https://www.researchgate.net/profile/Edip-Taskesen/publication/373559793/figure/fig1/AS:11431281197522606@1697015598032/The-basic-principle-of-concentrated-solar-power-CSP-systems-8.png.

For example, integration of CSP with biofuel or biogas has been proposed as a way of ensuring the production of energy during periods of low availability of the sun. The feasibility studies, as discussed by Nahdi et al. (2021), describe how the integration of biofuels can reduce fossil fuel dependency and improve sustainability. These combined systems can not only mitigate intermittency issues but also leverage the existing fossil-fuel power plant infrastructure, hence making it easier to transition to renewable energy resources. As highlighted in recent research, the optimization of such integrated systems can lead to considerable improvements in water and energy output, doubling the benefit with lower operating costs, especially in arid regions. Furthermore, exploration of innovative materials and intelligent technologies to further enhance CSP systems has further raised their potential applications. The creation of nanofluids, discussed at length here, has brought noteworthy thermal efficiency enhancements, a crux to the feasibility of CSP plants. With enhanced heat exchange, nanofluids facilitate smaller, less expensive collectors with excellent performance. Furthermore, use of Industry 4.0 technologies like artificial intelligence and machine learning has revolutionized process operations in CSP plants. These technologies facilitate real-time monitoring and predictive maintenance, hence increasing system reliability and ultimately reducing the cost of operation. The convergence of these technological advancements indicates a revolutionary shift towards more efficient CSP systems and confirms their place in comprehensive solar power integration schemes (Figure 4.9).

4.2.3. Hybrid Solar Systems

The innovation of hybrid solar systems is a breakthrough in the integration of solar technologies and additional sources of power for higher efficiency and reliability. The systems involve photovoltaic panels combined with other energy storage mechanisms, such as batteries and hydrogen fuel cells, to create a stronger grid system. The benefits are manifold: hybrid systems are able to provide steady energy output regardless of solar availability, thereby bridging the intermittent nature of solar power. This is especially important in application such as telecommunications, where power supply reliability is paramount. Hybrid systems also reduce dependence on traditional fossil fuels by harnessing alternative energy sources, thereby enabling the introduction of a cleaner, more sustainable energy age. It is possible to enhance the operational efficiency of a hybrid solar system by incorporating other forms of energy technology. Using electrolytic hydrogen technologies alongside solar power, for instance, can create a creative energy storage and management system. The reason for this is that surplus sun energy can be retained as hydrogen in the process and used as a fuel source when there is minimal sun exposure. Evidence indicates that this integration is able to achieve a drastic reduction in the requirement of battery storage and cessation of wasted renewable energy. With the integration of hydrogen technology and solar systems, hybrid configurations are able to provide a reliable and constant source of energy for off-grid use, hence justifying their role as part of the greater plan for renewable energy (Bustamante et al., 2015). Furthermore, the potential for energy efficiency of hybrid solar systems extends beyond the boundaries of mere energy storage technologies. Emerging trends in energy harvesting technologies, such as flexible solar cells that are produced on wearable substrates, present opportunities for innovative applications in smart textiles. These materials can harvest solar energy while continuing to be useful as smart fabric interfaces for communication networks. As has been observed in previous studies, the use of high-efficiency power management and energy harvesting systems makes it possible to have uninterrupted energy scavenging and, consequently, maximizes the total energy efficiency of artificial as well as solar light sources. Maximizing the use of alternative forms of energy, hybrid solar systems enhance the user experience and also maximize sustainable power solutions in diverse environments (Lemey et al., 2015) (Figure 4.10).

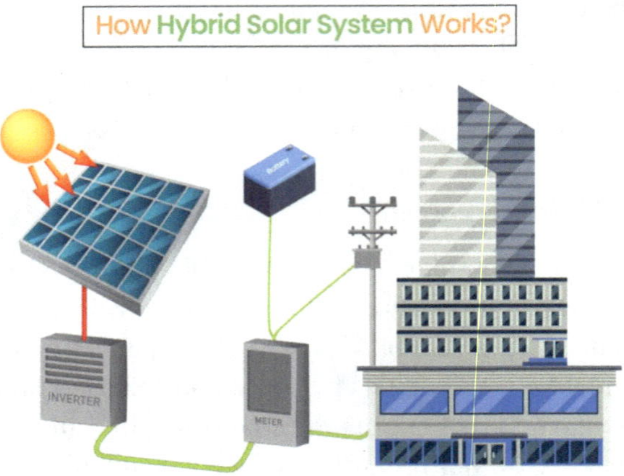

Figure 4.10. *How does a hybrid solar system work?*

Source: https://www.novergysolar.com/wp-content/uploads/2024/06/How-Hybrid-Solar-System-Works1-1024x683.webp.

4.3. BENEFITS OF SOLAR POWER INTEGRATION

The integration of solar energy into the current energy system has various benefits, greatly enhancing energy sustainability and security. Solar energy is a pivotal component in reducing dependence on fossil fuels, which are finite and contribute to environmental pollution. Solar technology use promotes cleaner air quality and lower greenhouse gas emissions, which are essential in countering climate change. Through energy source diversification, solar power integration minimizes the energy supply disruptions' risks, and hence, there is enhanced energy independence. This transformation not only stimulates national energy policy, but also international climate objectives provided by organizations like the International Energy Agency. The growing reliance on solar energy translates into a robust infrastructure capable of storing renewable resources, thus making consumers and industries benefit from stable and predictable energy prices in the long term, hence creating economic sustainability and resilience. Besides the environmental gains, integration of solar power also indicates significant economic benefits. With technological progress bringing about increased efficiency and decline in the cost of solar panels, individual and corporate energy users both benefit from the financial aspect in a better manner. Solar power has an effect on jobs with different sectors related to the creation, installation, and maintenance of the technology increasing at a higher rate. A recent study indicated a multitude of advantages that are associated with electric vehicles and vehicle-to-grid, with the setting that economy savings, emission reduction, and renewable energy installation rank high on this list of advantages. Further to this, the innovative spirit associated with the use of solar power contributes not only to increased energy efficiency but also to community investment, whereby the neighborhood economy flourishes through the support of green technologies.

Figure 4.11. *Benefits of solar energy.*

Source: https://www.greenlivinganswers.com/blog/wp-content/uploads/2024/08/solar-energy-benefits.webp.

The cumulative economic value solidifies solar powers position as a cornerstone in the emerging new energy sector, fueling sustainable growth. Long-term sustainability depends heavily on continued adaptation of energy technology, and this renders solar power integration at all levels imperative. An integrated approach to energy transition involves challenges like rapid deployment of technology, social acceptance, and innovation. Solar integration can thus bring about a transition toward decentralized energy systems, promoting communities to use local resources with an opportunity for participation and involvement. With fine-grained energy technologies like solar energy in the forefront, societies will be impacted through successful attainment of climate goals with universal access to clean power. This aligns with current research indicating the viability of understanding complex technological elements in improving efficacy and effectiveness in energy deployment. Solar power integration, therefore, is not only a technical upgrade; it is a comprehensive shift towards green energy practice that improves economic, social, and environmental resilience in the face of global adversity (Bento et al., 2020; Kester et al., 2018) (Figure 4.11).

4.3.1. Environmental Advantages

Solar energy utilization has vast environmental advantages beyond the conservation of greenhouse gases; it promotes sustainable usage of energy. Solar power systems, as a replacement for fossil fuels, which are finite and cause climate change, offer a lesser carbon dioxide emission that influences climate change directly. The utilization of solar energy is vital since traditional energy sectors are responsible for emitting vast amounts of CO_2 globally. Where energy efficiency measures are constrained, as in the case of heritage buildings, solar power provides a solution by merging energy supply technology with minimum structural intrusion. Rather than altering the building's architecture, the integration of solar solutions promotes sustainability without sacrificing heritage, addressing the pressing necessity of preserving such locations while reducing their carbon impact (Matteo et al., 2017). Besides, solar power allows for the creation of new and improved technologies to enhance its environmental benefits. One such example is microbial solar cells (MSCs) that are an emerging area of research that brings together biotechnology and renewable energy. By utilizing photosynthetic organisms for converting solar energy into usable electricity, the systems not only supply energy but also optimize the ecosystem. The new technology of MSC-powered green roofs shows the potential of combining urban infrastructure with environmental sustainability, illustrating how solar technology can be used to improve air quality and as a key urban cooling factor. Such twin use reinforces the notion that solar energy innovation is far more than application-only products, promoting a differentiated agenda of environmental benefits (Buisman et al., 2010). Furthermore, de-centralization provided through solar power generation enhances environmental resilience, particularly in communities that are vulnerable to climate disturbances. By facilitating energy generation where it is required, solar systems reduce the need for extensive transmission infrastructure, which is characteristically associated with energy losses as well as additional emissions. Not only does this local approach reduce the environmental impact of energy transport, but it also enables communities to become reliant less on dirty energy sources. As nations incorporate solar energy into their energy mix, they create a cushion against volatile fossil fuel markets and supply disruptions, enhancing energy security. Ultimately, embracing solar power through integration not only provides near-term environmental dividends but also lays the foundation for a sustainable and secure energy future (Figure 4.12).

Figure 4.12. *Environmental benefits of solar energy.*

Source: https://www.encstore.com/assets/blogs/1650368737-5-environmental-benefits-of-solar-energy.jpg.

4.3.2. Economic Impacts and Job Creation

Solar integration impacts the economy broadly, particularly in employment. With growth in the demand for renewable sources of energy, the solar industry has witnessed an unprecedented surge in employment opportunities. Recent data show solar energy job growth repeatedly overtaking many traditional industries, leading to a revolution in labor patterns. This shift is not only crucial to economic sustainability but also to the sustenance of local economies devastated by the fossil fuel economy. The installation and deployment of solar technologies require a combination of capabilities, ranging from project management and engineering to manufacturing and construction. This, in turn, generates a lot of entry-level and quality employment, creating a workforce fueled by innovation and sustainability. Such an economic transition not only favors the individual but also creates long-term economic viability and community resilience. Moreover, the net economic gain of integrating solar power is more than the provision of jobs to encompass a broader array of technological investments. The International Energy Agency indicates that of 45 major energy technologies required to achieve global climate objectives, 38 need substantial cost and performance improvement to be successful (Bento et al., 2020). This puts the economic potential advantages that motivate industries and investors to invest in solar power and allied technologies. By investing in the solar industry, the stakeholders can drive systemic change, boost technological developments, and even more so enhance job creation. Additionally, the danger of lock-in to the conventional energy systems is reduced as there are more granular and decentralized energy solutions available. Hence, systematic investment in the solar infrastructure changes not merely the landscapes of jobs but also the economic overall viability of the communities aiming for long-term sustainability. Apart from the immediate economic impacts, solar integration offers ample opportunities for impact investing that can lead to broader economic development. Investors are seeking to diversify their portfolios with sustainable assets as financial return and social good can be achieved simultaneously. Looking at policy decisions in conjunction with growing impact investing in clean technology can release a lot of money into solar projects (Thornley et al., 2012). Such investments do not only bring in money but also the framework necessary to propel the deployment of solar technologies at an accelerated rate, thereby enhancing job creation and community engagement. Governments and institutions adopting impact investing strategies can make the economic dividend of solar power more straightforward, ensuring a robust and agile employment market in spite of evolving energy requirements. Overall, the mutual dependency of the economic impacts and job potential in the field of solar power integration is a critical aspect of a more sustainable energy future (Figure 4.13).

Figure 4.13. *The economic impact of solar power.*

Source: https://cdn.prod.website-files.com/65b217df08a6f1f6aba3d4ce/681b31f37 18e6cfba9c6bf43_pasted-image-0--8--compressed-1.jpeg.

4.3.3. Energy Independence and Security

Energy security and independence form fundamental aspects of contemporary energy policy, especially with the integration of solar energy into the current energy mix. As nations seek to reduce reliance on imported fossil fuels, solar power offers an achievable means for securing greater levels of energy independence. Leverage of untapped solar resources allows nations to lower reliance on externality-generated market volatilities and geopolitical tensions that usually follow imports. The integration of solar power not only diversifies the energy mix but also enhances the strength of the energy network. This is particularly relevant in diversified generation systems like Western Australia, which has several non-interconnected and interconnected power systems (McHenry et al., 2010). Their advancement makes local energy production possible, but also regional energy cooperation and security via shared renewable sources. Moreover, energy autonomy is intertwined with national security in a natural way since the energy supply chain of a nation essentially establishes its stability and prosperity. Through the onset of climate change challenges, the substitution of conventional energy sources with renewable sources like solar power is a necessity both environmentally and in seeking to secure energy futures. As the European Union's bid to reform energy markets in south Mediterranean countries implies, regulatory systems can trigger the use of renewables in such emerging economies (Cambini et al., 2014). By putting in place an effective governance that allows the assimilation of solar technology, nations can improve energy independence and reduce exposure to world oil and gas price shocks. The structural transformation is essential to ensure cleaner energy proposals and assist nations in their journey to energy independence through sustainable means. Aside from the environmental advantages, energy independence through solar integration can also promote economic benefits, which include creating employment opportunities and technological innovation. With solar power systems becoming increasingly common, new opportunities in the local labor market are created, hence stimulating the local economy. The possibility of creating new employment opportunities in manufacturing, installation, and maintenance of solar technology cannot be overemphasized, especially in areas where traditional industries like oil and gas are on the decline. Through solar investments, economies not only gain more energy security but also gain bargaining power in the global market of renewable energy technologies. With such trends continuing to evolve, the nexus between independence and security in the context of energy will become more crucial in shaping energy policy and investment strategy. Therefore, integrating solar power offers a multi-faceted solution to achieving a secure, independent, and sustainable energy future (Figure 4.14).

Figure 4.14. *Energy independence.*

Source: https://smartenergygap.com/wp-content/uploads/2023/06/Energy-Independence.webp.

4.4. CHALLENGES IN SOLAR POWER INTEGRATION

The integration of solar energy into the existing energy infrastructure is subject to a number of challenges that are intrinsic to its variable and intermittent nature. Solar energy generation is intrinsic to sunlight, leading to fluctuations of availability that pose difficulties for the grid to handle. The variability necessitates a stronger and more flexible grid infrastructure in order to handle fluctuations of supply and demand. Apart from that, as has been realized in the quest for grid flexibility, challenges in incorporating renewable power sources, like the sun, can be exacerbated by poor forecasting equipment and limited access to the electricity market. The capacity of grids to absorb efficiently these uncertainties is most critical, and countries with varying levels of preparedness in grid flexibility have varying challenges in achieving renewable energy targets under reliability in power supply (Huang et al., 2019). One more significant challenge of solar power integration is the need to invest in superior technological options and modernize infrastructure. The transition to a cleaner energy system not only calls for increased grid flexibility but also investments in smart grid technologies enabling real-time monitoring and control of electrical currents. These upgrades are needed since they can improve the reliability of the grid and support the increasing share of renewable generation. Moreover, the postponed use of these technologies can also be a critical barrier, relying on fossil fuels and obstructing efforts to counteract climate change. New strategies, such as including solar microgeneration on low-power IT products, show how technology can lead to improved energy use and utilization, thereby helping in mitigating some of the strains on the grid during peak demand hours (Counsell et al., 2009). Regional cooperation is central to the addressing of challenges in integrating solar power. With nations setting sights high towards renewables, international cooperation has the ability to enhance knowledge sharing and utilization of resources. The mutual efforts can include establishment of joint renewable energy grids to facilitate greater distribution of solar power, in addition to joint efforts towards technological advancement. Such cooperation would not only benefit the improvement of grid flexibility but also to the common issues such as the lack of forecasting systems and irregular access to electricity markets. As the case of ASEAN shows, countries can benefit by grouping themselves into cooperative frameworks that focus their goals and capitals towards more sustainable energy systems (Huang et al., 2019). By embracing collaborative methods, nations are in the best position to respond to the inherent challenges associated with incorporating solar energy into their energy mix (Figure 4.15).

Figure 4.15. *Challenges in solar power integration.*

Source: https://www.vtacexports.com/media/magefan_blog/solar_and_grid.jpg.

4.4.1. Intermittency and Reliability Issues

As the demand for the utilization of solar power increases, the issue of intermittency presents itself as a significant obstacle to its widespread utilization. Solar power is not constant; its supply depends on location, timing, and weather. This fluctuation may yield periods of inadequate solar production relative to needs, thereby resulting in possible grid instability. The welfare implications of these disruptions are considerable; it has been estimated that the unpredictable intermittency of the sun's energy can produce a welfare loss equal to 3% of the average cost of solar energy generation in some context, such as an electric utility for the southeastern region of Arizona. These findings point towards the necessity of devising solutions that enhance reliability while promoting the growth of solar power in the overall energy matrix (Gowrisankaran et al., 2025). Addressing reliability issues related to solar power involves a two-way approach considering technological improvements and policy interventions. Diversifying the source of energy in a grid is one potential solution. By utilizing a mix of renewable and non-renewable sources of energy, the grid becomes robust against potential disturbances from any single source of energy. Evidence from the UK Electricity System shows that greater long-term diversity in the energy source facilitates the mitigation of reliability problems inherent with the intermittency of renewable energy sources like solar and wind (Butler et al., 2025). Such diversification enhances security of supply but also supports low-carbon aspirations, investing in a safer and more sustainable future without compromising reliability. In order to further mitigate the intermittency and reliability issues involved in solar power generation, energy storage technologies are equally critical. These technologies, including lithium-ion batteries, pumped hydro storage, and emerging solutions, facilitate energy capture and storage during times of high solar generation. The stored energy is then utilized when generation is low, thus eliminating the supply spikes characteristic of solar power. The addition of these storage solutions can significantly enhance the stability of the grid, lessening the reliability problems that usually come with renewable energy shift. But maintaining a balance between investments in energy storage and the generation capacity is still crucial to maximize welfare gains, especially as higher levels of solar penetration are the target in the future energy system (Gowrisankaran et al., 2025).

4.4.2. Infrastructure and Grid Integration

The integration of solar power into the existing infrastructure is a challenge as well as an opportunity to the process of modernization of the energy system. The grid infrastructure needs reinforcement and accommodation with growing demand for electricity and the decline in the consumption of fossil fuels. Solar photovoltaic (PV) systems, while being environmentally friendly, are inherently variable, and thus, a problem with grid stability due to insufficient strength and inertia. To address these, the utilities must alter their tactics, employing devices such as synchronous condensers (SynCons) in order to enhance grid resilience. These changes not only increase short-circuit capacity, but also voltage and frequency regulation, key aspects of maintaining an affordable power supply. The shift towards a solar-dominant grid thus needs adequate planning and investment in infrastructure with the capacity to accommodate renewable energy's idiosyncrasies of energy generation and consumption. Furthermore, the need for innovation is underscored by the increased complexity introduced through high penetrations of renewable energy. The management process of voltage profiles in distribution systems is most problematic, especially as the prevalence of electric vehicles and intermittent renewable resources rises. Current practices of Voltage/VAR Optimization and Control (VVOC) are prone to fail in addressing this evolving landscape, and a solution for the emerging gaps in current technologies is a priority. Better VVOC techniques are essential in enabling better voltage management and further facilitating effective integration of solar PV systems into the grid. Addressing the above research gaps will be of great significance in an effort to unlock the full potential of new technologies that can facilitate an enhanced integrated use of renewable sources of energy in the distribution network, ultimately resulting in a stable and efficient grid operation. In nations like Saudi Arabia, with high potential for solar energy, an intensive infrastructure development must be undertaken to facilitate proper integration of solar power. Complete replacement of conventional generators with solar PV in utilities like those owned and operated by the Saudi Electricity Company (SEC) demands innovative ways of addressing weakened grid resilience. Using measures such as the adjustment of protection systems to manage reduced fault currents is a direct function of enhanced stability and restoration following a disturbance. The application of performance metrics such as the short-circuit ratio (SCR) as a measure of strength can guide such efforts by directing attention to the effectiveness of measures such as SynCons. Together as a unit, a robust infrastructure that can evolve in tandem with advances in solar technology will be the driving force behind a transition to renewable energy, ensuring grid integration efficient and resilient in meeting future energy demands.

4.4.3. Financial and Investment Barriers

The transition to solar power integration is faced with significant financial and investment hurdles that constrain its widespread adoption. To note, one of the obstacles is the immense initial capital investment in the installation of solar technology. Although the long-term benefits of solar energy, such as reducing electricity bills and eco-friendliness, have been well documented, the initial investment tends to deter prospective adopters, especially in developing nations. Furthermore, funding vehicles can be unevenly accessible across various geographies and therefore contribute to differences in the adoption of solar. Governments and private capital need to collaborate to create favorable finance conditions that minimize entry points and stimulate investment. Without these basics in the provision of funds for solar undertakings, the green energy objectives that have been established under programs such as those launched by ASEAN in 2015 would remain elusive and compound the existing fossil fuel reliance in the energy sector even further (Huang et al., 2019). In addition to the complexities of accessing initial capital, complexities around the financial instruments used in the establishment of solar projects can also pose significant hindrances. Investors are often unclear regarding the risks involved in investments in solar energy, including unstable energy output due to occasional sun exposure and potential policy modifications affecting profitability. Uncertainty can contribute to reluctance in further investment in solar technology and infrastructure. To resolve these challenges, it is also crucial to develop innovative funding mechanisms such as public-private partnerships or green bonds that can help ensure greater stability and predictability for the investor. Through establishing more transparent investment channels and promoting the long-term payback of renewable energy, stakeholders can trigger a more favorable setting to secure financial capital towards solar power integration (Bento et al., 2020). Further, the lack of consistent regulatory systems has a tendency to increase financial barriers to solar energy projects. Different areas may have different regulations and policies for renewable energy, something that generates uncertainty and higher compliance costs for potential investors. Standardization of regulations to provide solar energy integration is critical in increasing investor confidence and supporting larger-scale projects. Effective policy measures can reverse the risks of investment and promote long-term investment in alternative sources. Improving grid flexibility, as ASEAN's recent analysis highlights, is crucial to be able to manage the technical demands of solar power. Redressing these investment and regulatory challenges not only encourages the transition to solar energy but also serves to advance more universal climate goals through advancing a stable and predictable environment encouraging the large-scale finance into clean energy technology (Huang et al., 2019).

4.5. TECHNOLOGICAL INNOVATIONS IN SOLAR POWER

Technological innovation in solar power has extensively transformed the renewable energy industry to make it more efficient and deployable. Some of the most significant achievements are the development of highly efficient photovoltaic (PV) cells with more than 25% conversion efficiencies. Such cells employ sophisticated technologies such as bifacial solar panels and perovskite layers to optimize their ability to capture sunlight from various directions and under varying conditions. Furthermore, the advancement of manufacturing technologies has lowered the costs of solar production, making solar technology more accessible and affordable to consumers and industries alike. As battery storage technologies continue to improve, they enhance solar installations to provide stable energy sources even during the absence of sunlight, thereby addressing one of the primary impediments in solar integration. This synergy is crucial in enabling the efficiency and stability of solar energy systems, making them more viable as a primary source of energy in contemporary energy mixes. Growing smart grid technology is another column in the area of solar innovation, with ease of integration with existing energy systems. Advanced communication technologies are employed in smart grids to enable real-time monitoring and control of energy production and consumption. This allows solar energy to be optimized and programmed to be utilized efficiently throughout the grid. Additionally, solar inverters that change solar electricity to be used on the grid have been enhanced to aid in grid stability by providing reactive power and frequency management. The technology facilitates greater responsiveness to supply and demand fluctuations, reducing the risk of blackouts. As utilities increasingly incorporate distributed solar power into their business models, smart grid technologies open the door to a more flexible and secure energy system that addresses renewable energy integration goals and an age of sustainable energy. Alternative energy market forces have also played a strong role in influencing solar technology advancements, particularly through pressures within the market and regulatory changes. Clean energy policies have driven competitive business actions, provoking quicker development in solar technologies. As Bhuyan et al. (2025) identifies, alternative energy has undergone a revolutionary change since the 1970s, as a reaction to outside factors such as shifts in oil prices and subsidies from the government. These pressures have urged firms to invest in research and development, creating a variety of technologies catering to various needs of energy. Besides this, substantial financial investments, and collaborative efforts by stakeholders like governments, private corporations, and research organizations have been the cause of the accelerating advancements in solar technology. This dynamic ecological niche is maintained by the twin drivers of technological innovation and shrewd market changes, which ensure that solar energy is ever an integral component of the planet's energy shift.

4.5.1. Advances in Solar Panel Efficiency

The solar panel efficiency scene has dramatically evolved in the last decade with a shift towards greener sources of energy. Scientists and engineers have designed advanced photovoltaic material to enhance the energy absorption and conversion rate, which is closely related to overall system performance. For instance, technologies such as multi-junction solar cells, by utilizing layers of different semiconductor materials to capture a broader spectrum of sunlight, have achieved efficiencies above 40%. This is quite a distance from traditional silicon panels with a capacity to generate about 20% efficiency. Advances in materials science with the advent of perovskite solar cells provide the potential for greater efficiency at lower production costs. By increasing solar panel efficiency, we will have the capability to maximize the integration of solar power into existing energy systems, hastening a new revolution of renewable energy utilization. Furthermore, the implications of increased solar panel efficiency extend beyond technical achievement; they have the potential to significantly influence climate policy and worldwide economic systems. With the United Nations calling for a cut in greenhouse gas emissions by radically, technological advancements in the solar sector can go a long way in averting climate change. Recent estimates have revealed that the technologies projected to become cost competitive over the next decade will enable rapid decarbonization of the world's energy economy, which will fight against the looming danger of CO_2 emissions (Flavin, 2008).

Highly efficient solar panels optimize energy yields while also facilitating low-cost installation of renewable energy to the point of allowing communities to transition from fossil fuels. This shift not only supports global climate targets but also spurs economic growth by opening new industries and employment opportunities in clean energy technologies. Furthermore, incorporating such sophisticated solar products into existing infrastructures has multiple advantages. With the expansion of electric vehicles and increased emphasis on smart grid technology in full swing, high-efficiency solar cells are capable of providing synergistic advantages in that they may be employed to enhance value in energy systems developed for sustainability. The technologies developed through part of the military solar power system, as presented in previous projects, show the adaptability of high-efficiency solar technology in terms of different missions and energy needs (Barthelemy et al., 2025). Not only will these technologies meet current energy demands but also lay the groundwork for future energy security. With the implementation of advanced solar technologies, nations can redefine their energy profiles, reduce reliance on foreign fuels, and foster energy autonomy, which is vital to economic stability as well as environmental sustainability. Overall, the ongoing research and development of solar panel efficiency offer a promising path in the global pursuit of tapping solar power in an effective way.

4.5.2. Energy Storage Solutions

The integration of solar power into the existing power grid is hindered by unique requirements that are attainable with novel energy storage technologies. The latter are central to mitigating the intermittency of solar-powered power generation, which depends on weather patterns as well as on daily sunlight exposure patterns. Using various technologies of storage such as lithium-ion batteries, pumped hydro storage, and newer technologies such as flow batteries, energy can be stored in surplus at times of peak sunshine and supplied during peak-demand or low-generating hours. Such balancing of supply and demand not only augments grid reliability but also helps in more penetration of renewable sources of energy and thus contributes to the attainment of climate objectives. In addition, with the advancement of energy storage technology in efficiency and cost, it performs increasingly well to optimize solar power integration at scale to provide a more resilient and sustainable energy infrastructure (Bento et al., 2020). Further, the financial viability of energy storage technology is also assisted by major technology improvements and increased investment in renewable resources. A diversified approach to energy storage—both centralized and distributed solutions—will maximize the benefits of solar power integration. For instance, distributed energy storage devices can be installed in residential homes and commercial buildings to allow home and business owners to use excess solar power at a later time. This decentralized architecture not only bestows autonomy on consumers but also reduces the stress on the central grid, testifying to the complementary nature of solar energy and storage systems. Additionally, studies indicate that regions with progressive policies for decentralized renewable technologies enjoy improved energy security and reduced greenhouse gas emissions and thus confirm the vital significance of energy storage for gaining viable energy solutions (Kammen et al., 2019). With mounting demand for clean energy options globally, efficient energy storage opportunities become a sine qua non in overcoming the bottlenecks of integrating solar power into the energy industry. Policy makers are required to give highest priority to activities leading to research and development of energy storage technologies, as well as funding models for their adoption. Such commitment is particularly crucial in developing economies, where rapid growth in energy consumption creates a spectrum of challenges and opportunities. For example, countries like Bangladesh are beginning to realize renewable energy potential, where geospatial analysis indicates significant utility-scale solar development potential. Such reports underscore the importance of strategic planning that reconciles land-use with renewable energy production, thus underscoring further how new energy storage technologies can enable a sustainable power generation and distribution plan (Kammen et al., 2019).

4.5.3. Smart Grid Technologies

The integration of solar power into the existing energy infrastructure is very difficult, primarily due to the old infrastructure of the traditional power grids. Smart grid systems offer innovative solutions to facilitate this integration, offering cures for the voltage regulation and power quality problems that are caused by the intermittent nature of solar power generation. As identified, the traditional grid network in areas like South Africa was not able to handle distributed generation at high capacity, and it will create challenges as solar photovoltaic installations rise. With the implementation of smart grid technologies, utilities are able to enhance their ability to handle these challenges in an effective manner, and thus it is possible to shift to renewable energy sources without incurring massive costs of network reinforcement. Thus, smart grids can transform existing infrastructures into more adaptive networks, responsive to existing renewable energy demands and significantly improving the resilience and reliability of power distribution networks (Parbhoo et al., 2015). The advent of information and communication technologies (ICT) has redefined the operation of power systems, allowing for integration with smarter grids that can utilize renewable sources of energy such as solar and wind. Traditional models of centralized power production are increasingly less appropriate as supply and demand patterns shift with rising decentralization of energy. Smart grid technologies incorporated into the system not only increase energy efficiency but also provide an environmentally cleaner means to deliver electricity. High-resolution data gathering and processing are enabled by advanced ICT, which enhances operational efficiencies and allows grid operators to monitor real-time conditions. Smart grid technology is crucial in dealing with the issue of complexity that arises from high levels of solar power penetration. Furthermore, the emphasis on the use of sustainable practices is resonant with the need to constrain climate change globally, and therefore, smart grids are a critical component of modern energy policy (Li et al., 2010). The integration of smart grid technologies into solar integration framework mechanisms is crucial for creating a more secure future for energy. These technologies facilitate improved load management and demand response processes, reducing the risk of outages and optimizing energy use. With growing distributed generation, robust communication infrastructures are a necessity. These infrastructures are the cornerstone of the sustainability of new operational modes, relying on real-time observation and control of energy flows. For instance, risk-limiting dispatch processes can evolve through effective smart grid technology deployment. By dynamically balancing supply and demand, these systems not only enhance the stability of the grid but also ensure environmental sustainability through the judicious use of available renewable energy resources. Consequently, smart grid technologies interfacing with solar power integration has a lot to offer in terms of energy efficiency and sustainability (Li et al., 2010; Parbhoo et al., 2015).

4.6. POLICY AND REGULATORY FRAMEWORK

The policy and regulatory framework for solar power integration is critical in facilitating the attainment of renewable energy targets as well as grid flexibility.

With countries planning to increase their use of renewable energy, as is the situation of ASEAN projecting increasing this from 13% to 23% by 2025, grid issues become apparent. The integration of intermittent power supplies such as solar requires a high degree of sophistication in the understanding of grid flexibility, which is the ability of the electric grid to react to changing supply and demand situations. By looking at grid reliability, access to the power market, and the diversity of renewable supplies, policymakers can identify vulnerabilities that pose obstacles to successful integration. Enhancing these aspects of grid flexibility is a vital step for ASEAN nations and others to achieve ambitious renewable energy objectives while maintaining overall system costs as low as possible (Huang et al., 2019). In Brazil, the policy environment regarding energy is filled with opportunities and challenges for solar power integration. With investments expected to hit 0.5 trillion U.S. dollars in the next 10 years, important choices will have a significant impact on the country's economic competitiveness and environmental sustainability. Policymakers must balance securing carbon-based infrastructures with developing efforts that would propel Brazil as a low-carbon economy leader. This setting demands energy and climate strategy reactions that are simultaneously, synergistic between the two fields, offering a harmonious policy strategy to reduce greenhouse gas emissions especially in major sectors such as transport, industry, and power generation. Synergic strategy feasibility, which promotes solar integration along with achieving climate goals, highlights the importance of visionary and sound regulation (Lucon et al., 2015). Regional regulatory frameworks focus on the importance of regional cooperation and shared policies to facilitate effective integration of solar energy. For ASEAN, the need for higher grid flexibility is a cross-cutting concern across member states, which calls for concerted action to address problems such as inadequate forecasting systems and limited access to the electricity market. A cooperative strategy can facilitate the exchange of technologies, resources, and best practices that not only increase regional energy security but also assist in achieving greater penetrations of renewable energy sources like solar power. Similarly, the Brazilian experience of energy policy underlines the necessity of a broad, coordinated strategy combining environmental commitments with economic objectives. Having policies in hand that will have good conditions to integrate solar energy can ultimately mean sustainable growth and energy resilience on a regional scale (Huang et al., 2019; Lucon et al., 2015).

4.6.1. Government Incentives and Subsidies

Government incentives and subsidies play a very core role in the integration of solar power technologies in various economies. These budgetary-driven policies not only relieve the significant initial cost burden of solar installations but also create a market competition and innovation-friendly environment. For instance, countries like Brazil are confronted with large policy decisions that will principally define their energy future. The national investment policy can lead to a dramatic shift towards renewable energy sources, thereby catalyzing economic competitiveness against the background of international climate commitment (Lucon et al., 2015). This shift is an example of how state action can stimulate sectoral growth and technology adoption, ultimately enhancing the core role of renewable energy in combating climate change and promoting sustainable development. Similarly, China's ambitious renewable energy plan is the quintessential case of the effect of government incentive on bringing solar power into the mainstream. With increased installed capacity and an abundance of supportive policies, China has come to be at the forefront of renewable energy technologies globally. The 2005 Renewable Energy Law imposing subsidies in terms of a nominal rate charged on electricity consumers is one such example illustrating how economic mechanisms facilitate the scaling up of renewable sources of energy (Campbell et al., 2010). Such policies are not only lowering the cost burden on consumers but also making solar projects feasible. With systems of regulation and financial incentives, the government of China promotes innovation and solar power installation, thus speeding up progress towards aimed renewable energy objectives while acting to address pressing economic and environmental needs. In addition, the economics of government subsidy play an enormous role in affecting the supply and demand dynamics of the solar power market. Subsidies on one hand make consumers cheaper to solar technology, hence broadening the clientele willing to spend on solar energy solutions. On the other hand, the incentives energize firms to become more efficient and reduce production costs, ultimately leading to market enlargement. As seen in China and Brazil, a blending of subsidy targeting and firm policy can stimulate the adoption of solar power, with both environmental and economic benefits. With the developing nations striving to achieve global greenhouse gas reduction levels, government incentives and subsidy will remain important drivers of sustainable displacement from fossil fuels (Campbell et al., 2010; Lucon et al., 2015).

4.6.2. International Agreements and Commitments

The assimilation of solar power into the global energy mix needs a firm foundation of international agreements and obligations. These agreements serve as platforms for collaboration between countries aiming to boost the utilization of renewable energy, counteract global warming, and attain sustainable development. For instance, initiatives like the Paris Agreement are aimed at the role of countries in setting and achieving their renewable energy targets and consequently creating a cooperative burden towards carbon diminution. Additionally, the proliferation of regional accords, as in those established by ASEAN, illustrates a strategic way of harmonizing energy policies between countries. These facilities not only facilitate knowledge transfer and technology adoption but also foster a political environment that promotes the adoption of solar power despite the numerous challenges, driving nations towards a sustainable energy future. Having an understanding of the unusual challenges that accompany the use of renewable energy sources like the sun, global agreements tend to focus on addressing specific technical and infrastructure demands. The ASEAN framework recognizes grid flexibility as a key resource for addressing the uncertainties of solar energy supply through intermittent sources. Based on the study by Huang et al. (2019), nations in the ASEAN region have different levels of preparedness when it comes to their electricity infrastructure, thus requiring specific regional strategies to improve the resilience of grids. By sharing best practices and investing in technologies that increase grid flexibility, member states are better positioned to accommodate increasing levels of renewable energy penetration. This cooperation also makes possible a more efficient use of resources, enhancing energy security at lower aggregate costs for the transition to a renewable energy-dominated system. Regional and international agreements also often create the framework for innovative financing mechanisms to drive solar power initiatives. For example, contributions by institutions like the Austrian Development Agency are key to assisting research and implementation activities that enhance energy infrastructure in less developed nations, according to Novakovic et al. (2019). Such investments assist in ensuring that those countries with limited resources can be engaged in the global transition towards solar energy utilization. Such collaborations among governments, private sector players, and research institutions have a synergistic effect, supporting efforts towards a shared goal of realizing renewable energy targets. In the end, the success of global deals in making solar power integration possible also relies on more than ambitions to reach for. It should also enable pragmatic solutions to the different capabilities and needs of the participating nations.

4.6.3. Local and State-Level Initiatives

Local and state initiatives have a key function to contribute to the promotion of solar power integration on a regional basis. Such initiatives normally find expression in policy drivers favoring the exploitation of renewable energy in the form of tax credits, rebates, and grants to homeowners and businesses adopting the use of solar technology. States, for instance, have implemented

Renewable Portfolio Standards (RPS) mandating part of the power to be sourced from renewable goods. This kind of legislation not only boosts growth in the solar industry but also leads to the employment of local residents. Moreover, effective outreach activities foster public awareness of the potential benefits of solar power, thereby boosting public support and engagement. By fostering heightened awareness and participation by communities, demand for solar installations arises, ultimately reinforcing the importance of local initiative in establishing a clean-energy future. Other than financial incentives, local governments are increasingly taking an active role in financing infrastructure for solar energy. This includes the utilization of solar microgrids, which are geographically localized networks of interconnected energy resources that can be stand-alone or grouped together. These systems enhance energy resiliency through the potential of communities to have power despite disruptions or peak demand. According to recent studies, these microgrids find most utility in zones of disaster risk, where they can provide much-needed stability and serve localized energy needs. With favorable federal and local policies supporting them, microgrid plans are being conceptualized in conjunction with state energy policy and market conventions. This ensures that they not only enhance the energy landscape at the local level but also contribute toward higher ideals of sustainability (Bao et al., 2018). Local and state initiatives also focus on education and cooperative strategies for establishing a culture of sustainability. School, business, and community organization initiatives work to provide training in solar technology and energy-saving methodologies. Through college workshops and collaborative efforts with local colleges, these initiatives empower individuals to apply solar solutions within their own lives and then impact others within their community. In addition, municipalities can partner with states to develop joint projects for developing overall solar development plans that are compatible with the needs and resources of the place. Such approaches not only enable an integrated method of installing solar power but also provide fair access to renewable energy to various communities. By enhancing awareness and coordination, state and local initiatives thus serve to be important drivers of the successful introduction of solar energy at a larger level.

SUMMARY

Incorporation of solar power into modern energy systems is a multifaceted approach to fighting the effects of climate change and energy shortages. As the globe shifts towards cleaner lifestyles, solar power is particularly well-placed since it can be utilized in large quantities, from single residential rooftops to industrial-level solar farms. Its successful integration not only lowers fossil fuel reliance but enhances energy security as well. By employing new technologies such as photovoltaic shading systems, buildings can conserve significant amounts of energy while also providing an effect of cooling at the same time. This double benefit provides a more economically viable option for energy production, evident in many research studies that record the cost-saving capability and improved payback terms for such installations (Bloem et al., 2005). Furthermore, the incorporation of solar microgeneration technology is indicative of the trend towards decentralized energy generation that can transform the way we engage with power consumption, particularly in technologically advanced settings. Utilization of standards like Power over Ethernet to provide energy in an effective way enhances the feasibility of solar solutions for urban centers where traditional grid infrastructure may be inferior or out of date. The ability to supply low-power products with renewable energy not only proves the versatility of solar solutions but also underscores their relevance in the current media and IT markets, which increasingly emphasize sustainability. The technological leap in energy management through this convergence is a critical steppingstone toward achieving the broader objectives of energy efficiency and reduced operating costs (Counsell et al., 2009). Finally, the path to complete integration of solar energy is set with issues and opportunities that deserve close examination by policymakers, corporate leaders, and academics alike. By leveraging innovations such as photovoltaic shading systems and microgeneration technologies, we can create a more resilient energy system that is centered on sustainability and efficiency. Finally, the implementation of solar power integration not only advances green goals but also drives economic growth via job creation and technological advancements. Inter-stakeholder collaboration to cultivate effective incentives and policies will be instrumental in driving this transformation to make solar power become a pillar of our energy source in the future. With solar power's potential yet to be fully revealed, sustained research, education, and community awareness will become the key to reaping its maximum benefits for humanity and the world. The integration of solar power into existing energy infrastructures is a fundamental shift in both infrastructural and technological paradigms. Among the key to successful adoption are drivers of performance that are associated with solar energy production and distribution. The models of analysis applied to the examination of various satellite subsystem requirements can also be applied to solar power systems, demonstrating how such models are dependent on cost, risk, and efficiency in operations. These parameters, as outlined by Pidgeon et al. (2025), serve as principles of instruction for system design and technology selection. By careful consideration of these performance indicators, stakeholders can make informed decisions on the technologies to implement towards maximum energy output and integration, leading to higher reliability and sustainability of solar energy systems. In the case of solar power integration, the complexity involved in modeling various parameters cannot be overemphasized. As the energy landscape expands to contain more variables-everything from solar cell efficiency ratings to more general environmental factors-the need for efficient modeling becomes critical. Dimensionality reduction methods, such as finding active subspaces, have been demonstrated to significantly alleviate the difficulties of dealing with multiple input parameters. A one-diode solar cell model, as cited by Campanelli et al. (2015), showed that the majority of predictive capability exists in a one-dimensional active subspace. This finding enables effective parameter studies and optimizations that are useful to the process of decision-making on scaling and implementing solar technology in current infrastructures. Besides, transition to solar power demands a symbiotic connection with energy storage and grid management technology advancements. Integration enables solar energy, as influenced by variable weather conditions and time, to be stored effectively and dispatched accordingly. With advancements in analytics and modeling technologies, as mentioned in the previous section, the energy providers are able to design a more responsive and robust grid system. While the technology is evolving in solar, along with its integration with current infrastructure, the stakeholders must understand the evolving dynamics, which are the key to realizing the potential of renewable sources. This sustained focus on both flexible technology and robust modeling will be key to future success in solar energy integration.

REFERENCES

1. Abdel Gelil, I., Aguayo, F., Boyle, G., et al. (2012). *Policies for Capacity Development.* https://core.ac.uk/download/9836421.pdf (accessed on 30th September 2025).
2. Al Nahdi, W. A., & Ali, M. I. H. (2021). Feasibility study for water-electricity cogeneration using integrated system of concentrated solar power and biofuel as renewable energy. *Renewable Energy*, *178*, 137–148. https://core.ac.uk/download/482000557.pdf (accessed on 30th September 2025).
3. Albukhari, W. M. R. (2025). *Enhancing Power System Strength and Inertia in the Saudi Electric Grid for Solar Power Integration.* https://core.ac.uk/download/660473212.pdf (accessed on 30th September 2025).
4. Arghandeh, R., Brown, M., Cibulka, L., et al. (2013). *Distribution System Voltage Management and Optimization for Integration of Renewables and Electric Vehicles: Research Gap Analysis.* https://core.ac.uk/download/288432751.pdf (accessed on 30th September 2025).
5. Arrighi, G., Boyle, G., Bradsher, K., Diao, Y., Du, J., Yeung, G., et al. (2012). Geographical interdependence, international trade and economic dynamics: The Chinese and German solar energy industries. *Energy Policy*, *48*, 462–471. https://core.ac.uk/download/9551789.pdf (accessed on 30th September 2025).
6. Bao, Y., Feng, W., Jin, M., et al. (2018). A review of microgrid development in the United States–A decade of progress on policies, demonstrations, controls, and software tools. *Applied Energy*, *228*, 1332–1349. https://core.ac.uk/download/323310623.pdf (accessed on 30th September 2025).
7. Barthelemy, R. R., Mahefkey, E. T., & Massie, L. D. (2025). Military needs for orbital power. *Acta Astronautica*, *196*, 521–530. https://core.ac.uk/download/pdf/42872072.pdf (accessed on 30th September 2025).
8. Becerra Villanueva, J. A., Calderón, B., Carvalho, M., Chacartegui, et al. (2019). Transcritical carbon dioxide charge-discharge energy storage with integration of solar energy. *Energy Conversion and Management*, *195*, 1353–1365. https://core.ac.uk/download/322845447.pdf (accessed on 30th September 2025).
9. Beluco, A., Canales, F. A., Guezgouz, M., et al. (2019). A review on the complementarity of renewable energy sources: Concept, metrics, application and future research directions. *Renewable and Sustainable Energy Reviews*, *112*, 725–747. https://core.ac.uk/download/286642706.pdf (accessed on 30th September 2025).

10. Bento, N., De Stercke, S., Grübler, A., Healey, et al. (2020). Granular technologies to accelerate decarbonization. *Science, 368*(6486), 649–654. https://core.ac.uk/download/323057192.pdf (accessed on 30th September 2025).
11. Bhuyan, R., & Mytelka, L. K. (2025). *Changing Configuration of Alternative Energy Systems*. https://core.ac.uk/download/pdf/6941347.pdf (accessed on 30th September 2025).
12. Blain, A. W., Dowell, C. D., Leisawitz, D. T., Benford, D. J., Young, E. T., Kooi, J. W., Mather, J. C., et al. (2005). Mission concept for the Single Aperture Far-Infrared (SAFIR) Observatory. *Astrophysics and Space Science, 298*(2), 343–349. http://arxiv.org/abs/astro-ph/0508578 (accessed on 30th September 2025).
13. Bloem, J. J., Colli, A., & Strachan, P. (2005). Evaluation of PV technology implementation in the building sector. *Renewable Energy, 30*(2), 289–303. https://core.ac.uk/download/9018964.pdf (accessed on 30th September 2025).
14. Blum, D. H., Gehbauer, C., Lee, E. S., et al. (2020). An assessment of the load modifying potential of model predictive controlled dynamic facades within the California context. *Energy and Buildings, 224*, 110243. https://core.ac.uk/download/287622187.pdf (accessed on 30th September 2025).
15. Bouznit, M., del Pablo-Romero, M. P., & Sánchez Braza, A. (2020). Measures to promote renewable energy for electricity generation in Algeria. *Renewable Energy, 146*, 2823–2835. https://core.ac.uk/download/299806337.pdf (accessed on 30th September 2025).
16. Brazier, F. M. T., Voulis, N., & Warnier, M. (2017). Impact of service sector loads on renewable resource integration. *Energy Policy, 105*, 618–629. http://arxiv.org/abs/1605.09667 (accessed on 30th September 2025).
17. Buisman, C. J. N., Hamelers, H. V. M., Helder, M., et al. (2010). Microbial solar cells: Applying photosynthetic and electrochemically active organisms. *Trends in Biotechnology, 28*(11), 566–574. https://core.ac.uk/download/30436016.pdf (accessed on 30th September 2025).
18. Bustamante, H., Newborough, M., & Scamman, D. (2015). Hybrid hydrogen-battery systems for renewable off-grid telecom power. *Renewable Energy, 76*, 561–569. https://core.ac.uk/download/79499881.pdf (accessed on 30th September 2025).
19. Butler, L., Grubb, M., & Sinden, G. (2025). Diversity and security in UK electricity generation: The influence of low carbon objectives. *Energy Policy, 83*, 221–234. https://core.ac.uk/download/pdf/7093211.pdf (accessed on 30th September 2025).

20. Callaghan, W., & McDonald, R. (2025). *Flat-Plate Solar Array Project. Volume 1: Executive Summary*. https://core.ac.uk/download/pdf/42837491.pdf (accessed on 30th September 2025).
21. Cambini, C., Franzi, D., Eberlein, B., Levy, B., Schimmelfennig, F., et al. (2014). Assessing the EU pressure for rules change: Perception by southern Mediterranean energy regulators. *Energy Policy*, *69*, 57–65. https://core.ac.uk/download/pdf/11432127.pdf (accessed on 30th September 2025).
22. Campanelli, M., Constantine, P. G., & Zaharatos, B. (2015). Discovering an active subspace in a single-diode solar cell model. *Journal of Renewable and Sustainable Energy*, *7*(4), 041705. http://arxiv.org/abs/1406.7607 (accessed on 30th September 2025).
23. Campbell, R. J. (2010). *China and the United States—A Comparison of Green Energy Programs and Policies*. https://core.ac.uk/download/5130065.pdf (accessed on 30th September 2025).
24. Counsell, J. M., Haughton-Brown, C., & Tuff, S. (2009). Renewable power for lean desktops in media applications. *Energy Efficiency*, *2*(3), 245–252. https://core.ac.uk/download/9022822.pdf (accessed on 30th September 2025).
25. Di Matteo, U., & Nastasi, B. (2017). Innovative use of hydrogen in energy retrofitting of listed buildings. *Energy Procedia*, *133*, 313–322. https://core.ac.uk/download/141694922.pdf (accessed on 30th September 2025).
26. Flavin, C. (2008). *Low-Carbon Energy: A Roadmap*. https://core.ac.uk/download/71340916.pdf (accessed on 30th September 2025).
27. Gowrisankaran, G., Samano, M., & Reynolds, S. S. (2025). Intermittency and the value of renewable energy. *Journal of Political Economy*, *133*(5), 1147–1190. https://core.ac.uk/download/pdf/6496966.pdf (accessed on 30th September 2025).
28. Huang, Y. W., Kammen, D. M., & Kittner, N. (2019). ASEAN grid flexibility: Preparedness for grid integration of renewable energy. *Energy Policy*, *128*, 711–718. https://core.ac.uk/download/323310932.pdf (accessed on 30th September 2025).
29. Kammen, D. M., Shiraishi, K., & Shirley, R. G. (2019). Geospatial multi-criteria analysis for identifying high priority clean energy investment opportunities: A case study on land-use conflict in Bangladesh. *Energy for Sustainable Development*, *50*, 104–114. https://core.ac.uk/download/323310931.pdf (accessed on 30th September 2025).
30. Kester, J., Noel, L., Sovacool, B. K., Zarazua de Rubens, G., et al. (2018). Beyond emissions and economics: Rethinking the co-benefits of Nordic electric vehicles (EVs) and vehicle-to-grid (V2G). *Journal of Cleaner*

Production, *190*, 807–818. https://core.ac.uk/download/161102773.pdf (accessed on 30th September 2025).

31. Koh, J., & Stremke, S. (2009). Sustainable energy transition: Properties and constraints of regenerative energy systems with respect to spatial planning and design. *Futures*, *41*(1), 13–21. https://core.ac.uk/download/pdf/29249814.pdf (accessed on 30th September 2025).
32. Lemey, S., & Rogier, H. (2015). Substrate integrated waveguide textile antennas as energy harvesting platforms. *Sensors*, *15*(7), 15626–15645. https://core.ac.uk/download/55782750.pdf (accessed on 30th September 2025).
33. Li, V. O. K., Wu, F. F., & Zhong, J. (2010). Communication requirements for risk-limiting dispatch in smart grid. *Proceedings of the IEEE*, *99*(1), 194–205. https://core.ac.uk/download/37951214.pdf (accessed on 30th September 2025).
34. Lucon, O., Fransen, T., & Romeiro, V. (2015). *Bridging the Gap Between Energy and Climate Policies in Brazil: Policy Options to Reduce Energy-Related GHG Emissions*. https://core.ac.uk/download/75782972.pdf (accessed on 30th September 2025).
35. McHenry, M. P. (2010). Diverse, remote and innovative–Prospects for a globally unique electricity network and market in Western Australia. *Energy Policy*, *38*(3), 1221–1230. https://core.ac.uk/download/11237237.pdf (accessed on 30th September 2025).
36. Millstein, D., & Wiser, R. (2020). Evaluating the economic return to public wind energy research and development in the United States. *Nature Energy*, *5*(1), 49–55. https://core.ac.uk/download/286361704.pdf (accessed on 30th September 2025).
37. N/A. (2025). *Technology for Large Space Systems: A Special Bibliography with Indexes (Supplement 04)*. https://core.ac.uk/download/pdf/42860983.pdf (accessed on 30th September 2025).
38. Novakovic, T., Rammel, C., Voshmgir, S., et al. (2019). *Sustainable Development Report: Blockchain, the Web3 & the SDGs*. https://core.ac.uk/download/286370731.pdf (accessed on 30th September 2025).
39. Oliver, R., Tillotson, J., & Toomey, A. (2009). Bioengineered textiles and nonwovens–The convergence of bio-miniaturisation and electroactive conductive polymers for assistive healthcare, portable power and design-led wearable technology. *Journal of the Textile Institute*, *100*(5), 407–419. https://core.ac.uk/download/9587153.pdf (accessed on 30th September 2025).

40. Parbhoo, N. (2015). *An Investigation into The Use of Smart Grid Technologies for the Northern Cape Province of South Africa, Where Extensive Solar Generation Is Planned in A Constrained Network*. https://core.ac.uk/download/39676175.pdf (accessed on 30th September 2025).
41. Petrollese, M., Braggio, M., Peer, M. S., Orrù, P. F., & Melesse, T. Y. (2025). Next-generation CSP: The synergy of nanofluids and Industry 4.0 for sustainable solar energy management. *Renewable Energy*, *200*, 1450–1463. https://core.ac.uk/download/657073628.pdf (accessed on 30th September 2025).
42. Pidgeon, D., Price, K. M., & Tsao, A. (2025). Mass and power modelling of communication satellites. *Acta Astronautical*, *197*, 531–540. https://core.ac.uk/download/pdf/42812001.pdf (accessed on 30th September 2025).
43. Thornley, B., Wood, D., & Grace, K. (2012). *Impact at Scale: Policy Innovation for Institutional Investment with Social and Environmental Benefit*. https://core.ac.uk/download/71358489.pdf (accessed on 30th September 2025).
44. Tuohy, P. G. (2009). Simulation and BIM in building design, commissioning and operation: A comparison with the microelectronics industry. *Automation in Construction*, *18*(2), 210–215. https://core.ac.uk/download/9032439.pdf (accessed on 30th September 2025).

CHAPTER 5

WIND POWER INTEGRATION

Contents

INTRODUCTION

The integration of wind power into the mainstream energy infrastructure is a crucial frontier in the global shift towards sustainable energy sources. As the need for clean, renewable energy becomes increasingly pressing, wind power has been projected as an answer due to its widespread availability and minimal environmental footprint. Nevertheless, the on-again, off-again nature of wind energy generation presents daunting challenges to demand-supply balancing, thereby making efficient integration worthy reasons. Emphasizing the exigency for a flexible grid, the operators need to deal with the unpredictability inherent in wind power, employing pioneering technologies and operational strategies to render the grid secure. This demand for flexibility is further exacerbated by ASEAN's plan to increase its renewable energy share, which illustrates the challenge of incorporating intermittent resources into an existing energy system (Huang et al., 2019).

Figure 5.1. *Wind power integration in Egypt.*

Source: https://www.ea-energianalyse.dk/wp-content/uploads/2020/02/sunrise-4317151_1920.jpg.

The contribution of electric mobility systems to wind power integration should not be underestimated. The advent of electric vehicles (EVs) presents opportunities alongside challenges in electrical grid stability. As outlined in a recent workshop on coupling electric mobility with grid infrastructure, technology has taken considerable strides beyond our understanding of how to deal with the interplay between increasing electricity demand from electric vehicles and clean-energy supply (Cassandras et al., 2019). Balancing EV charging with times of peak wind generation can help relieve some of the grid pressure, enabling improved utilization of renewable sources. This is, nonetheless, dependent on further research and collaboration between stakeholders to dispel uncertainties and optimize the capacity of infrastructure appropriately. Integration of wind power in the future will be best accomplished via a strategy that encompasses technological innovation, policy frameworks, and cross-sectorial collaboration. Institutions and the energy supply industry must play the key role in establishing regulatory environments conducive to investing in renewable technology and grid improvements. Concurrently, popular acceptance and popular involvement are necessary to gaining approval for the transition to wind power and electric vehicle systems. With pressure mounting to meet climate goals, the stakeholders must move rapidly to create and implement solutions that match renewable resources with the infrastructure ability of the existing grid. Investment in grid flexibility and system resilience, as well as interregional coordination, as highlighted in the ASEAN scenario, remains paramount to guarantee that the transition toward wind power is not only cost-effective but cost-beneficial in the long run (Huang et al., 2019) (Figure 5.1).

Definition of Wind Power

Wind energy is the technique of harnessing air movement induced by wind to produce energy, primarily electricity. Two are the most widely utilized technologies in this sector: horizontal-axis wind turbines (HAWTs) and vertical-axis wind turbines (VAWTs). HAWTs are characterized by their effectiveness and conventional design, with blades moving about a horizontal axis. On the other hand, VAWTs operate on the vertical axis and are usually installed in urban settings since they are capable of capturing wind from any direction. Wind energy is distinct among energy sources since its generation does not create direct emissions and is founded on an abundant and renewable natural resource. The integration of wind power into the energy infrastructure requires an understanding of its variability in relation to climatic forces, which presents challenges for grid reliability and stability. The integration of wind power into traditional energy systems requires resolution of its variability and intermittency. Weather and climate play a significant role in the amount of energy generated by wind turbines during a specific period, creating variability that could render supply management challenging. To counter this, hybrid energy systems integrating wind with other forms of renewable energy sources such as solar or hydroelectricity are gaining popularity. Among the proposed solutions to match demand and supply is one that utilizes the complementarity of various energy sources using a single power plant, as noted in recent research. An example is the co-location of wind and solar technologies, which can offer a more constant energy supply by having maximum generation during mixed weather conditions, hence improving overall efficiency and reliability (Beluco et al., 2019). Furthermore, energy storage technologies play a critical role in facilitating wind energy integration into energy systems. Because the output of wind energy is not necessarily synchronized with electricity demand, efficient storage mechanisms are developed to store excess energy during periods of high generation for use at a later date. Methods like pumped-hydro and emerging technologies revolving around thermal energy systems have been proposed as viable solutions for addressing this issue (Villanueva et al., 2019). New methods are capable of storing energy when production exceeds use and releasing it when the network requires more power, increasing the reliability of renewable energy systems. Lastly, the successful integration of wind energy requires a comprehensive approach that converges technological creativity, master planning, and energy dynamics to create an equilibrated, efficient, and sustainable energy future (Figure 5.2).

Figure 5.2. *Wind turbines.*

Source: https://www.windsystemsmag.com/wp-content/uploads/2019/10/1019-CW-I1.jpg.

Importance of Renewable Energy Sources

The move towards renewable energy sources is essential in facing the looming issues associated with climate change, energy security, and financial sustainability. Wind energy, in this case, has emerged as a key driver in the global shift towards cleaner energy systems. As governments and institutions wake up to the extreme environmental impacts of fossil fuels, tapping into wind power becomes ever more evident. Wind energy offers a sustainable solution that is not just abundant but also cost-effective, with technology and prices improving over time. This shift is central to mitigating the release of greenhouse gases because wind power produces electricity without any direct carbon emissions. By incorporating wind power within the overall energy provision system, we are one step closer to achieving a decarbonized economy, with sustainability and planetary stewardship being the core focus. While the environmental benefits of wind power are significant, its incorporation into the existing power systems with success is marred by a variety of challenges, especially relating to energy storage and transmission. Its intermittent nature of production necessitates the setting up of effective energy storage systems to ensure the stability of the supply. The use of transcritical carbon dioxide charge and discharge operations in combination with thermal energy storage systems is one new approach to it. This technology eliminates the non-dispatchability problem of wind power using the capability to store inexpensive energy during off-peak periods for utilization during peak demand. The findings of recent studies indicate that with solar power as an input too, the efficiency levels of these systems can be over 60% and hence a cost-effective option for broad-scale integration of energy. Thus, it is important to develop these technologies for the future viability of wind power in the renewable energy industry (Villanueva et al., 2019). A review of regenerative energy systems highlights the transformational potential of integrating renewable technologies like wind into our built environment. Comparative analysis shows that regenerative systems enhance environmental quality and help develop more resilient infrastructure. For instance, those districts that pursue a regenerative approach are more efficient in energy and less dependent on fossil fuels. In a case study in South Limburg, the Netherlands, most important attributes, and weaknesses of these systems have been evaluated, emphasizing the importance of spatial planning to ensure effective energy integration. Such an admiration is important for policymakers and urban planners who would like to design sustainable communities that maximize the energy efficiency of renewable energy. Ultimately, the inclusion of wind energy, accompanied by advances in energy storage and the guidelines of regenerative design, is a key trajectory toward a sustainable energy future (Koh et al., 2009) (Figure 5.3).

Figure 5.3. *The importance of renewable energy in combating climate change.*

Source: https://c402277.ssl.cf1.rackcdn.com/photos/8740/images/original/wind-mill_optimized.jpg?1429192877.

Overview of Wind Power Integration Challenges

The integration of wind energy into conventional electrical grids is confronted with a wide range of issues related to the natural variability and intermittency of wind power output. Unlike conventional energy supplies with the ability to generate a consistent output, wind generation varies with changes in wind direction and speed. This variability requires advanced forecasting techniques and flexible grid management practices. In addition, the rising penetration of wind power can cause voltage variations and impact distribution transformer operation, particularly in low voltage distribution systems. Research has shown that as the integration of wind power increases, these issues become more acute, and simulation models must be used to analyze power flow behavior and harmonic emissions, particularly at the high voltage (HV) and low voltage (LV) distribution level (Ali et al., 2013).

Dynamic voltage stability is another significant aspect of wind power integration, which poses significant challenges to the electrical network. As their contribution towards the energy supply increases with wind sources, their impact on voltage stability has to be thoroughly researched. Wind farms require reactive power control strategies to maintain voltage at acceptable levels, especially during intermittent wind patterns. Poor control of reactive power can lead to voltage collapse and undermine overall system stability. By developing sophisticated dynamic wind farm models, researchers are in a position to simulate various grid perturbations, such as faults and frequency and reactive power demand variations.

Using these models, grid operators can implement countermeasures, enhance dynamic voltage stability, and ensure stable electricity supply as renewable energy technologies evolve (Ferreira et al., 2015). In addition to voltage stability, wind power integration poses concerns regarding grid infrastructure and technology adaptation. Most of the available grid transmission and distribution infrastructures were designed for centralized fossil fuel-based generation, creating incompatibilities because distributed wind generation is increasing. Upgrading of grid infrastructure is a capital-intensive process and may involve the creation of smart grid technologies to enhance monitoring, control, and automatic functions.

Moreover, the interaction between wind energy and traditional systems complicates the control of power flow and grid stability. Thus, addressing these integration challenges demands collaborative work among utility organizations, policy makers, and researchers to ensure a secure power system capable of addressing a vast range of renewable energy sources, such as wind energy in this scenario.

5.1. HISTORICAL CONTEXT OF WIND POWER

From the history of wind power, it indicates a change from conventional application to the incorporation of contemporary times into energy systems. Wind energies use started with ancient societies in grinding grains and moving water, as far back as the Persians and Europeans built windmills going as early as 1000 A.D. The Industrial Revolution was a milestone insofar as technological advancement made possible the development of larger, more efficient wind turbines. It was not until late in the 20th century, however, that wind power gained momentum as a reliable alternative source of power because of the oil crises and increasing environmental concern. This tour de force in history points to the importance of societal and governmental support in propelling innovation, which provided the foundation for the re-emergence of wind power in modern energy policies aimed at integrating renewable energy sources into national energy grids. As demand for sustainable energy solutions grew in the latter part of the 20th century and later, the majority of nations began investing in wind power installations.

Figure 5.4. *History of wind power.*

Source: https://www.windustry.com/image/windmills.jpg.

The development of coherent government policies and incentives played a crucial role in creating this industry. In America, for example, the extension of the Production Tax Credit (PTC) and state renewable portfolio standards significantly stimulated investment in wind energy projects. Similarly, on the European continent, ambitious energy policies promoted the establishment of onshore and offshore wind farms, leading to a colossal increase in world installed wind capacity. Internationally, as nations transitioned to reduce greenhouse gas emissions, wind power arrived not just as a necessary component in the diversification of energy but also as frontier technology that signaled synergy between private and public sectors in addressing climate change challenges, which is increasingly relevant to today's energy transition discourse. The economic implications that may result from the integration of wind power into the energy markets cannot be overlooked. Experiments have identified that although variable renewable energy (VRE) sources like wind are causing difficulties to traditional baseload power generation, they are also contributing to the development of a dynamic energy market. For instance, the effect of VREs on market pricing reflects a considerable shift in the supply and demand curve of energy, which is leading to declining wholesale electricity prices according to historical data. However, some regions with limited transmission capacity have experienced more significant impacts, such as off-peak negative pricing. Therefore, the decision today to invest heavily in renewable infrastructure in nations like Brazil can decelerate or accelerate the growth of a low-carbon economy. By acknowledging the economic value and historical precedent for wind energy, stakeholders can better plan for the integration of renewable sources of power in existing systems (Botterud et al., 2020; Lucon et al., 2015) (Figure 5.4).

5.1.1. Early Uses of Wind Energy

Its ancient utilization goes back thousands of years and was so diverse in different civilizations and regions. The first ever recorded utilization is in ancient Persia, where windmills were employed to grind cereals and provide water irrigation, which reflects a new method to harness natural resources for agricultural cultivation needs. These early windmills featured vertical sails rotating about a vertical shaft, a setup which laid the foundation for later innovations in wind power. This method of energy harnessing focused on the potential for wind to be a reliable source of energy, and before long it was used in other forms elsewhere in the globe. The evolution from rudimentary windmills to more sophisticated models followed a trajectory that would ultimately be realized in modern wind power systems, demonstrating an early interest in the utility of renewable power to sustain human activity. As wind power continued to evolve, it found applications across Europe during the Middle Ages. The innovation of horizontal-axis windmills was a tremendous step, particularly in nations like the Netherlands that became synonymous with wind power on the basis of its low-lying topography and windy climate. They worked primarily for grain milling and drainage of water from lowlands, with an ability to adapt to local environmental demands. Moreover, the introduction of wind technology into agriculture showed an integral link between farm output and renewable energy. This not only increased crop yields but also helped to grow local economies. By the 19th century, the windmill had become a common sight in rural towns, promoting a culture of innovation that would encourage future generations to research and invest in renewable energy technologies. The importance of early wind energy uses goes far beyond simply technological development; it marks an increased sensitivity to the demand for renewable energy resources. Throughout most past instances, civilizations were highly dependent on wind power not only for farming but also as a means to drive ships over the seas. Wind was utilized to facilitate maritime navigation, demonstrating the diverse character of wind power and its probable future role in future energy systems. The quest for complementary energy sources, as seen in contemporary research, echoes these earlier practices of tapping diverse forms of energy. The integration of wind energy into other renewable sources, such as solar and hydroelectric power systems, is an illustration of how the historical practices can inform contemporary practices in energy production and transmission, underscoring the options for future innovations in the integration of wind power (Alcalá et al., 2019; Beluco et al., 2019) (Figure 5.5).

Figure 5.5. *A windmill near the town of Alkmaar, the Netherlands, known for its cheese fairs.*

Source: https://www.energyencyclopedia.com/inc/imgresize.php?img=%2Fvar%2Fwww%2Fenergyencyclopedia%2Fdata%2Fwww%2Fenergyencyclopedia.com%2Fdata%2Fweb%2Fimg-wind-energy%2Fthe-history-of-wind-power-utilization%2F3684000.jpg&method=fitbox&nw=1280&nh=1280&rotate=0&jpgquality=80&noresize_small=1&max_enlargement=2&outputformat=webp.

5.1.2. Development of Modern Wind Turbines

The evolution of modern wind turbines has been a major driver in integrating wind energy into the global energy mix. Over recent decades, improvements in technology and design have made efficiency and reliability increase significantly. As the need for clean energy to fight climate change has increased, wind turbines have not only become larger but more capable as well. New turbines have greater rotor diameters and larger capacity factors, which allow them to generate more electricity from identical resources of wind. This turbine technology development is directly comparable to the enhanced economic feasibility of wind power, where manufacturers can drive maximum output even in capacity-limited markets like the United States, where low-specific-power turbines are obviously advantageous (Bolinger et al., 2021). The economic impacts of turbine design are greater than merely efficiency; they involve the overall reduction in the levelized cost of energy (LCOE). In the push for increased rotors and lower specific power ratings, wind turbines are now a profitable return on investment for investors and developers. Megawatt-hour optimizing turbines will likely reduce energy costs while also reducing logistical challenges in terms of handling, constructing, transporting, and installing large blades.

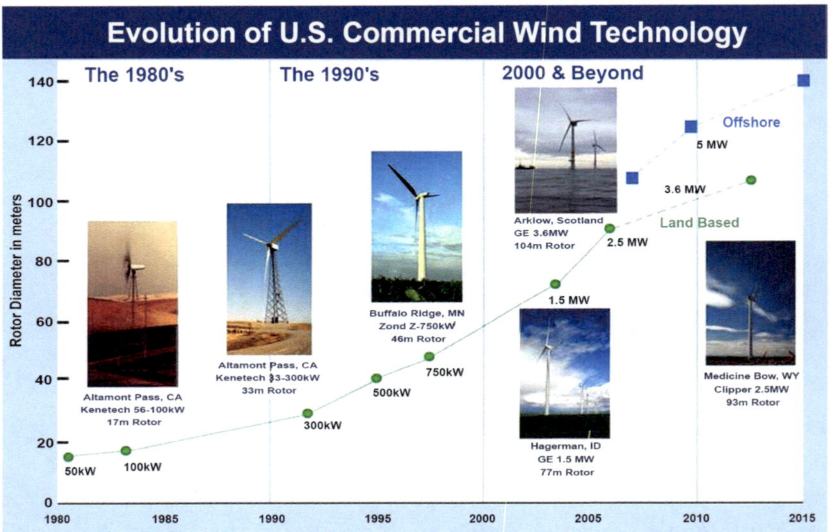

Figure 5.6. *The development path and size growth of wind turbines.*

Source: https://www.researchgate.net/profile/Paul-Veers/publication/253677103/figure/fig7/AS:668355971272715@1536359889456/The-development-path-and-size-growth-of-wind-turbines.jpg.

Additionally, as penetration of wind energy increases into the power grid, market value added by such advanced turbines becomes increasingly relevant, suggesting a sustained demand for the same in the future (Bolinger et al., 2021). Despite a thrilling technology quantum leap in wind turbine design, there is no smooth sailing ahead. There are still concerns regarding the environmental footprint of wind farms that can taint the growth in wind power capacity. Land use and landscape modification at local scales, wildlife effects, and noise pollution are typical themes in debates on wind power development. These conflicts will have to be balanced thoughtfully against the benefits of renewable energy, which can dramatically reduce carbon emissions and lower the world's dependence on fossil fuels. As the world realizes that climate change must be addressed today, overcoming these conflicts and finding sustainable answers will become crucial to making sure the world continues to develop and integrate wind power into the global energy base (Figure 5.6).

5.1.3. Growth of the Wind Power Industry

The renewable energy segment of wind power has seen record expansion over the past two decades, driven by increasing demands for renewable sources of power in the face of pressing climate and energy security issues. The sector's expansion comes largely from sustained government policy backing that ensures an investment-friendly climate for wind power. These policies encourage domestic supply of services and capital equipment, creating employment and economic growth. With the wind industry maturing, advances in technology and efficiency continue to reduce the cost of generating wind power, making it increasingly competitive with fossil fuels. Therefore, investors are likely to invest in wind power projects because they relate to their capacity to provide a significant contribution to the future of low carbon as well as to address energy supply requirements. International integration is crucial to the further growth of the wind power industry because countries increasingly participate in cross-border investment and trade. This integration is driven by the reality that investments in wind technology cross borders, enabling countries to collaborate and trade resources without obstacles. Internationally, countries can combine their strengths, with optimized supply chains and production processes. However, barriers of this integration are still a concern, with nontariff trade barriers as well as varying local industrial standards being some of the challenges. The overcoming of such hurdles is imperative to reach the potential of the wind power sector, as optimized practices and policies have the ability to enhance international trade flows and decrease the costs of wind energy schemes. Removal or lowering of such barriers is essential to reduce the growth prospect of the industry and expand its market penetration. Countries like Brazil demonstrate the investment needed for strategic decision-making to harness wind energy's potential alongside economic growth and climate targets. With energy infrastructure investments of around 0.5 trillion U.S. dollars expected, Brazil stands at a crossroads where it will continue carbon-based energy policies or move towards a sustainable low-carbon economy. The future direction of Brazil's energy landscape depends on the coordination of wind power growth with climate policies integrated together. By adopting a more integrated system of energy infrastructure that prioritizes renewable sources like wind energy, Brazil can be the leader in the global renewable energy market. This transformation is not only important to national energy security and economic competitiveness but also required so as to assist in advancing global attempts at reducing greenhouse gas emissions and slowing climate change and thereby ensuring a sustainable energy future for all.

5.2. TECHNOLOGICAL ADVANCES IN WIND POWER

The technology of wind energy has experienced impressive technological advancement in the last decades, which has catapulted a significant revolution in the manner in which renewable energy is being integrated into traditional systems. The innovation with respect to larger, more efficient turbines has enabled the utilization of wind energy on previously unprecedented scales, leading to higher total energy output. These technological innovations not just maximize the effectiveness of deriving electricity from wind but also minimize the cost per megawatt-hour of electricity produced. Moreover, modern wind turbines also have sophisticated computer systems that maximize efficiency in real-time by altering blade pitches to maximize efficiency based on wind conditions. This greater operational efficiency is essential to making wind power a serious competitor for traditional fossil fuels and allowing its adaptation into the broader energy industry to help meet the growing need for clean energy solutions. With nations making greater efforts to turn to renewable energy as a way to combat climate change, the regulatory climate is changing to keep up with advances in technology in wind power. Governments worldwide are incorporating policies which favor investment and innovation in clean energy technologies. Examples of these are funding, streamlined permitting processes, and enabling infrastructure investments which facilitate quicker deployment of wind farms. Of specific interest is the fact that alternative energy technologies, of which one is wind power, are assuming major components of national energy policies aimed at reducing greenhouse gas emissions. The synergy of technological advancement and supporting policy can create an effective market environment, enabling businesses to compete in the new clean energy market. The interdependence is important in continuing to enhance the capabilities of wind power and making it the major source of energy in coming decades, especially for regions with a high energy need. Furthermore, the integration of geospatial analysis into wind energy development is enhancing the strategic planning of energy initiatives to make sure that environmental issues keep pace with technological innovation. The use of tools such as MapRE enables the policymakers to identify optimal areas where the development of wind farms should take place with few land-use disputes. By assessing factors such as the availability of resources and proximity to transmission facilities, geospatial models facilitate decision-making by allowing intelligent choices that ensure maximum energy production while facilitating environmental protection. In Bangladesh, for instance, where land is limited, the selection of locations for renewable energy projects needs to be carefully planned not to conflict with agricultural and ecologically useful lands. This prudent planning enables the optimization of the potential of renewable energy so that technological innovation serves the interests of both economic prosperity and environmental conservation, thus supporting the argument for the incorporation of wind power within a portfolio of diversified energy (Bhuyan et al., 2025; Kammen et al., 2019) (Figure 5.7).

Figure 5.7. *Technological advancements and innovations in wind energy in Latin America.*

Source: https://media.licdn.com/dms/image/v2/D4D12AQGDAilAFgTujg/article-cover_image-shrink_720_1280/article-cover_image-shrink_720_1280/0/1716219546188?e=2147483647&v=beta&t=v5__10LIdNCuTchgaFh2KP6YpdrQmVMXyAWI0zWM2f8.

5.2.1. Innovations in Turbine Design

The evolution of wind turbine design has been instrumental in ensuring optimization of efficiency and scalability when integrating wind power. The recent evolution has been focused on optimizing blade aerodynamics to its maximum, which directly means more energy harvesting. The innovation of longer but lighter blades, for instance, allows turbines to collect wind energy efficiently even at low wind speeds. These innovations not only increase overall energy production but also allow turbines to be installed in more varied environments, even city environments. These kinds of innovation are needed because countries are looking to increase their share of renewable energy. From an agreed framework proposed to evaluate innovations in wind technology, the understanding of turbine advancements will help policymakers prioritize their innovation objectives and the expenditures towards them so that investment is directed towards technologies with most likely possibilities of diffusion to national energy grids (Dinica et al., 2009). Further, the integration of smart technology into the turbines' design is a quantum leap. Now, turbines with advanced monitoring technology can modulate their activities in real time based on changing wind patterns and grid demands. Such responsiveness enhances reliability and efficiency, overcoming one of the greatest challenges with wind power—its volatility.

Figure 5.8. *The aerodynamics of efficiency: Innovations in wind turbine design.*

Source: https://windcycle.energy/wp-content/uploads/2023/09/Aerodynamics-of-Efficiency-1.jpg.

Moreover, smart turbines can plan maintenance for ideal times through predictive analytics, reducing downtime and maintenance costs. The impact of these advances reaches beyond the one turbine operation; it promotes a more integrated approach to wind power in the energy system as a whole. Initiatives such as the Special Initiative on Offshore Wind demonstrate strategic partnerships aimed at leveraging such technological advancements to facilitate offshore wind mass installation, underpinning the need for innovative turbine design to allow for a robust renewable energy infrastructure (Ozkan et al., 2016). Lastly, the push for sustainable materials in turbine technology cannot be overlooked when discussing innovations in the technology. As the environmental footprint of energy technology comes under scrutiny, recycling materials and environmentally friendly ones used in making turbines is becoming an essential trend. For example, bio-composite utilization in blade making not only reduces the carbon footprint but offers performance levels that are as good as the use of traditional materials. This innovation is hand-in-hand with the rising pressures of sustainability in energy networks, espousing a circular economy model for wind energy. Moreover, these advancements also advance the feasibility of large-scale integration of wind energy into existing power systems. Addressing performance challenges and environmental concerns alike, these advancements are at the heart of advancing public trust and facilitating the acceptance of wind energy as a mainstream energy supplier (Figure 5.8).

5.2.2. Improvements in Energy Storage Solutions

The ongoing development towards renewable energy sources, particularly that of wind power integration, calls for considerable advancements in energy storage technologies. Effective storage technologies are essential in responding to the variability of renewable energy output. Developments in recent energy technologies have offered various options of storage, such as batteries, thermal storage, and fuel cells, able to manage supply in the absence of wind resources. For example, the combination of batteries and thermal storage facilities enables a more robust and flexible grid to provide round-the-clock power supply from renewables. The capability to store electrical and thermal energy can maximize energy efficiency and consumption, making more sustainable and reliable power systems. The potential of achieving high levels of self-consumption of renewable energy can be demonstrated in studies that quantify the efficiency of the technologies in a real system, complementing the significance of innovative storage solutions in wind power integration (Blumberga et al., 2024). Moreover, the concept of ubiquitous energy storage, as portrayed in new publications, emphasizes storage integration at all scales of the electricity grid. This approach recognizes the growing demand for storage capacity driven by the rising usage of distributed generation, particularly from wind and sun resources. Various storage technologies such as lithium-ion batteries and newcomers are in the process of being examined for applications and economic feasibility in the energy system. The common challenge is that some storage technologies are not economically feasible. Empirical evidence indicates that while electrical storage is not necessarily economically appealing, thermal storage presents economically viable options, especially if deployed in conjunction with renewable energy sources of generation (Ault et al., 2005). The above discussion not only highlights the importance of innovation technology but also necessitates evaluation of regulatory frameworks facilitating the deployment of such technologies. With increasing demand for renewable energy, advancements in energy storage technologies will continue to play a key role to address challenges related to integrating wind power. Innovative energy storage solutions can go a long way in improving energy system reliability and efficiency by allowing improved synchronization of demand and supply dynamics. Higher energy storage capacity allows for the utilization of excess energy generated with higher wind levels, which can be supplied during lean generation periods. Through a combination of sophisticated storage technologies—ranging from chemical, mechanical, to thermal systems—energy communities are able to transition towards cleaner operations. This holistic approach not only assists in reducing the need for fossil fuels but hastens the application of renewable energy devices at scale. Ultimately, as ongoing research and development endeavors continue to enhance storage alternatives, their integration into wind power systems will be critical to charting a sustainable energy future (Blumberga et al., 2024) (Figure 5.9).

Figure 5.9. *Advanced energy storage solutions powering the future.*

Source: https://www.soleosenergy.com/wp-content/uploads/2025/03/Blog-Post-Image-2048x2048.png.

5.2.3. Integration of Smart Grid Technology

There is a need to use smart grid technology to facilitate the integration of wind power into the energy mix without any hitches. Smart grids are designed to optimize the efficiency, reliability, and sustainability of the supply of electricity, which is crucial in addressing the intermittency nature of wind power. Advanced metering infrastructure, real-time data analysis, and automated fault detection enable energy to be managed more responsively. This sensitivity enables grid managers to match time-of-day power consumption with the unpredictable nature of wind generation. Furthermore, smart grids can optimally coordinate the interconnection of distributed generation, as witnessed in various case studies from the United States and other nations illustrating disparate POA implemented in managing flows of energy, ultimately reducing the specter of curtailment and investment imbalance among stakeholders (N/A, 2013). Besides, smart grid technology makes it possible to have microgrid systems, offering a decentralized power production model that can ensure enhanced integration of wind power. Microgrids make it possible to have local power production, which may be wind, solar, and other renewable resources, thus ensuring resilience and stability for communities.

Figure 5.10. *Renewable energy (RE) integration and smart grids.*

Source: https://www.eprmagazine.com/wp-content/uploads/2024/06/10.jpg.

As the U.S. has realized the benefits of using such a system, various federal and local policies have been there to encourage the installation of microgrids, especially in integrating renewable energy sources (Bao et al., 2018). Smart grid and microgrid technologies facilitate smoother transitions in terms of energy supply while allowing for effective dispatch of renewable energy inputs even in times of intermittent supply. Such an integrated approach towards renewable energy incorporation not only improves the security of energy supply but also promotes large-scale adoption of greenhouse gas reduction schemes. Additionally, the benefits of smart grid technologies incorporation extend beyond connectivity and operational efficiency. More grid intelligence supports demand response programs, whereby consumers can adjust their usage pattern based on real-time availability and prices. These initiatives are critical to maximizing the utilization of wind power, which, although abundant, sometimes outpaces demand on heavy output days. Using smart technology, grid managers are able to interface effectively with homes and businesses, calling for energy-conserving habits or rewarding use during these peak output periods. Therefore, this convergence of consumer behavior and renewable energy supply makes a more responsive and engaging power system, eventually leading to improved economic outcomes and reduced carbon footprint in the power sector. The collective advance of smart grid technology and its deployment illustrate the path towards not only wind power addition but also leveraging its function as part of the entire energy system (Figure 5.10).

5.3. ECONOMIC IMPACTS OF WIND POWER INTEGRATION

Incorporation of wind energy into the energy system has profound economic impacts, aside from electricity generation. Firstly, establishment of wind farms contributes substantially to the economy of specific regions in terms of job opportunities in the construction, operation, and maintenance sectors. It is proven by studies that communities that host wind energy projects experience a boost in job opportunities locally, which enhances people's livelihood in those communities. In addition, wind farms stimulate market activity by encouraging local material and service demands, thus benefiting supply chains, and causing diversification of the economy. Patents of innovative wind tower designs, such as the precast concrete towers minimizing environmental and social impacts and optimized for efficiency, highlight improvements that can lead to future cost savings (Aguado et al., 2017). Overall, these forces reinforce the core role of wind energy in regional economic resilience. In addition, the economic impacts of wind integration extend across the entire energy market, particularly with respect to pricing and energy expenses. Wind energy, as a form of variable renewable power (VRE), contributes to reducing wholesale electricity prices during times of excess generation. This effect is primarily because it has the possibility to replace more expensive thermal generation and decrease the capacity factors of traditional baseload generators. By enhancing competition for electricity, VRE contribution can decrease consumers' cost, which serves as one primary economic incentive for single households and companies. Research indicates that while the collective effect of VRE on market prices has been modest relative to others like natural gas prices, its effect is bound to increase with increasing levels of penetration (Botterud et al., 2020). This points towards the transformative power of the energy sector as wind power integration becomes prevalent. The long-term viability of incorporating wind power is equally dependent on solving economic concerns regarding up-front capital expenditures and infrastructure modifications. The cost of initial investment in setting up wind facilities is high and, in most instances, demands large sums of financing alongside risk analyses that discourage potential investors. Yet, technology evolves, and economies of scale connected to bigger projects have been discovered to increasingly reduce such costs, making wind power more competitive. Moreover, the need to expand transmission infrastructure to handle random inputs from wind power can be an enormous economic cost. However, investment in the upgrades needed makes it possible to integrate the renewable energy and also increase energy resilience and reliability, which results in a net economic gain to society while it addresses critical sustainability goals (Aguado et al., 2017). Thus, meeting these challenges lies at the core of realizing the full economic potential of wind power integration.

5.3.1. Cost Analysis of Wind Energy versus Fossil Fuels

The economics of wind energy over fossil fuel is a compelling case for transition to greener energy platforms. Fossil fuel reliance has been characterized in the past by lower initial costs in terms of current infrastructure and technology. But these internalized expenses—derived from health impacts, ecosystem destruction, and volatile fuel prices—are typically not included in traditional economics. Increased technology for wind has seen tremendous cost savings in capital expenditures as well as operating costs over the past few years, and wind energy has become competitive, if not preferable, compared to fossil fuels. The expectation of future price volatility for fossil fuels under stringent climate controls also worsens the fiscal situation, which more and more is shifting towards investments in alternative forms of renewable energy like wind power, as the efficiency of smart grid technologies combined with wind power enhances and the potential for maximum energy distribution and usage rises. The future potential for wind power hinges not just on its blatant price but also on an increasing realization of long-term sustainability and resource availability. Wind power technologies are supported by a mounting body of evidence demonstrating opportunities for forceful decarbonization—a feat fossil fuels are unable to accomplish through either intrinsic transformation of technology or otherwise. As detailed by research, enhanced technologies, and gains in the efficiency of energy systems can allow for a shift wherein the world's energy needs are satisfactorily met with minimal reliance upon fossil fuels, without unaffordable cost additions in energy services (Flavin, 2008). This is a mirror of the high potential of wind energy to emerge as one of the dominant forms of energy, which may take away the economic competitiveness of fossil fuels in developed and developing nations as well. Furthermore, investments in renewable energy are being made more appealing through their dual role of enhancing socio-economic disparity and reducing global environmental challenges, providing an integrated solution to modernizing the energy system. Further, an effective policy framework will play a key role in realizing the highest Economic value of wind energy at the expense of fossil fuels. Policy makers are encouraged to search for policies that stimulate the adoption of renewable technologies, driving innovation with the promise of affordability through government support programs. Investment in wind energy infrastructure, besides playing an important role in mitigating greenhouse gas emissions, also drives economic growth through the creation of employment in developing technology sectors (Lucon et al., 2015). With countries like Brazil and others at the crossroads of defining important energy policy directions, fulfilling such with climate targets offers a colossal opportunity to reshape their economic trajectory in the global economy. The incorporation of wind energy into established energy systems presents a clean and renewable answer that ultimately speaks to the costs and benefits of energy spending, propelling the move toward a more sustainable and resistant future.

5.3.2. Job Creation in the Renewable Energy Sector

The transition to renewable energy, particularly in the area of wind power, holds significant promise for employment creation, a critical element in the economy today. As government policies become supportive of boosting demand for wind energy, local supply chains for capital goods and services are developing well. These policies provide the ground base for employment creation for groups engaged in wind energy activities, from manufacturing to installation and maintenance. If structural incentives are in place, such as carbon pricing that allows wind power to be made more competitive, the demand for turbines and complementary equipment increases considerably. Such a structure not only improves local economies but also achieves long-term market stability that eventually converges the interests of a range of different stakeholders in advancing wind power integration (Kirkegaard et al., 2025). Furthermore, localizing the wind sector transcends the role of job creation in an immediate sense to building a sustainable local base for production. In regions like South Africa, where there is high prioritization of integrating renewable energy resources, building offshore wind capacity can stimulate economic growth while reducing carbon emissions. The unique geographical features, such as expansive coastal regions, enhance the feasibility of wind power projects, making them eligible for investment and job creation. However, harmonization of wind power with the existing grid is not a problem-free process that has to be addressed methodically, ranging from technology requirements to policy settings. As policymakers and business executives struggle to ease these issues, the likelihood of job creation and economic stimulation in the renewable energy sector rises significantly (De

La et al., 2011). Moreover, enhancing opportunities for employment in the wind sector necessitates a strategic approach in cross-border investment and trade. International integration of the wind industry, however, provides that local labor markets are able to benefit from external cooperation and expertise, as well as obtain non-tariff barrier circumvention which could otherwise hinder growth. De-stimulation of such barriers as various national industrial standards and local content requirements thus becomes pivotal to enhancing investment attractiveness. Focus on cross-border investment over domestic production alone can expand job creation activities by maximizing current technologies and expertise within the international economy. Here, establishing robust trade policies that enable international collaboration is essential for unlocking the vast wind industry potential while ensuring that local workforces are benefiting from the resulting economic gains (Kirkegaard et al., 2025).

5.3.3. Economic Incentives and Government Policies

Government policy and the provision of economic incentives play a large role in the integration of wind energy into existing energy infrastructure. These are needed to create a robust renewable energy industry. Governments around the world employ a variety of financial mechanisms, such as tax credits, rebates, and subsidies, to encourage the development of wind energy installations and reduce upfront investment barriers. These incentives, in addition to stimulating private sector investments, offer a guarantee to stakeholders in terms of long-run returns. A notable example is the U.S. federal production tax credit (PTC), whose substantial impact has been to boost wind energy capacities. Such policies not only render wind energy competitive with traditional fossil fuels but also reconcile economic development with clean environmental policies, ultimately reducing greenhouse gas emissions and enabling a lower-carbon economy, as evident from the longer-term implications depicted in cases like Brazil's energy policy landscape (Lucon et al., 2015). Additionally, the policies of the government have a critical role to play in establishing regulatory systems that contribute to hindering or facilitating integration of wind energy. Zoning, permitting, and grid access policies significantly influence the size and pace of wind energy developments. In some instances, streamlined permitting and favorable zoning policies enhance project deployment, thus enhancing the overall share of renewables in the energy mix. Conversely, excessive regulation can be used as an entry barrier, deterring investment. An excellent prime example exists with the promotion of microgrid technology in the U.S., as promotional policies have driven industry growth and local energy resilience. Through an assessment of these policy settings, policymakers can determine areas of possible harmonization of wind power development with other energy goals in order to pursue the socio-economic benefits of renewable energy integration (Bao et al., 2018). Ultimately, effective wind power integration relies in large part on the prudent alignment of economic incentives with social and environmental purposes. Policy makers ought to consider very seriously not just how incentives can promote renewable energy but how they can promote social justice and energy equity as well. For instance, investments in wind energy produce job opportunities for renewable energy industries benefiting local communities and regional economic growth. Thus, economic incentives should also pay special attention to workforce training programs and local outreach efforts to ensure local participation in wind energy development. The challenge is designing policies that do not just encourage investment but incorporate the overall direction of energy transition, such as with Brazil's changing energy policies working to reconcile economic and environmental needs (Lucon et al., 2015). The transition towards wind power can thus be an inclusive strategy towards sustainable development.

5.4. ENVIRONMENTAL BENEFITS OF WIND POWER

The environmental benefit of wind power is significant, i.e., reduction of greenhouse gas (GHG) emissions. As countries work towards the adoption of sustainable power networks, integration of wind energy is a viable alternative towards preventing climate change. Wind power generates electricity without emitting carbon dioxide, a major cause of global warming. By using natural winds, this renewable energy source lowers the use of fossil fuels. Wind power adoption is not a theoretical benefit; numerous studies indicate that intensified wind power generation can lead to steep reductions in national emissions as a whole. These reductions are particularly significant in countries where governments have strong renewable energy policies in place, implying that judicious investment in wind infrastructure can lead to tangible environmental change along with better public health impacts due to cleaner air quality. Moreover, using wind power contributes to the stability of the power grid, often complementing other sources of renewable energy such as solar and hydroelectric power. The volatility of wind is improved by combined use with hybrid systems or storage to avoid the potential for curtailment when generation is higher than the grid needs. In Portugal, for instance, studies show that plug-in electric vehicle usage can contribute to improving the wind energy performance and stability, effectively using it during off-peak hours, and minimizing wasted potential production. This two-faceted effort not only optimizes the utilization of energy but also that of the grids' flexibility to adjust to varying demands. By incorporating such technologies into wind energy projects, countries are able to supply a better source of energy while simultaneously addressing environmental concerns associated with the use of fossil fuels (Carvalho et al., 2012). Besides lowering emissions and grid stability, wind power is vital in saving water usage and aquatic resources. Traditional technologies for producing energy, particularly fossil fuels and nuclear power, consume huge amounts of water for cooling and processing, which can trigger drastic ecological imbalances. Alternatively, wind turbines do not require water to generate electricity, thereby allowing for more efficient water resource management. With countries striving to secure their water supply with climate variability, wind power becomes a sustainable alternative. Strategic wind power deployment is also able to maintain biodiversity by reducing the loss of habitat associated with the production and burning of fossil fuels. Thus, by leading with wind power, nations are able to provide several environmental benefits, resulting in a greener, cleaner energy future (Lucon et al., 2015) (Figure 5.11).

Figure 5.11. *The environmental benefits of wind power.*

Source: https://www.turbinesinfo.com/wp-content/uploads/2023/01/The-Environmental-Benefits-of-Wind-Power-Reducing-Greenhouse-Gas-Emissions-and-Air-Pollution_turbinesinfo-1200x750.jpg.

5.4.1. Reduction of Greenhouse Gas (GHG) Emissions

With the urgency in dealing with climate change increasing, greenhouse gas (GHG) emissions reduction has become of paramount importance across different industries, above all the energy sectors. Wind power integration is among the viable alternatives in achieving sizeable reductions in power generation emissions. Through the process of converting wind kinetic energy into electricity with minimal or zero GHG emissions, wind turbines directly lead to cleaner energy supplies. Transitioning towards wind energy not only decreases carbon-heavy fossil fuel reliance but also leaves nations in a stronger position to meet their climate objectives. This policy shift is crucial for countries like Brazil, whose investments in energy infrastructure can either solidify carbon-intensive technologies or force the innovation of low-carbon economies, as discovered in recent analysis of Brazil's GHG emitting energy sectors (Lucon et al., 2015). Moreover, the proper integration of wind power into the existing energy supply system can ensure a multisided decrease of GHG emissions by increasing the proportion of renewable energy. The development of high-tech technology has made wind power more practical and cost-effective, encouraging a quick deployment of wind farms.

Figure 5.12. *Reduction of greenhouse gas emissions.*

Source: https://i0.wp.com/sciencemediahub.eu/wp-content/uploads/2021/07/55-reduction-of-CO2-gases-by-2030-Its-a-Herculean-effort.jpg?resize=1024%2C768&ssl=1.

The emphasis on renewables, or wind and geothermal generation of power, is vital, as exposed in the Carbon Emissions Pinch Analysis (CEPA) method, which identifies the huge challenges and strategies in reducing emissions in various sectors of energy (Atkins et al., 2010). As nations strive toward ambitious greenhouse gas emissions cuts, wind power can be a cornerstone of national energy strategy, fostering the development of a more resilient and cleaner energy system. However, incorporating wind into the grid to achieve significant reductions in GHG emissions requires concerted policies and robust investment in infrastructure. Moreover, managing variability in renewable energy output is tough challenges that must be addressed in a systematic way. Governments and stakeholders must convene to enhance grid flexibility and storage capacity, ensuring the safe supply of electricity generated from wind. Effective energy policies, as proposed in Brazil's case, should incorporate an integrated approach that accord equal preference to climate and energy goals, identifying major abatement opportunities to enhance GHG reduction actions (Lucon et al., 2015). By creating a comprehensive framework that promotes wind energy, nations are not only able to reduce emissions but also increase economic development, facilitate innovation, and improve total energy security (Figure 5.12).

5.4.2. Impact on Biodiversity and Ecosystems

The incorporation of wind energy is a multifaceted series of issues and opportunities for ecosystem health and biodiversity. Although the shift away from fossil fuels to renewable energy resources like wind turbines is necessary for slowing global climate change, there are significant environmental considerations that need to be taken into account. Wind farms will alter natural habitats within the local wildlife, especially for birds and bats who will sometimes fly into the turbine blades. This potential effect is concerning in the context of population reduction in vulnerable species and ecosystems' overall well-being. A comprehensive understanding of these dynamics necessitates analysis that is multidisciplinary, as outlined in the model that adopts Economics, Ecology, and Technology for evaluating implications of land use shifts resulting from wind power development (Bachev et al., 2013). Besides, development of wind power facilities has the capability to alter land and marine ecosystems by reforming natural environments and impacting environmental mechanisms. Such alteration may lead to changes in species composition and perturbations in food webs. Successful integration of wind energy with the preservation of biological diversity demands the implementation of sophisticated environmental management strategies. These approaches need to consider the direct consequences of turbine installation and also the cumulative impact of more widespread human activity on regional ecosystems. Developing an integrated assessment framework would allow these interactions to be disaggregated, and more efficient governance and stakeholder engagement (Austen et al., 2016). Ignoring these variables ensures that the ecological benefit of wind power can be overbalanced by unwanted impacts. Lastly, the encouragement of biodiversity in the incorporation of wind energy requires dynamic conservation intervention and sustainable handling. This is achieved through the analysis of the health of ecosystems against pressure from energy infrastructure, as well as restoration where it is required. Interdisciplinary measures, such as insights drawn from ecology, technology, and socio-economic factors, can enhance the outcomes of conservation. By allocating highest priority to the health of ecosystem and biodiversity during wind energy project planning phases, stakeholders are able to offset the need for renewable energy with the need to conserve ecological integrity. It is an approach that is critical for the promotion of sustainable development that benefits human communities as well as natural ecosystems upon which they depend (Bachev et al., 2013) (Figure 5.13).

Figure 5.13. *When wind energy is developed in suitable locations using appropriate technologies, there is arguably no need for biodiversity conflict.*

Source: https://www.nsenergybusiness.com/wp-content/uploads/sites/4/2022/01/shutterstock_1476824375.jpg.

5.4.3. Comparison with Other Renewable Energy Sources

The contrast of wind power with other renewable sources of energy is essential to its potential contribution in an energy future that is sustainable. Wind power contrasts with solar energy, with a heavy reliance on direct sunshine. Wind power can, nevertheless, generate electricity even on cloudy days. The two technologies do, however, suffer from intermittency and variability; hence, grid flexibility becomes essential in managing these fluctuations. Studies have shown that areas with well-established grid infrastructures achieve higher levels of success in incorporating renewable sources into their mix, with far-reaching effects on the energy composition. For example, as ASEAN targets raising its share of renewable energy, it needs to advance grid infrastructure to integrate wind and solar energies seamlessly with other renewables such as hydroelectric and geothermal power, which can offer steadier performances. This diversity guarantees the availability and stability of resources while being cost-effective in pricing energy, as reported by Huang et al. (2019). Comparing wind and bioenergy, integration into the current energy infrastructure shows varying characteristics and viability for continued use. Wind energy generates less carbon emissions in operation than bioenergy, which frequently requires fuel burning. However, bioenergy is able to provide a more consistent supply of energy since it is capable of being stored and used at whim. This function makes bioenergy a suitable ally for wind power, allowing utility operators to balance demand and supply. Knowing the capacity of every source is important when the city is pushing for strong renewable energy targets. The consumption of energy by service industries, such as commercial enterprises, also has to be addressed through convergent energy models. These can improve the credibility of integrating renewable resources into cities, reporting comparative advantages of wind supplemented with bioenergy and other technologies, as observed by Brazier et al. (2017). Lastly, the successful integration of wind power into a broader renewable energy mix highlights its merits and complements other sources of energy well. For instance, hydropower can provide high base load output, smoothing out wind electricity's variability. While wind power has historically been seen as variable and unreliable, its capability to generate clean electricity vastly outweighs its shortcomings when compared with fossil fuels. Furthermore, high-tech forecasting tools and grid access, as outlined by Huang et al. (2019), are essential to counterbalance the natural volatility inherent with wind energy. As nations pursue grid upgrades and renewable objectives aggressively, it would be necessary to understand such dynamics so as to facilitate higher penetrations of wind power without compromising stability and reliability in the general energy infrastructure. Combining these renewable technologies thus develops a diversified and robust energy portfolio, which is vital to attain a sustainable energy future.

5.5. CHALLENGES OF WIND POWER INTEGRATION

Integration of wind energy into existing power systems is plagued by issues that must be appreciated and addressed to unlock its potential in the world energy pool. The most significant among these issues is variability and intermittency of wind power generation. Wind energy, unlike conventional power sources, fluctuates with variations in environmental conditions, thereby creating uncertainty in electricity supply. This volatility places tremendous stress on grid managers to provide consistent supply of power in real time to meet varying demand. The increase in the share of wind power in many geographies requires higher grid flexibility and improved forecasting systems to manage these oscillations effectively. As it is noted, while wind power is currently one of the least expensive low-carbon sources of electricity, it is still faced with technical, economic, and social challenges in being fully integrated into the larger energy systems that require innovative solutions (Bucha et al., 2025). The planning and permitting process also generally serves as immediate barriers to the rapid deployment of wind power. In some countries, red tape is causing serious delays, inhibiting the development of new wind farms. Go-to places and rapid-track permits are novel solutions which have been put forward in countries such as the EU but whose long-term efficacy and social acceptability remain uncertain. Processes must be efficient in the sense that slowed permitting can postpone investment in wind energy facilities, inhibiting national and regional renewable energy targets. As emphasized in recent studies, despite existing solutions, the present complexities in regulation frameworks emphasize the necessity for comprehensive policy reforms aimed at promoting renewable energy integration and winning public support (Bucha et al., 2025). At the heart of addressing wind power integration issues is to ensure maximum grid flexibility within existing electrical networks. Countries like those in the ASEAN group have experienced the need to get the most out of their portion of renewable energy, keeping in mind the character of variability from intermittent sources. The availability of grid flexibility analysis tools accounts for better existing capacity analysis, demonstrating the absence of uniform readiness among nations for increased penetration of renewables. Higher prediction technology and greater access to electricity markets are known as essentials in reducing issues with fluctuating energy supplies. With ASEAN countries gearing up renewable energy targets, inter-collaborative action and support will be instrumental in overcoming the listed shortcomings, thereby maximizing the share of wind power in cleaner, greener energy systems (Figure 5.14).

Figure 5.14. *Addressing the challenges of wind energy.*

Source:https://www.greenlivinganswers.com/blog/wp-content/uploads/2024/08/wind-energy-challenges.webp.

5.5.1. Intermittency and Reliability Issues

Merging wind power into existing electricity networks also presents various challenges, primarily in terms of intermittency and reliability. Wind power takes on an inherently variable nature based on fluctuating wind patterns, which tend to be uncertain and irregular in behavior. Such variability creates issues for grid operators to achieve supply and demand balance at all times, ensuring that electricity supply remains in proportion with consumption. The Nordics provide the best case of such challenges, where a significant share of their experts identify the integration of renewables as the most significant challenge facing the achievement of low-carbon electricity goals, along with an observation that managing intermittency is very essential in providing a reliable energy supply (Kester et al., 2018).

This requires improved grid management methods and technological advancement in reducing the natural uncertainty associated with wind turbine returns. Reliability issues related to the integration of wind power similarly emphasize the imperative need for varied energy sources. Supply of electricity security is not merely an issue of managing the variability of wind power but also of optimizing the combined fuel mix in the grid. Current estimates indicate that utilization of several energy sources would enhance the system's resilience to interruption from any single generation strategy (Butler et al., 2025). In addition, incorporating energy storage facilities may help bridge supply-demand during low wind periods and thus ensure system reliability. Minimizing reliance on fossil fuels and shifting to an energy system with high penetration of renewables does require, however, an overall perspective in evaluating trade-offs between low carbon objectives and electricity supply operation security.

Such considerations necessitate rethink and revision of current policies and practices with regard to energy diversity and grid operation. Finally, handling the intermittency and reliability issues of integrating wind power is vital to achieving sustainable energy programs. Success in addressing these issues through initiatives will travel far in the widespread acceptability and use of wind energy. As illustrated from the Nordic example, integrating renewables has considerable impacts on electricity reliability and total carbon emissions of energy systems (Kester et al., 2018). Thus, policy guidelines and research should continue to address adaptive strategy development in the form of flexible grid operations, enhanced forecasting tools, and public engagement in energy planning. Through such an emphasis on these strategies, the power sector can transition step-by-step towards more efficient and dependable integration of wind power and managing the risks associated with intermittency.

5.5.2. Infrastructure and Grid Compatibility

Conversion of wind energy into existing electrical grids involves extensive analysis of infrastructures and grid compatibility. Traditional power systems utilize a uni-directional flow of information and energy, with a tendency to create inefficiencies and challenges in the management of dynamic inputs from renewable sources like wind. To overcome these shortcomings, the shift to Smart Grids (SGs) has become unavoidable. SGs also facilitate a two-way energy flow, promoting better communication between consumers and providers. Enhanced connectivity is necessary for variable energy sources integration due to the fact that real-time monitoring and regulation of energy production and use become feasible. Utilization of the Internet of Things (IoT) in SGs enhances the functionality of the system significantly through devices that enable monitoring and regulation of energy flows through all phases of the system. This grid infrastructure upgrading is critical to achieve maximum wind power integration and ensure a secure supply of energy (Copeland et al., 2019). In addition, the interaction of wind power systems with grid infrastructure raises major issues of power quality and electromagnetic compatibility (EMC). Power variability of wind power output affects voltage, causes flicker, and introduces harmonics into the power grid and consequently degrades power quality in general.

Mitigating the challenges requires a collective effort in voltage continuity and safety standards in the distribution networks. Variability and intermittency inherent in wind energy sources highlight the importance of establishing strict standards and procedures for maintaining power quality. Enhancing grid compatibility should include comprehensive monitoring programs for the detection and correction of problems in voltage fluctuation and electromagnetic interference that are crucial to efficient energy supply. This ongoing discourse regarding power quality (PQ) and EMC is essential to upcoming energy networks with an emphasis on integration of renewables (Desmet et al., 2015). Finally, wind power infrastructure compatibility with the grid entails constructing innovative solutions to enhance reliability and resilience. Extensions to existing transmission and distribution networks are urgently necessary to provide the particular challenges imposed by wind integration. Grid modernization investment not only facilitates the integration of renewables but also enhances the overall infrastructure to withstand outages and disruption. In addition, ongoing research and collective action across industries can drive innovation of new technologies and best practices that enhance grid resilience. As seen with upcoming IoT applications in Smart Grids, the future grid of electricity will have to be directed towards adaptive and responsive capabilities so that wind energy can be integrated efficiently. This will allow the infrastructure to meet growing energy demands while accommodating increased variability from renewable sources, eventually resulting in an energy future that is sustainable (Copeland et al., 2019).

5.5.3. Public Perception and Community Acceptance

Public attitudes and social acceptance are significant factors in the successful integration of wind energy into national power grids. The large-scale deployment of wind turbines typically evokes fear about potential impacts on surrounding environments, landscapes, and communities. It is critical to overcome these perceptions through an interdisciplinary framework consisting of quantitative and qualitative analysis. For instance, community attitude surveys can give well-informed indication of what matters could be of concern to residents regarding wind facilities, including concerns of noise, visual impact, or damage to local wildlife. This kind of information can assist stakeholders in addressing concerns effectively, ultimately contributing to higher levels of community participation and support for wind projects. As has been demonstrated in similar research, social license to operate needs to be created through transparent communication and involvement of the local community for ensuring acceptance of renewable energy projects (Ipsos, 2025). In addressing the issue of public perception, targeted information campaigns need to be employed that highlight the advantages of wind power. These efforts must seek to furnish factual information regarding the economic, environmental, and health benefits of wind power, debunking misconceptions that are typically attached to its use. For instance, using community forums and workshops can help foster grounds for dialogue between residents and industry players, allowing room for flow of information as well as relaxation of fears. Further, incorporating the perspectives of local communities into planning can lead to creative solutions that align with their values and preferences, thereby producing more goodwill towards wind energy projects. Creative strategies for building stakeholder knowledge, particularly among the general public, have proven effective in building support and acceptance for renewable energy projects (Prins et al., 2009). The key to the success of incorporating wind power lies in winning the trust and cooperation of stakeholders, including policymakers, business executives, and communities. This is only possible through a commitment to ongoing communication and responsive action to local concerns. For example, instituting a system of feedback mechanisms where citizens can articulate their opinions at each phase of project development can significantly enhance perceptions of ownership and acceptance. Besides, emphasis on issues such as the generation of local jobs and economic development can help reconcile wind energy projects with community hopes. As research still indicates, proactive interventions that engage the public and address non-technical barriers contribute to the overall acceptance of wind power projects, again emphasizing the importance of integrating carefully community opinions within the renewable energy framework (Ipsos, 2025; Prins et al., 2009).

5.6. CASE STUDIES OF SUCCESSFUL WIND POWER INTEGRATION

Wind power is integrated into existing energy systems in most instances by new technologies and approaches that enhance flexibility and grid dependability. By examining a variety of case studies, it is clear that smart grid technology application is crucial in facilitating greater levels of renewable energy integration, including wind energy. Microgrids as an integral component of smart grids allow for the integration of diverse distributed energy resources, thereby reducing the need for extensive distribution system upgrade. A case study demonstrates an operational effectiveness of microgrids that significantly improved reliability indices through optimized charging of critical loads, establishing that enhanced wind power integration does not always result from increased capacity but by making the entire energy system more robust (Ansari et al., 2017). Therefore, strategic microgrid technology integration becomes a best practice in current wind integration operations. Another equally important element of successful wind integration is the assessment of grid flexibility. Flexibility is paramount, especially in regions where energy needs fluctuate wildly, which is still the challenge even when wind power is intermittent. A good example among ASEAN countries illustrates this principle because their goal to increase renewable sources means they need a solid grid management policy. The region has implemented a flexibility assessment tool that quantifies a range of indicators, such as load profile ramp capacity and diversity of renewable energy generation. This analysis identifies the various degrees of readiness in member states and emphasizes the necessity for collaboration at the regional level. Through encouraging cooperation and sharing best practices, countries as a whole can enhance their grid flexibility to accommodate higher volumes of wind power, thereby facilitating more ambitious targets for renewable energy (Huang et al., 2019). Wind power integration is not merely the outcome of technological innovation but also includes a broad policy framework for fostering sustainable energy development. Successful case studies demonstrate the roles of governments and stakeholders in creating support environments for the deployment of wind energy. Regulations and incentives for investment within renewable energy facilities are vital in building investor confidence, thereby leading to capacity and grid integration development. As an example, having regulatory systems tackle issues such as market access and forecasting errors is able to lessen integration challenges considerably. Emphasizing collaboration among private and public sectors can promote adoption of policies that strengthen grids and make them efficient in operations. Strategic policy introductions are therefore crucial in driving successful wind power integration and ensuring the potential of wind energy to be maximized towards a sustainable tomorrow.

5.6.1. Examples from Europe

The use of wind power in Europe is the best example of harnessing renewable sources of power to meet growing energy demands. In nations like Denmark and Germany, the share of wind energy in contributing to electricity generation has increased significantly, illustrating how wind systems have been successfully upscaled. Notably, power source hybridization has been adopted as a policy to reduce the natural uncertainty of wind power. By incorporating wind into solar and hydroelectric power, countries can achieve greater stability in the supply of their energy. Such systems are benefited by the complementarity nature of these renewable sources, with solar power supplementing wind generation during low-wind periods, thus ensuring maximum energy yield and a consistent energy grid (Beluco et al., 2019). This approach highlights the imperative need for systematic planning in energy generation to counter the volatility associated with renewable sources. European nations have also advanced considerably in terms of integrating multi-megawatt wind farms into electricity networks, thereby enhancing their operating capabilities. The ongoing projects and research in states such as Spain also indicate advances in the wind turbine technology, which facilitate better interaction with transmission systems. Nowadays wind farms can handle operations such as voltage and reactive power control in a way that they can facilitate the stability of the electricity grid. This shows why there is a need to couple intelligent technology with renewable power sources in order to enhance efficiency and reliability in network performance. These collective steps by grid operators and technology developers serve the essential role of demonstrating how wind energy can be not only an energy source but also a balancer in the energy network (Broe et al., 2025). Further, the European experiment with wind energy integration teaches us something about the necessity of policy frameworks and welcoming infrastructures to foster renewable energy. Countries that adopt stringent regulatory standards have promoted ecosystems that enable innovative projects to flourish. In such an ecosystem, power output management systems that enable real-time monitoring and adjustment of power output are instrumental. As integration efforts increase, research continues to explain the complementarity between various forms of renewable energy sources, setting the stage for improvements in energy management methods in the future. These models will be pivotal to enhancing grid reliability as well as minimizing dependence on fossil fuel. With Europe still leading the wind power integration agenda, the benefits achieved through these models will certainly shape international action towards sustainable energy goals (Beluco et al., 2019) (Figure 5.15).

Figure 5.15. *European countries smashing records for wind power generation.*

Source: https://www.fuergy.com/media/pages/blog/european-countries-smashing-records-for-wind-power-generation-wind-power-v/e9d68bb19c-1712731791/european-countries-smashing-records-for-wind-power-generation-wind-power-v.webp.

5.6.2. Innovations in the United States

Wind use in America is an articulation of passion for energy development, aimed at addressing economic as well as environmental issues. Among the notable efforts is the Special Initiative on Offshore Wind, which is housed at the University of Delawares College of Earth, Ocean, and Environment. This project plays a crucial role in preserving offshore wind energy developments by offering expertise and sharing crucial information across industries from government to industry. Through participatory stakeholder involvement, the program offers comprehensive understanding needed to successfully install offshore wind systems. Such collaborative work not only goes a long way in enhancing the capacity of the country's wind power but also in addressing some of its most pressing energy requirements, emphasizing strategic partnerships in guaranteeing sustainable energy solutions (Ozkan et al., 2016). Recent research emphasizes that policy-led innovation is key to the development of renewable energy technologies, including wind power. A detailed analysis proves that various innovation mechanisms, primarily support from policies, can make the integration of wind power more effective in the United States. Notably, those states with deregulated power markets and relatively low barriers to entry are likely to witness a significant growth in renewable energy innovations.

Such interaction among environmental policies and the performance of innovation proves that purposive regulatory regimes can facilitate technological progression in wind power. It has been found that well-designed policies, especially when complemented with considerable public R&D expenditure, have a strong effect on stimulating green patent development, which is crucial in the transformation towards cleaner energy systems (Nicolli et al., 2025). In addition, the composition of a nation's knowledge base and market structure determine the pace of wind energy innovation. Research has established that income inequality sources have the ability to slow down technological progress in renewable energy, but heterogeneous bases of knowledge have the ability to generate more innovation outputs. Aspiring policies, along with product market liberalization, is what creates an enabling environment for the generation of green patents. The complementarity is what necessitates not only regulatory support but also the broader socioeconomic environment towards achieving maximum use of wind power technologies. Given that the United States is still investing in wind energy research and development, it will be imperative to comprehend these dynamics in implementing effective strategies that drive both economic growth and energy sustainability (Nicolli et al., 2025) (Figure 5.16).

Figure 5.16. *Wind turbines near palm springs, California.*

Source: https://image.cnbcfm.com/api/v1/image/106216240-1572537090795gettyimages-1139164884.jpeg?v=1572537148&w=1480&h=833&ffmt=webp&vtcrop=y.

5.6.3. Lessons Learned from Global Initiatives

The integration of wind power into world energy infrastructure learns significant lessons from global projects. The most significant one of them is the importance of adaptive systems of governance, which have the capacity to deal with the dynamic challenges of renewable energy deployment in a timely manner. Successful projects, such as those realized in Europe, illustrate how cooperative regimes among governments, corporations, and citizens can foster the adoption and accommodation of wind power. Such networks enable stakeholders to share resources, technologies, and best practices, hence building a more robust energy paradigm. With the development of wind power projects, the importance of adaptive policies that take into account uncertain variables such as technological advancement and public perception becomes evident. Such adaptability remains crucial to combating institutional inertia in combating climate change issues, as noted in the study of climate governance (Meadowcroft et al., 2025). Furthermore, experience from global wind power projects indicates the significance of fracturing opposition from interests invested in dominant energy systems. These stakeholders often exhibit unwillingness to embrace alternatives to dominant practices, and this represents a key barrier to effective climate mitigation. Projects able to overcome this struggle often employ strategies such as the activation of opponents, the provision of participation incentives, and the redefinition of public expectations. For instance, the emergence of novel economic bases of power in sustainable sectors has enabled the shift toward embracing renewable energy. With governments seeking to stimulate interest perception change, coalition building is required to maintain momentum towards lofty wind power targets. Such actions are novel means of responding to the underlying conflict of power and interest inherent in climate change policy (Meadowcroft et al., 2025). Lastly, global action is focusing on creating education and awareness campaigns to establish public acceptability for wind power integration. These campaigns are pivotal in shaping people's attitudes and dispelling myths regarding wind energy's reliability and efficiency. Good programs demonstrate that informing stakeholders about wind energy's economic and environmental benefits can make a notable shift in people's attitude and facilitate adoption of policies and development of projects. Coupled with cooperative mechanisms for climate management, education initiatives can inculcate a sense of community ownership of wind energy projects, pushing implementation further. As countries struggle through the complexities of integrating renewable sources of energy, insights from these programs will be invaluable in driving the global transition to sustainable energy systems, ultimately contributing to mitigating climate change (Meadowcroft et al., 2025) and making a more resilient future possible.

SUMMARY

The incorporation of wind power into existing energy systems is characterized by both phenomenal possibilities and large-scale difficulties. Effective integration of renewable energy sources, but particularly wind, requires advanced modeling techniques capturing the unique urban energy demand patterns in an effective manner. As discussed, models which traditionally focus on household consumption fail to respond to service sector activity significantly. The need for representative demand profiles, recognized in recent studies, is central to developing realistic estimates of the integration potential for renewables. Through building surrogate service sector demand profiles, urban energy models can enable more reliable prediction, thereby supporting improved grid planning and operation. The findings point out that including the service sector can give much greater estimates of integrated potential for renewable resources, thus the necessity of an integrated approach in future studies on wind power deployment (Brazier et al., 2017). Further, with increasing growth in wind generation, maintaining grid stability and reliability is a priority. Technological development of wind turbines, particularly Doubly-Fed Induction Generators (DFIG), facilitates the process by isolating rotor inertia from the energy system. The technological revolution enables the provision of rapid frequency response, which is vital in offsetting the effects of increased wind penetration. In new research, a control technique was proposed that enhances the capability of DFIGs to deliver inertial response and responds directly to problems of system balance whenever energy supply is variant. The technique is in a way that rotor velocities can be maximized while ensuring safe operation, thereby paving the way for stronger integration of wind energy into the grid. These advancements identify the requirement of novel solutions that not only efficiently harvest wind energy but also ensure the integrity of the electrical grid (Tomsovic et al., 2017). Altogether, effective integration of wind power relies on a multi-variable solution that covers advanced modeling, technological advancement, and adaptive operating practices. The interplay between urban energy loads and renewable sources is used to emphasize the necessity of integrated planning covering a broad array of considerations beyond residential use. As the inclusion of service sector profiles is demonstrated through empirical analysis to greatly enhance the integratability of renewables, technological advancements such as those surrounding DFIGs provide valuable support toward maintaining grid stability. These developments highlight that the wind power integration of the future relies on the development of better forecasting models and continuous technology improvement for wind turbines. By the use of these methodologies, stakeholders are able to handle effectively the complexity of modern energy systems and with promoting a sustainable transition toward renewables (Brazier et al., 2017; Tomsovic et al., 2017).

The integration of wind power into the current power infrastructure has some opportunities and challenges that need rigorous examination. With the greater emphasis on clean energy, particularly in nations such as ASEAN, there is a clear objective to accelerate the share of renewables in the energy mix, with particular goals to reach double from 13% to 23% by 2025. This change places extra attention on the importance of grid flexibility, or the capability of the electricity grid to adapt to fluctuating electricity demand and supply. Dependence on the successful integration of wind power is heavily reliant on grid operators' ability to manage this volatility because intermittent wind patterns have the potential to lead to supply uncertainties. Consequently, ASEAN nations exhibit various stages of preparedness for such, and therefore a need arises for an assessment framework that examines grid flexibility in terms of multiple indicators and identifies areas of improvement in terms of operational capacity and forecasting systems (Huang et al., 2019). Reducing the variability of wind generation requires innovation in power system operation. An economic dispatch model facilitated by adaptive robust optimization provides an appropriate methodology to counteract the implications resulting from high penetration of wind power. The model utilizes dynamic uncertainty sets to reduce the time and location variability of wind generation. Closing this gap with predictive statistical tools improves the economic dispatch procedure to achieve higher cost reduction and system reliability. Large-scale computational simulations in practice have yielded encouraging outcomes, proving that the above-introduced framework not only improves over current models but also allows for rational decision-making under uncertain energy supplies. Thus, this technique is a significant advancement in the ongoing pursuit of integrating wind power in a sustainable way into the power grid (Lorca et al., 2014). Moreover, efficient integration of wind power calls for, along with technical solutions, regional cooperation with the goal of improving access to electricity markets and forecasting (Lorca et al., 2014). ASEAN member states, as varied in their levels of readiness, need to put greater emphasis on grid flexibility development to allow for larger ratios of renewable energy generation effectively. Through collective cooperation and collaboration among members, information exchange on best practices, and upgrading of grid infrastructure enhancement, the region is poised to adapt and overcome key vulnerabilities holding back highest wind power integration. In addition, as demand for clean energy sources rises, strong regional systems have the potential to facilitate shared technological advancement in deployment and regulation convergence. Such a holistic approach will eventually help ASEAN achieve its renewable energy targets while ensuring a safe and affordable supply of energy to its member nations (Huang et al., 2019).

REFERENCES

1. Abdel Gelil, I., Aguayo, F., Boyle, G., et al. (2012). *Policies for Capacity Development*. https://core.ac.uk/download/9836421.pdf (accessed on 30th September 2025).
2. Aguado de Cea, A., Armengou, J., de la Fuente Antequera, A., et al. (2017). Towards the sustainability in the design of wind towers. *Renewable Energy*, *106*, 204–216. https://core.ac.uk/download/87657604.pdf (accessed on 30th September 2025).
3. Alcalá, G., Garrido, J., Hernandez-Escobedo, Q., et al. (2019). Wind power cogeneration to reduce peak electricity demand in Mexican states along the Gulf of Mexico. *Energy Reports*, *5*, 669–677. https://core.ac.uk/download/222767697.pdf (accessed on 30th September 2025).
4. Ali, S., Oo, A. M. T., Shafiullah, G. M., et al. (2013). Influences of wind energy integration into the distribution network. *Renewable and Sustainable Energy Reviews*, *24*, 355–364. https://core.ac.uk/download/190972688.pdf (accessed on 30th September 2025).
5. Ansari, O. A., Chung, C. Y., & Safar, N. (2017). Reliability assessment of microgrid with renewable generation and prioritized loads. *International Journal of Electrical Power & Energy Systems*, *93*, 395–405. http://arxiv.org/abs/1709.07970 (accessed on 30th September 2025).
6. Atkins, M. J., Morrison, A. S., & Walmsley, M. R. W. (2010). Carbon Emissions Pinch Analysis (CEPA) for emissions reduction in the New Zealand electricity sector. *Energy Policy*, *38*(11), 6570–6580. https://core.ac.uk/download/29197470.pdf (accessed on 30th September 2025).
7. Ault, G. W., Currie, R. A. F., Foote, C. E. T., et al. (2005). Ubiquitous energy storage. *Energy*, *30*(13), 2410–2419. https://core.ac.uk/download/9034038.pdf (accessed on 30th September 2025).
8. Austen, M., Berg, T., Borja, A., et al. (2016). Editorial: Bridging the gap between policy and science in assessing the health status of marine ecosystems. *Marine Policy*, *68*, 1–3. https://core.ac.uk/download/151161503.pdf (accessed on 30th September 2025).
9. Bachev, H. (2013). Natural resources conservation management and strategies in agriculture. *Journal of Environmental Management and Tourism*, *4*(1), 20–31. https://core.ac.uk/download/12036314.pdf (accessed on 30th September 2025).
10. Bao, Y., Feng, W., Jin, M., Liu, et al. (2018). A review of microgrid development in the United States–A decade of progress on policies, demonstrations, controls, and software tools. *Applied Energy*, *228*, 1332–

1349. https://core.ac.uk/download/323310623.pdf (accessed on 30th September 2025).

11. Becerra Villanueva, J. A., Calderón, B., Carvalho, M., et al. (2019). Trans critical carbon dioxide charge-discharge energy storage with integration of solar energy. *Energy Conversion and Management*, *195*, 1353–1365. https://core.ac.uk/download/322845447.pdf (accessed on 30th September 2025).
12. Beluco, A., Canales, F. A., Guezgouz, M., Jurasz, et al. (2019). A review on the complementarity of renewable energy sources: Concept, metrics, application and future research directions. *Renewable and Sustainable Energy Reviews*, *112*, 725–747. https://core.ac.uk/download/286642706.pdf (accessed on 30th September 2025).
13. Bhuyan, R., & Mytelka, L. K. (2025). Changing configuration of alternative energy systems. https://core.ac.uk/download/pdf/6941347.pdf (accessed on 30th September 2025).
14. Blumberga, A., Blumberga, D., Bohvalovs, Ģ., Pakere, L., et al. (2024). Transition towards positive energy district: A case study from Latvia. https://core.ac.uk/download/617057184.pdf (accessed on 30th September 2025).
15. Bolinger, M., Hammond, R., Hoen, B., Lantz, E., et al. (2021). Opportunities for and challenges to further reductions in the "specific power" rating of wind turbines installed in the United States. https://core.ac.uk/download/287622883.pdf (accessed on 30th September 2025).
16. Botterud, A., Levin, T., Mills, A. D., Seel, J., et al. (2020). Impacts of variable renewable energy on wholesale markets and generating assets in the United States: A review of expectations and evidence. https://core.ac.uk/download/305124749.pdf (accessed on 30th September 2025).
17. Brazier, F. M. T., Voulis, N., & Warnier, M. (2017). Impact of service sector loads on renewable resource integration. http://arxiv.org/abs/1605.09667 (accessed on 30th September 2025).
18. Broe, A. M. de, Burges, K., & Feijóo Lorenzo, A. E. (2025). Advanced power control in a wind farm network. https://core.ac.uk/download/657150270.pdf (accessed on 30th September 2025).
19. Bucha, M., Burdack, A., Burgherr, P., Camargo, J. R., et al. (2025). System impacts of wind energy developments: Key research challenges and opportunities. https://core.ac.uk/download/644524495.pdf (accessed on 30th September 2025).

20. Butler, L., Grubb, M., & Sinden, G. (2025). Diversity and security in UK electricity generation: The influence of low carbon objectives. https://core.ac.uk/download/pdf/7093211.pdf (accessed on 30th September 2025).
21. Carvalho, E., Faias, S. M. R., Sousa, J. A. M., Ventim-Neves, M., et al. (2012). Is the electric vehicle a solution for the wind power integration in the Portuguese power system? https://core.ac.uk/download/47134545.pdf (accessed on 30th September 2025).
22. Cassandras, C. G., Fox-Penner, P., & Kelly, H. (2019). Final report: Workshop on: Integrating electric mobility systems with the grid infrastructure. https://open.bu.edu/bitstream/2144/39613/1/SloanWshopFinalReport.pdf (accessed on 30th September 2025).
23. Copeland, R., Crespi, N., Rehmani, M. H., Saleem, Y., et al. (2019). Internet of Things-aided smart grid: Technologies, architectures, applications, prototypes, and future research directions. http://arxiv.org/abs/1704.08977 (accessed on 30th September 2025).
24. De la Bat, J. G. (2011). Development of a system for testing grid-connected permanent magnet wind generators. https://core.ac.uk/download/185417581.pdf (accessed on 30th September 2025).
25. Desmet, J., Heimbach, B., & Renner, H. (2015). Power quality and electromagnetic compatibility: Special report, session 2. https://core.ac.uk/download/55822716.pdf (accessed on 30th September 2025).
26. Dinica, V. (2009). Wind technology: A framework for the evaluation of innovations' impacts on the diffusion potential. https://core.ac.uk/download/pdf/11482434.pdf (accessed on 30th September 2025).
27. Flavin, C. (2008). Low-carbon energy: A roadmap. https://core.ac.uk/download/71340916.pdf (accessed on 30th September 2025).
28. Huang, Y. W., Kammen, D. M., & Kittner, N. (2019). ASEAN grid flexibility: Preparedness for grid integration of renewable energy. https://core.ac.uk/download/323310932.pdf (accessed on 30th September 2025).
29. Ipsos. (2025). Establishing the social licence to operate large scale solar facilities in Australia: Insights from social research for industry. https://core.ac.uk/download/pdf/30672015.pdf (accessed on 30th September 2025).
30. Kammen, D. M., Shiraishi, K., & Shirley, R. G. (2019). Geospatial multi-criteria analysis for identifying high priority clean energy investment opportunities: A case study on land-use conflict in Bangladesh. https://core.ac.uk/download/323310931.pdf (accessed on 30th September 2025).

31. Kester, J., Noel, L., Sovacool, B. K., Zarazua de Rubens, G., et al. (2018). Expert perceptions of low-carbon transitions: Investigating the challenges of electricity decarbonisation in the Nordic region. https://core.ac.uk/download/146490613.pdf (accessed on 30th September 2025).

32. Kirkegaard, J. F., Weischer, L., & Hanemann, T. (2025). It should be a breeze: Harnessing the potential of open trade and investment flows in the wind energy industry. https://core.ac.uk/download/pdf/6486668.pdf (accessed on 30th September 2025).

33. Koh, J., & Stremke, S. (2009). Sustainable energy transition: Properties and constraints of regenerative energy systems with respect to spatial planning and design. https://core.ac.uk/download/pdf/29249814.pdf (accessed on 30th September 2025).

34. Lorca, A., & Sun, A. (2014). Adaptive robust optimization with dynamic uncertainty sets for multi-period economic dispatch under significant wind. http://arxiv.org/abs/1409.2936 (accessed on 30th September 2025).

35. Lucon, O., Fransen, T., & Romeiro, V. (2015). Bridging the gap between energy and climate policies in Brazil: Policy options to reduce energy-related GHG emissions. https://core.ac.uk/download/75782972.pdf (accessed on 30th September 2025).

36. Machado Ferreira, C. M., Maciel Barbosa, F. P., & Monteiro Pereira, R. M. (2015). Reactive power management of a wind farm to prevent voltage collapse of an electric power system. https://core.ac.uk/download/143395680.pdf (accessed on 30th September 2025).

37. Meadowcroft, J. (2025). Climate change governance. https://core.ac.uk/download/pdf/6305225.pdf (accessed on 30th September 2025).

38. Monsees, J., & Ohl, C. (2025). Sustainable land use against the background of a growing wind power industry. https://core.ac.uk/download/pdf/6670980.pdf (accessed on 30th September 2025).

39. N/A. (2013). Understanding best practice regarding interruptible connections for wind generation: Lessons from national and international experience. https://core.ac.uk/download/42337156.pdf (accessed on 30th September 2025).

40. Nicolli, F., Vona, F., & Nesta, L. (2025). Determinants of renewable energy innovation: Environmental policies vs. market regulation. https://core.ac.uk/download/pdf/6441011.pdf (accessed on 30th September 2025).

41. Ozkan, D., McClellan, S., & Kempton, W. (2016). Massachusetts offshore wind future cost study. https://core.ac.uk/download/75785617.pdf (accessed on 30th September 2025).

42. Prins, W., & Thornley, P. (2009). Barriers to European bioenergy expansion. https://core.ac.uk/download/55739198.pdf (accessed on 30th September 2025).

43. Tomsovic, K., & Wang, S. (2017). Quantitative control approach for wind turbine generators to provide fast frequency response with guarantee of rotor security. http://arxiv.org/abs/1712.06190 (accessed on 30th September 2025).

CHAPTER 6

HYDROPOWER AND SMALL-HYDRO INTEGRATION

Contents

INTRODUCTION

The significance of hydropower as a clean energy source has gained traction in recent years, largely due to increasing worldwide energy requirements and the need to mitigate the impacts of climate change. While nations strive to ease their dependence on fossil fuels and transition towards cleaner alternatives, hydropower offers an already-developed technology with the promise of reliable power generation while reducing greenhouse gas emissions.

This chapter addresses the intricacies of hydropower and its integration into small-hydro systems, a subclass that has the potential to greatly complement conventional large-scale operations. The integration of these systems is as much a strategy for augmenting energy supply as it is against offsetting the variability challenges of other renewable systems, like wind and solar, with a focus on hybrid energy solutions (Beluco et al., 2019). Emphasizing the significance of hybridization in energy production, it is apparent that small-hydro systems play a vital role to provide maximum power reliability and availability. By employing the characteristic temporal and spatial features of different renewable energy sources, these systems are able to meet supply and demand fluctuations effectively.

Figure 6.1. *Hydroelectric power plant.*

Source: https://energytheory.com/wp-content/uploads/2023/12/hydroelectric-power-plant.jpg.

The concept of complementarity is central to this argument, as synergistic grouping of wind, solar, and hydropower can stabilize power production and align it more with the consumers' demand. Research has proven that the integration of the three sources leads to improved overall performance and efficiency, showing how an energy system in modern times needs a diversified portfolio (Beluco et al., 2019). In energy production planning in the future, the integration of small-hydro with large hydropower schemes becomes a critical component of sustainable planning for energy. With improving technology and policy support for renewables, the prospect for more small-hydro plant installations improves. Not only do the systems increase the resilience of the grid, but they also enhance local development by the provision of clean power in inaccessible areas. Besides, the combined system of such energy systems also demands more research on innovative means of how to achieve their peak performance, thus contributing to a greener energy future. This integration cannot be exaggerated, considering that it is a strategic step towards addressing combined challenges of energy demand and environmental conservation, opening the way for integrative development in renewable energy facilities (Figure 6.1).

Definition of Hydropower

Hydropower, a vital and renewable energy sector, transforms the potential and kinetic energies of flowing or standing water into electric power. Giant dams typically create reservoirs whose water exits through turbines to generate electrical energy from moving water. The sophistication of hydropower, however, extends beyond such a standard system in considering the small-hydropower (SHP) systems. SHP plants, or facilities that generate up to 10 megawatts of power, are a more contained and environmentally sound option than large installations. This distinction is significant, as SHP systems can be installed in various settings, effectively meeting energy requirements while reducing environmental strain. With increasing reliance on renewables, knowledge of these various applications of hydropower becomes essential for energy sustainability projects globally. In addition to providing a renewable source of energy, hydropower is also significant in water management and agriculture. The flexibility of operation of small-hydropower plants presents simultaneous benefits like irrigation and flood control and is hence particularly well-suited to nations like Bangladesh. In Bangladesh, for example, SHP systems are able to provide cost-effective alternatives to energy at minimal environmental costs. These crops are not only energy producers but also a method of enhancing agriculture yields and efficient management of water resources. The dual role of hydropower as an energy source as well as a system for managing water helps to highlight its role in sustainable development in various forms, indicating the need to seek innovative ways to exploit the resource sustainably, as discussed by Joseph et al. (2006). Furthermore, the hybridization of SHP with other renewable energy sources enhances the overall efficiency and dependability of energy systems. The complementarities of different renewable resources, such as solar and wind power, provide a viable method of addressing the difficulties resulting from their inherent temporal and spatial variability. The method facilitates more dependable power generation, which offsets the typical oscillations in renewable energy systems. Implementing such integrated approaches not only achieves stability in the energy supply but also fulfills the overall environmental goals by reducing dependence on fossil fuels. Accordingly, literature reports a growing interest in appreciating such synergies, which can lead to new solutions to fulfill energy demand, as referenced in the case of the existing research in the complementarity among renewable energies in power generation systems, as stated by Beluco et al. (2019) (Figure 6.2).

Figure 6.2. *How hydropower works?*

Source: https://www.greenlivinganswers.com/blog/wp-content/uploads/2024/08/hydropower-mechanism.webp.

Importance of Renewable Energy Sources

The transition to renewable energy is no fleeting fancy; it is a call to action towards sustainable development and climate change mitigation. As the global energy demand grows, the consumption of fossil fuels has propelled the emission of greenhouse gases to record levels, whose role in global warming cannot be exaggerated. In this context, renewable energy, particularly hydropower and small-hydro systems, comes across as an honest alternative. Hydropower is among the oldest and most mature renewable energy technologies, able to supply large-scale energy requirements while having low environmental effects. Additionally, the integration of small-hydro plants can increase local energy autonomy and robustness, thus lowering community dependence on centralized fossil fuel systems. By strategically investing in these technologies, countries can cut carbon emissions by quite a significant amount and emerge as world leaders in the transition to a low-carbon economy, as with ongoing discussions in Brazil regarding energy investment (Lucon et al., 2015). The importance of renewable energy sources is that they can provide economic competitiveness and improve social welfare. Renewable technologies are embraced by nations that are able to expect investments that serve energy needs as well as create opportunities for employment in new sectors. For instance, the development of hydropower plants tends to foster opportunities for employment in the surrounding area through construction and maintenance, hence boosting regional economies. Cleaner energy also enhances health outcomes in terms of removal of air pollution and associated health risks. The policymakers need to understand that the deployment of small-hydro systems as part of the energy mix can be a pillar of energy security and meeting international climate commitments. There is an entire strategy, as noted in recent policy appraisals, that places emphasis on having a strong regulatory platform that supports investment in sustainable energy schemes (Athena et al., 2010). Lastly, the incorporation of renewable energy technologies, notably hydropower, within national energy policies aligns with overarching global climate and energy goals. The imperative with which climate change has to be dealt with necessitates a redesign of existing energy infrastructures in order to trigger investments in green technologies that decarbonize energy supplies. Countries like Brazil are faced with critical choices regarding their energy futures; today's decisions can either bet them into a path of high carbon or push them into a low-carbon economy. Successful achievement of small-hydro integration not only mitigates the effects of climate but also promotes sustainable environmental behavior by saving water resources and promoting biodiversity. Therefore, policy, investment, and convergence of technology in renewable energy is the key feature for triggering an economic setting that gives equal significance to environmental integrity and social justice concerning climate problems (Lucon et al., 2015).

Overview of Small-Hydro Integration

In the current renewable energy landscape, small-hydro integration is a vital component harnessing local water streams to generate power in a green way. Unlike large hydropower projects, which are usually put through rigorous environmental and social assessment, small-hydro plants typically demand smaller initial investments, shorter development periods, and lower environmental impact. These power plants can be strategically located at sites of limited adverse effects on local ecologies and thus conserve biodiversity while addressing energy needs. Small-hydro plants are a flexible alternative that can accommodate variable loads and form part of aggregated renewable mixes. This flexibility is particularly valuable in countering the variable character of wind and solar power, underlining the significance of diversified generation in modern power grids (Beluco et al., 2019). Moreover, integration of small-hydro plants into local power systems is a step toward decentralized energy systems, allowing for local energy autonomy and reliability. This model conserves fossil fuels and, moreover, enhances grid stability through local generation. Intercombination of small-hydropower with other renewable technologies such as wind power and solar power enables energy producers to create hybrid systems, optimizing efficiency and minimizing dependence on any one source. Hybrid systems exploit the complementarity between alternative energy sources by smoothing out their temporal and spatial variability. Research in these integration steps shows an increasing trend towards maximization of energy generation from different geographical sites, hence paving the way to new practices and research on hydropower integration (Elverhøi et al., 2025). As there is an increased interest in small-hydro integration, stakeholders are faced with numerous challenges and opportunities brought about by infrastructure development and investment decisions. Investors and policymakers need to examine the feasibility of refurbishing existing plants compared to constructing new ones, particularly in the background of unstable electricity prices and regulatory policies. Real options valuation techniques allow investors to make better decisions based on the uncertainties present in energy markets. Small-hydro integration in existing grids has a significant impact on energy security, besides providing economic advantages to nearby communities. Integrated models addressing technical as well as financial concerns of small-hydro schemes offer a totality of perception about their role in the new energy paradigm, thus facilitating the process of informed decision-making in the hydropower sector (Elverhøi et al., 2025) (Figure 6.3).

Figure 6.3. *An example of a small-hydropower plant, Sveta Petka.*

Source: https://upload.wikimedia.org/wikipedia/commons/1/14/Small_hydro_power_plant_%22Sveta_Petka%22.IMG_0013.jpg.

6.1. HISTORICAL CONTEXT OF HYDROPOWER

The historical background of hydropower presents a revolutionary force in electricity production, with origins dating back to ancient times. An early illustration comes in the form of water wheels that existed in Greco-Roman civilizations and utilized flowing water for performing mechanical activities like grinding grain. The Industrial Revolution provided the impetus for the development of hydropower, as it became possible to produce electricity in larger systems. It was the period of transition from decentralized power sources to concentrated power generation, and gargantuan feats like the Hoover Dam embodied human resolve in harnessing natural resources. Technological developments in hydropower technology not only catered to the needs of society to become more efficient and predictable but also paved the path for subsequent generations to explore its feasibility as a renewable energy source. The persistent quest for innovation in hydropower is proof of human sensitivity to energy requirements and environmental concerns.

Figure 6.4. *Historical development of hydropower.*

Source: https://www.greenlivinganswers.com/blog/wp-content/uploads/2024/08/hydropower-history.webp.

Throughout the 20th century, hydropower emerged as a dominant player in the international energy sector, particularly in industrialized nations. Incorporating hydropower into national grids provided a clean and reliable source of energy that promoted industrialization and urbanization. They spent heavily on big dams and reservoirs that controlled river flows and offered energy storage during peak demand times. However, this rapid acceleration had environmental and social impacts, raising doubts regarding the sustainability of such projects. Despite these doubts, hydropower remains a viable source to achieve energy security and reduce greenhouse gas emissions. Contemporary arguments now center on incorporating small-hydropower schemes, which can mitigate some of the adverse environmental effects associated with large dams but still provide critical energy solutions for local communities. Recent interest in small-hydro integration is being driven by an emerging recognition of potential benefits from decentralized power schemes. By using local water resources, small-hydropower schemes have the ability to provide electricity to remote villages that are normally neglected by traditional energy supply infrastructure. Small-hydropower schemes promote environmental protection because they will generally have a smaller ecological footprint compared to large schemes. Furthermore, the utilization of small-hydropower conforms with new worldwide trends towards decentralized energy technology and sustainable development. While nations are grappling with the effects of climate change and energy transition, they should encourage small-hydro integration. Methods used to spur this sector, such as those outlined in Colombia's renewable energy programs, can be adopted as a model for how to apply hydropower technology to meet present-day energy demands effectively without disturbing the environment (Deeb et al., 2025; Inyang et al., 2025) (Figure 6.4).

6.1.1. Development of Hydropower Technology

Technological advancement of hydropower has been an integral part of global energy plans, particularly with the increasing concern of climate change and reliance on fossil fuels. In the hope that energy consumption would increase by leaps and bounds over the next two decades, there has been a realization to shift towards greener sources of energy. The Fukushima Daiichi nuclear meltdown caused this realization to dawn on people, highlighting the risks involved in the use of nuclear energy. Stakeholders now consider hydropower as a sustainable source which can minimize greenhouse emissions and help with energy security. In utilizing the water resources efficiently, hydropower not only provides a stable power supply but also helps with grid flexibility, as required in balancing other renewable sources. These technologies play a crucial role in balancing the twin challenges of energy use and environmental stewardship, further reinforcing the role of hydropower in the broader debate of deploying renewable energy resources (Abolhosseini et al., 2014). Hydropower technologies have also opened up the suitability of small-hydro schemes, which are now increasingly viewed as a means of contributing to local energy supply. Unlike large hydropower facilities that may disturb ecosystems, small-hydro schemes may be added with minimal environmental impact, promoting rural sustainable growth. Small-hydropower plants may further enhance energy efficiency and reliability, even in local conditions. Such a model motivates the community to invest in decentralized energy infrastructure, reducing transmission losses and enhancing energy resilience. The application of sophisticated technologies, such as better turbine design and computerized observation systems, also enhances the efficiency and feasibility of small-hydropower systems, ultimately resulting in greater adoption in geographical areas. All these developments align with the worldwide drive towards reducing carbon emissions and promoting sustainable energy resources (Hampl et al., 2014). Moreover, the ongoing evolution of hydropower technology is accompanied by a robust system of assessing the disruptive potential of hydropower in the energy industry. Grounded in theories of disruptive innovation, experts are better positioned to study the consequences of emerging hydropower technologies in redefining existing energy systems. Such systematic examination not only defines future trends in renewable technology but also informs strategic decision-making on the part of energy providers and policy makers. Examining the dynamics of different nations, for instance, Austria and Germany, the model presents regional variations and the particular issues they face in integrating hydropower. It is important to realize these points while formulating policies that allow the development of clean energy sources in a competitive environment. With hydropower continuing to develop, its interconnection with other renewable energies will be pivotal to the realization of sustainable energy objectives and the diminution of dependence upon less ecologically sound sources (Hampl et al., 2014).

6.1.2. Key Milestones in Hydropower History

Over time, the use of hydropower has seen milestones that have influenced practices of energy production and sustainability. One significant one was in the late 19th century when the first hydroelectric power plant was built in Appleton, Wisconsin, in 1882, showing that water flow could be converted into electricity. This pioneering feat paved the way for larger hydroelectric projects, including the construction of Hoover Dam in the 1930s, which not only provided massive amounts of electricity to the areas nearby but also ensured flood control and irrigation. These developments have further established the role of hydropower as it assists in transforming the energy situation with a consistent and renewable source of energy that satisfies expanding population demands while concurrently addressing environmental problems. Hydropower technology developments have also seen key developments with the introduction of small-hydro schemes, especially in the latter half of the 20th century. With emerging technologies in turbine and generator, regions with small streams and rivers can now be harnessed for hydroelectric power, allowing people to utilize locally available resources in energy production. This is particularly significant as small-hydro schemes have the ability to be more adaptable to the local environment and can offer tailored energy solutions that support sustainability. The potential of decentralized energy applications through small-hydro facilities is the key to energy resilience where large-scale projects may be impractical. The widespread use of the systems in place renders the integration of hydropower into multipurpose energy mixes a central interest. In the present day, Brazil is an interesting case study for the future of hydropower in the context of the interface between energy development and environmental conservation. The country's energy strategy points to the need for investment that can pave the way to a low-carbon future or extend carbon-based approaches, thus framing stakes on the growth of hydropower. With approximately 0.5 trillion U.S. dollars of investments in energy infrastructure on the horizon, decisions now will significantly affect Brazil's environmental contributions and climate goals. As highlighted through the dialogue on energy-based GHG emissions, focus on hydropower abatement opportunities is imperative to achieving climate and energy objectives simultaneously (Lucon et al., 2015). Such strategic alternatives embody the far-reaching implications of hydropower milestones toward advancing both human development and climate change mitigative practices in the context of a globalized world (Li, 2025) (Figure 6.5).

Figure 6.5. *The site of the first American hydroelectric dam, built on the Fox River at Appleton, Wisconsin, in 1882.*

Source: https://www.researchgate.net/profile/Mike-Snead/publication/359439748/figure/fig3/AS:11371588071383O8@1648131195521/The-site-of-the-first-American-hydroelectric-dam-built-on-the-Fox-River-at-Appleton.png.

6.1.3. Evolution of Small-Hydro Systems

Small-hydro system development dates back to the early application of waterpower for mechanical and electrical energy. Small-hydro systems were initially used for local mechanical applications, i.e., grinding grain or sawing wood. As technology evolved in the late 19th century, efforts turned towards generating electrical energy, particularly with the invention of efficient turbines and generators. This transition initiated small-hydroelectric stations that electrified local grids and energized rural communities. Their decentralized nature offered a geographically stable substitute for fossil fuels, improved energy access, and paved the way for a cleaner energy future. Therefore, this development demonstrates how small-hydro schemes have come to address the shifting energy needs of communities, and their place in the renewable energy story. Over the past decades, increasing demand for alternative clean energy sources has fueled the expansion of small-hydro schemes worldwide. These projects are recognized for being comparatively environmentally sound compared to giant hydroelectric plants, and therefore, they represent a desirable substitute for green power generation. As countries make efforts to meet their carbon reduction targets, small-hydro schemes contribute to a diversified energy supply and improve energy security. In addition, the development of turbine technology and material design has improved the efficiency and robustness of these schemes, enhancing their competitiveness against other forms of renewable energy. Policymakers recognize this potential and have encouraged investment in small-hydro plants by offering incentives and streamlined processes, symptomatic of a growing emphasis on the integration of small-scale renewable systems into national energy grids to enhance overall sustainability objectives. However, development of small-hydro systems entails numerous challenges that must be overcome if they are to remain successful. There are the parameters of hydropower potential as affected by the effect of climate change, which can have an effect on the reliability of energy production. Studies show that interannual and inter-basin variability of hydropower resources is affected by climatic oscillations such as El Niño. According to recent research, acquisition of these differences and their incorporation into the working arrangement of small-hydro schemes is very important to facilitate long-term sustainability and viability (Dieppois et al., 2019). Furthermore, bringing small-hydro into synchronization with the existing energy system requires cautious assessment of local ecological condition as well as stakeholder interests to avoid likely conflicts on the use of water. Through engaging with the challenges through diversified policy frameworks that recognize the dynamics of small-hydro systems, opportunities to make a contribution towards ushering in a resilient and sustainable energy future can be significantly expanded (Lucon et al., 2015).

6.2. MECHANISMS OF HYDROPOWER GENERATION

The generation of hydropower is fundamentally reliant on the kinetic energy of flowing or descending water being used to generate electrical energy. It begins with water taken from river basins or reservoirs, which is channeled to turbines.

Through the water's passage in the turbine, it transmutes the potential energy of the higher amount of water to mechanical energy. This mechanical energy is used to drive a generator, which subsequently converts it into electrical energy. These large-scale uses are primarily based on hydropower plants, which can significantly vary in scale but typically involve complex structures like dams. These systems not only provide good energy output but are also important regulators of river flows, ensuring water supply, and even alleviating flood risks, putting their importance within the broader ecosystem and economic energy production systems into perspective.

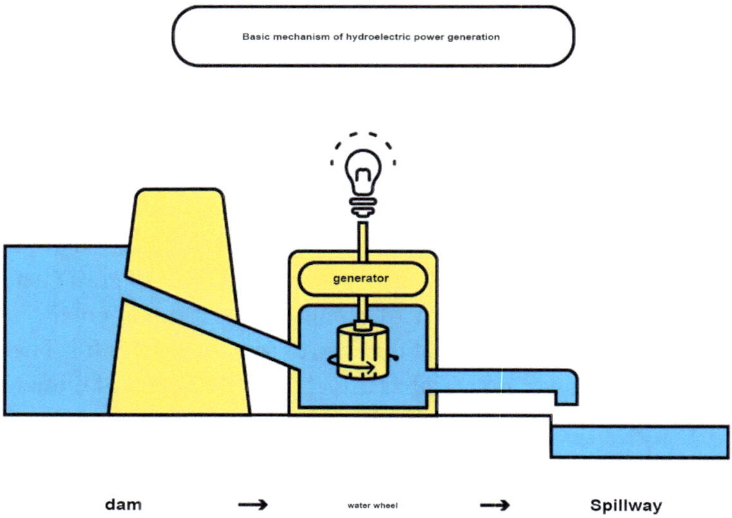

Figure 6.6. *Basic mechanism of hydroelectric power generation.*

Source: https://cdn.clipkit.co/tenants/1224/item_field_assets/assets/000/002/344/large/903f95ed-9b03-42ed-a6a6-e14ad8bb8aaf.png?1731916744.

Small-hydroelectric systems embody the principles of hydropower generation on a smaller scale. Using smaller amounts of water flow, these mini-systems are placed strategically in rural or far-off areas where comprehensive development is not feasible. Mechanisms are typically analogous: water flows through turbines and produces mechanical energy that drives generators. Small-hydro installations employ more efficient, less intrusive technologies with lower environmental impact and land use, however. Such a decentralized approach allows for flexibility in delivering local energy requirements, with communities enjoying enhanced energy independence. Furthermore, hybrid systems incorporating small-hydro and other renewable technologies have the potential to create hybrid systems that maximize resource utilization, which produces a reliable source of energy and minimizes greenhouse gas emissions to an absolute level, an important consideration for climate change. Also, as nations seek environmentally friendly sources of energy, the need for hydropower's mechanism becomes increasingly evident in bigger energy policies. The integration of hydropower into national energy policies is capable of minimizing reliance on fossil fuels, offering a reliable source of energy while addressing climate goals. For example, when Brazil is deciding on energy investments, prioritizing hydropower can place the country at the vanguard of low-carbon energy in the context of cross-pressures in other areas, including the industrial and transport sectors (Lucon et al., 2015). On the other hand, minor economies such as Colombia can deploy localized hydropower sources to more diversify their energy supply, catalyzing renewable energy and diversified energy portfolio development (Deeb et al., 2025). All these details determine the numerous benefits of including hydropower mechanisms and support more extension in energy programs globally (Figure 6.6).

6.2.1. Basic Principles of Hydropower Generation

The fundamental concepts of hydropower production rely on transforming the kinetic energy of flowing water into electricity. It usually begins by constructing a dam that creates a reservoir, increasing the head of the water—the vertical height from which waterfalls. The potential energy in the reservoir is released in measurable amounts to drive turbines. As water flows through the turbines, it makes them spin, converting the kinetic energy of the flowing water into mechanical energy. The mechanical energy is then utilized to produce electrical energy with the help of generators. Energy conversion efficiency is crucial in the process since this will determine the viability of hydropower as a source of renewable energy in the context of total energy requirements and environmental impact. Small-hydropower plants, commonly defined as installations that produce no more than 10 megawatts, have several advantages over conventional large hydropower plants, especially when it comes to their environmental impact. In comparison with large plants, small-hydro plants generally need less elaborate infrastructure and affect local ecosystems and communities less. Small-hydro plants can utilize micro watersheds without requiring gigantic dam constructions, making it easier for them to blend into existing landscapes in a more sustainable manner. Besides, the schemes promote local energy independence, allowing societies to generate their own power with a lower loss in transmission. Therefore, small-hydropower is a feasible option in regions where conventional hydroelectric plants would be met with social or environmental resistance due to their larger size and implications involved. However, implementing small-hydropower as part of larger energy plans demands an adequate understanding of regional environmental and social circumstances. For example, in Nepal, the hydropower industry has been operating with extremely high pressures to develop faster while being exposed to uncertainty of seismicity, among other natural hazards. Before the destructive earthquake that struck in April 2015, stakeholders in Nepal underplayed the geophysical weaknesses that have been accentuating hydropower development, leading to what is referred to as a socially organized denial of the problems. This harsh fact supports the necessity for full-risk assessments and stakeholder involvement to manage the intricacies represented in hydropower projects so that the pursuit of renewable energy does not undermine local livelihoods or accelerate vulnerabilities in vulnerable communities (Butler et al., 2017). These are considerations that are central because the globe continues to move towards hydropower as a potential solution to climate change and the promotion of sustainable energy options (Gelil et al., 2012) (Figure 6.7).

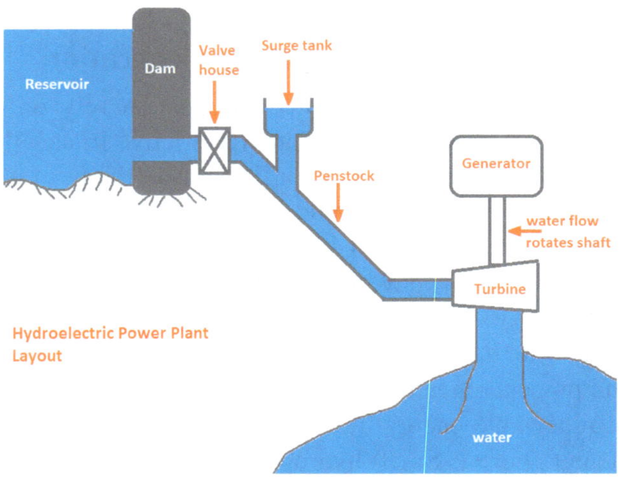

Figure 6.7. *Working principle of hydroelectric power plant.*

Source: https://yourelectricalguide.com/wp-content/uploads/2018/09/working-principle-of-hydroelectric-power-plant.png.

6.2.2. Types of Hydropower Plants

Hydropower stations are usually categorized into various types based on the operation principle and construction. Run-of-river, reservoir, and pumped storage stations are the most common and each has characteristics suited to certain geographical and environmental conditions. Run-of-river stations use rivers' natural flow without significant water storage on a large scale, with limited disturbance to ecology. On the other hand, reservoir plants store large amounts of water in a dam, which is an enclosed environment for energy generation during peak loads. Pumped storage facilities are specifically energy storage devices that move water from a higher point to a lower point, selling power during high demand by releasing stored water, while simultaneously pumping water up to the upper reservoir during low demand.

Figure 6.8. *Types of hydropower plants.*

Source: https://www.greenlivinganswers.com/blog/wp-content/uploads/2024/08/hydropower-plant-types.webp.

This plant diversity allows for a bespoke strategy in integrating hydropower, which is vital for supplementing other renewable resources in the energy sector globally. Incorporating different types of hydropower is instrumental in balancing the intermittency of renewable sources, especially where demand and supply are unreliable. By making use of complementary renewable energy such as solar and wind power along with hydropower, energy systems can balance output and be more consistent. For instance, solar and wind power variability can be balanced by making use of the predictable nature of hydroelectric power to deliver stable energy. Experiments proved that the blending of energy sources, e.g., solar-hydro or wind-hydro combinations, taps into the respective strength inherent in each technology, thus achieving greater overall energy production and minimizing greenhouse gas emissions. As the energy market develops, such hybrid systems are progressively gaining significance in realization of a robust and sustainable energy infrastructure, responding directly to the factors introduced by climate-driven variability in generation (Beluco et al., 2019). In addition, the complementarity of hydropower and other renewable resources underscores the necessity for long-term planning in energy systems. It has been discovered through studies that efficiency gains from integration of wind, solar, and hydro sources are substantial, particularly when evaluated across various time scales. As an example, solar-hydro generation pattern synchronization can be used to provide stable flow during low solar output hours. Synergy not only improves the reliability of energy supplies but also causes a reduction in fossil fuel dependency, thereby accelerating the shift towards more sustainable energy configurations (Beluco et al., 2019). As future energy policies increasingly center on the deployment of hybrid systems, it will be essential to understand the various types of hydropower plants and integration capacity in order to achieve an optimal and efficient renewable power mix (Figure 6.8).

6.2.3. Role of Turbines and Generators in Energy Conversion

The primary role of turbines and generators in energy transformation is perhaps best appreciated in the context of hydropower plants, where kinetic energy of flowing water is transformed into mechanical and subsequently electrical energy. Turbines provide the initial mechanical means of harnessing energy of flowing water through a series of blades that rotate to convert hydraulic energy into rotational energy. This is crucial to the overall performance of hydropower technology. As global energy consumption is increasingly reliant on renewable energy, how effective turbines are is most crucial. Advances in turbine material and design allow for greater energy capture, meeting the need to wean the world off fossil fuels, which have significant environmental impacts, as indicated in the background of growing energy requirements and climate change mitigation efforts (Abolhosseini et al., 2014). After kinetic energy has been converted into mechanical energy using turbines, the role of generators takes center stage in the energy conversion process. Generators translate mechanical to electrical energy, which can be fed into the electricity grid or utilized locally. The level of this energy conversion efficiency will be determined by the interaction between the performance of the turbines and their capacity to generate. Optimizing the joining of the two elements is not merely a case of maximizing energy harvesting but also ensuring the quality of the power generated to feed into modern power systems. Technological advancements in turbines and generators have gone a long way in promoting energy efficiency and output, especially for small-hydro schemes where system constraints demand energy-efficient operation with low maintenance (Boero et al., 2012). Additionally, the use of advanced technology in turbines and generators enables the installation of smart energy systems. Intelligent sensors and computer controls are utilized to optimize system operation and reduce the environmental footprint of hydropower plants. For instance, self-contained machines can monitor system efficiency and develop data that is utilized to plan maintenance schedules to avoid interrupting energy production. Moreover, energy scavenging technology is becoming more widely integrated into hydropower systems, wherein tiny turbines harness localized water flow to generate electricity, offering independent power for off-grid uses. This coordination not just underscores the versatilities of generator and turbine functions in energy conversion but also indicates their significance in the broader context of green energy solutions and community resilience in the face of environmental challenges (Boero et al., 2012) (Figure 6.9).

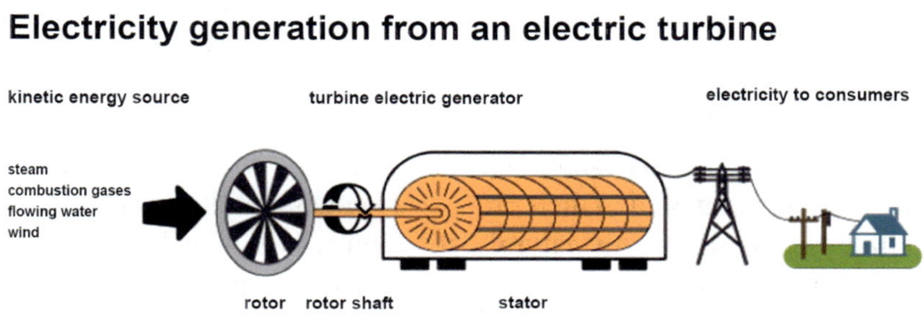

Figure 6.9. *Electricity generation from an electric turbine.*

Source: https://www.eia.gov/energyexplained/electricity/images/electricity-generation-turbine.png.

6.3. BENEFITS OF HYDROPOWER

The integration of hydropower into the energy mix is not simply a matter of producing electricity but presents a myriad of benefits. One of the most appealing benefits is the potential to generate low-cost energy, particularly when considered in the context of small-hydropower (SHP) plants. For countries like Bangladesh, power plants based on SHP offer an affordable alternative, not only delivering low-cost energy but also addressing essential services such as irrigation and flood control. These benefits are incurred without invoking heavy environmental costs, the typical characteristic of other energy sources. Thus, the installation of SHP plants positions them at the focal point in ensuring the management of the energy demands and socio-economic issues of developing nations, promoting both resilience and sustainability in energy use (Joseph et al., 2006). Furthermore, hydropower plants can be a stable and ongoing provider of energy that potentially makes up for the variability of other clean energy technologies. In the process of the global energy model shifting toward clean sources, the inclusion of hydropower can also serve as a buffer to natural wind or sunlight variability. By creating a hybrid power system that involves the use of over one source, i.e., solar and hydropower, the reliability of energy supply is highly enhanced. Hybridization is particularly beneficial in regions with irregular energy demand, offering a more stable and uniform supply of electricity. The strategic use of hydropower, therefore, alleviates not just energy shortages but also encourages a more resilient and adaptive energy system that stands prepared to attend to evolving consumption patterns (Beluco et al., 2019). In addition to financial and performance advantages, hydropower is environmentally compatible. Compared to fossil fuels, hydropower is able to generate electricity with minimal greenhouse gas emissions after the plant's setup. This aspect is only more critical as global efforts become more robust to combat climate change and move towards cleaner sources of energy. Furthermore, small-hydro power plants are likely to leave a smaller footprint on the environment compared to their large-scale counterparts, making them more compatible to integrate with natural ecosystems. Utilizing localized sources of water, SHP systems can support conservation efforts as well as safeguard neighboring ecosystems. Thus, the benefits of hydropower go beyond the production of energy to advocate for environmental stewardship, and it is a major player in the global movement toward sustainable energy strategies (Joseph et al., 2006) (Figure 6.10).

Figure 6.10. *Benefits of hydropower.*

Source: https://www.greenlivinganswers.com/blog/wp-content/uploads/2024/08/hydropower-benefits.webp.

6.3.1. Environmental Advantages

The ecological advantages of hydropower, particularly small-scale hydro integration, are great when compared to conventional fossil fuel electricity generation. Through the harnessing of the kinetic energy of flowing water, hydropower generates electricity with minimal greenhouse gas emissions, which contributes to significantly alleviating the impacts of climate change. Small-hydropower schemes that are intermittent enable scaling up in development, sometimes with reduced ecological footprint compared to large-scale hydropower developments. In addition, these systems promote the use of renewable sources, broadening the energy supply base and reducing dependence on nonrenewable. More specifically, small-hydropower can integrate well with local environments, offering advantages like maintaining river flows and supporting aquatic life. The environmental impacts are significantly lower in relation to those of conventional energy sources, leading to a cleaner and more sustainable solution for meeting energy demands in a clean and efficient manner, as advocated by emerging studies on the utilization of renewable energy (Beluco et al., 2019). Further, the integration of small-hydropower at the regional level enhances the resilience of the energy system, particularly in multi-diversified climatic regimes. The complementarity of hydropower with other renewable energy technologies, such as solar and wind, plays a critical role in ensuring a stable grid. Temporal complementarity allows hydropower to deal with the intermittency of solar and wind resources effectively. In cases of low solar and wind output due to adverse weather conditions, hydropower can act as a back-up.

Figure 6.11. *Environmental advantages of hydroelectric power.*

Source: https://graconllc.com/wp-content/uploads/2021/10/environmental-advantages-of-hydroelectric-power.png.

This two-energy system approach not only optimizes overall energy efficiency but also encourages on-site energy autonomy. Such interdependencies accentuate hydropower's environmental advantages as regions with these combined systems can effectively reduce their carbon footprints and limit the environmental impacts typically created by traditional fossil fuel reliance (Abolhosseini et al.,

2014). Moreover, small-hydropower installations often trigger local economies while promoting environmental stewardship. By generating clean energy at the site, these systems reduce the transportation of energy from distant fossil fuels and hence associated emissions and energy losses. Small-hydro plant operation and construction can also generate employment in host communities, supporting economic development that is environmentally friendly. Smaller infrastructural footprint compared to large dams for hydropower allows for less disruption of natural ecosystems, thereby ensuring systems are not disturbed and biodiversity is preserved. Small-hydropower is thus not only an impetus for the adoption of renewable power but also a tool of local ecosystem enrichment alongside empowering communities economically, presenting a twofold vision for human well-being and environmental protection (Beluco et al., 2019) (Figure 6.11).

6.3.2. Economic Impacts and Job Creation

The combination of hydropower and small-hydro technology has significant economic significance, both in terms of jobs and spurring local economies. With a growing demand for alternative renewable power sources, driven by rising fossil fuel prices and environmental issues, hydropower is an attractive option. Local labor is utilized in the construction, operation, and maintenance of small-hydro developments, creating employment that strengthens regional economies. Such projects, while redeveloping rural areas with secure job opportunities, also lead to ancillary industries and businesses, and therefore the overall economic resilience is enhanced. Such types of projects create a multi-dimensional economic scenario in which direct jobs through energy production are complimented with indirect job creation in connected industries, enhancing benefits throughout the community, and highlighting the importance of investment in renewable energy facilities (Abolhosseini et al., 2014). Moreover, the contribution of renewable energy to the fight against climate change can have a major economic imposition. Through the curtailment of reliance upon carbon-based sources of energy, nations not only assist in global emissions reduction but also become competitive in the emerging new low-carbon economy. Investment in hydropower infrastructure is crucial, as seen in the case of Brazil, which is planning enormous energy investments that will shape its economic map for generations to come. Strategic utilization of renewable energy infrastructure like small-hydro can facilitate tremendous capital inflow, amplify sustainable economic policies, and provide a solid basis for long-term growth. While Brazil struggles to reconcile economic competitiveness with environmental sustainability, the incorporation of hydropower technology could be a fulcrum for climate and economic development ambitions (Lucon et al., 2015). Economic benefits of hydropower extend beyond employment creation and investment; they extend to energy reliability and savings. By diversifying energy supplies, countries can reduce their vulnerability to market fluctuations in the prices of fossil fuels, leading to cheaper energy for consumers and businesses. Hydropower can be included in an energy mix that adds reliability while making energy independence possible. In addition, as more countries join the cause of curbing greenhouse gases, the shift towards renewable energy, reflected by hydropower schemes, provides entry to new avenues of economic development and prospective markets in the field of green technology. This change not only underlines global climate goals but also represents a transition towards sustainable economic development, ensuring that future generations will be left with a strong and successful economy founded on environmentally friendly practices (Abolhosseini et al., 2014; Lucon et al., 2015).

6.3.3. Reliability and Efficiency of Energy Production

Efficiency and reliability in power generation are essential when developing renewable energy systems, particularly with the integration of hydropower and mini hydro plants into the national grid. Renewable energy sources, as much as they are poised to reduce greenhouse gas emissions, are dominated by variability since they rely on weather conditions. Unpredictability of energy supply complicates the provision of a constant supply of energy and necessitates innovative measures to make them more reliable. Specifically, the use of different renewable sources—e.g., wind, solar, and hydro—can take advantage of each is complementing capacity to enhance the general system stability and efficiency. According to studies, different temporal scales are employed to ensure that energy output is stable even with variability in each source (Beluco et al., 2019). In addition, production of energy that is not only sustainable but also reliable must consider policies and investment offers that decide future energy infrastructure. Countries like Brazil are faced with a choice of how to invest huge amounts of capital into energy projects that either fund carbon-hungry technologies or low-carbon technologies. The investment has long-term implications not only for the reliability of energy but also for competitiveness, public health, and climate change mitigation. The interconnectedness of energy production efficiency and facilitating regulatory frameworks underscores the significance of integrative policies for guaranteeing both energy security and sustainability. The structure is pivotal in guaranteeing energy output reliability in tandem with respecting climate obligations, hence reconciling short-term energy needs with long-term environmental considerations (Lucon et al., 2015). Lastly, small-hydro system integration plays an important role in facilitating enhanced reliability and efficiency in energy production settings. Small-hydro systems can operate separately or in parallel with other big hydropower plants, exhibiting energy generation flexibility that could effectively react to local demands. Compared to conventional energy sources, small-hydro plants tend to have less operational costs and environmental influence, thus good options for sustainable development. Moreover, small-hydropower modularity facilitates incremental investment that can adapt to shifting energy needs with time. Through the incorporation of small-hydro in the overall energy mix, nations can achieve resilience, lower fossil fuel reliance, and improve grid reliability—ultimately propelling a more efficient and sustainable energy paradigm.

6.4. CHALLENGES IN HYDROPOWER DEVELOPMENT

Hydropower schemes construction is linked to a series of issues that can significantly restrain development within the sector. Despite its reputation as a clean energy supply, the large-scale construction and operation of hydropower cause interference with local habitats and agricultural activity. The resulting environmental pressures always generate social tensions as individuals protest against the damage to the environment that comes with dam construction schemes. Moreover, there are also heavy economic expenses involved in the creation of hydropower; countries are forced to traverse a massive capital investment, which may limit investment in other crucial infrastructure sectors. As highlighted by existing policy discussions, countries aiming to balance energy requirements against sustainability must make decisions that will influence investment in energy infrastructure in the future, with the potential to either pour fuel on carbon-intensive patterns or drive a low-carbon destiny (Lucon et al., 2015). In the case of Nepal, the intersection of geological risks with the urgent need for the generation of energy results in an exacerbating issue. With hydropower seen as a means towards increasing domestic power demands, it will overlook the relevant factors that include seismic activity. The passion to undertake hydropower development on the basis of likely economic growth as well as foreign investment can result in a phenomenon referred to as socially organized denial against environmental hazards. This disregard is born out of the twin pressures of guaranteeing energy supply and enticement of foreign investment, with actors demanding short-term economic dividends at the expense of long-term safety and sustainability. The result is that without proper geological surveys, projects end up endangering both the environment and local peoples, whose livelihoods are directly tied to the equilibrium of natural resources (Butler et al., 2017). Finally, the use of small-hydropower plants provides a promising method to overcome some of the issues that are particular to large hydropower schemes. Small projects generally cause fewer environmental impacts and have a lower initial capital requirement, enabling more localized energy production. Moreover, these plants can be made to coexist more harmoniously with local ecosystems, enabling sustainable development options that enrich communities without sacrificing necessary energy supplies. However, effective integration is not without difficulties, including regulatory factors, finance issues, and technology transfer. Through the overcoming of such challenges, stakeholders are able to assist in enabling a more sustainable energy sector that balances the need for economic development with environmental sustainability in hydropower schemes (Figure 6.12).

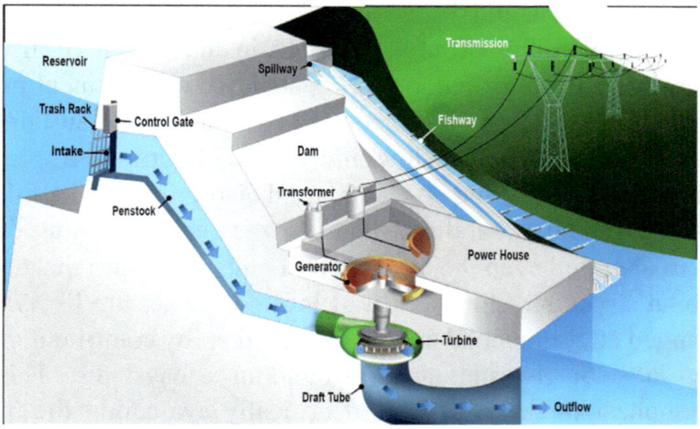

Figure 6.12. *Challenges in hydropower development.*

Source: https://media.licdn.com/dms/image/v2/D4D12AQHQSdtLQY0_zQ/article-cover_image-shrink_600_2000/article-cover_image-shrink_600_2000/0/1686839028977?e=2147483647&v=beta&t=B1BkLzAyDxZAfoSTszvuSJT2e5JbBwf0DFziv-DqGmQ.

6.4.1. Environmental Concerns and Ecosystem Disruption

Operation and development of hydropower dams significantly impact aquatic habitats of freshwaters, raising severe environmental concerns. As water is being moved from areas of excess to areas of deficit, which has been done traditionally and continues to be done through actions like interbasin water transfers (IBTs), the environmental balance of recipient river basins is disrupted. Although government policies in principle are designed to alleviate water shortages, the long-term effects on habitat destruction and biodiversity are significant. Rivers and aquatic ecosystems are harmed when massive water transfers occur without full scrutiny of their environmental costs. Consequently, such projects lead to the deterioration of vital freshwater ecosystems that provide essential services, ultimately exacerbating the very problems these projects aim to solve, as quoted in the analysis of IBTs in various regions (Pittock et al., 2009). Besides, hydrological risks associated with climate change exacerbate the problems resulting from hydropower projects, particularly in Southern African regions. The Zambezi River, for example, is confronting increasing hydrological uncertainty, threatening planned and proposed hydropower dams. With climate change becoming apparent in terms of more severe droughts and flooding, energy planners must account for such uncertainties. Current energy planning is short-sighted in neglecting the fact that unless adapted to hydrological change, newly proposed dams could be economically unviable or even dangerous, undermining the reliability of energy supply in the region (Beilfuss, 2012). There is therefore a need for incorporating adaptive management principles in ensuring that development of hydropower is sustainable to sustained climate change. Beyond altering water flow and quality, hydropower plants disrupt local environments by breaking up habitats and disrupting species migration. As rivers become dammed, natural sediment flow is cut off, causing downstream ecosystems to alter and even the extinction of aquatic species that are used to a specific habitat structure. Moreover, translocated species are likely to struggle with the changed environmental circumstances, thereby contributing to reduced biodiversity in both riverine and surrounding ecosystems. This ecologic disruption emphasizes the necessity of critically evaluating the hydropower development's environmental consequences. Anticipatory strategies, including environmental impact assessments and stakeholder engagement, need to be employed to minimize unwanted effects on ecosystems and enable harmonious coexistence between energy production and natural resource conservation.

6.4.2. Regulatory and Permitting Hurdles

Hydropower integration, particularly small-hydro installations, is faced with significant regulatory and permitting hurdles that can impede development and deployment. These typically arise from a sophisticated combination of federal, state, and local regulations that can vary extremely by jurisdiction. Developers may be required to navigate an environmental analysis maze, water rights negotiations, and National Environmental Policy Act (NEPA) regulation compliance. Complementing these are the involvement of multiple agencies in the permitting process, each with its own agenda. Not only does this increase the project time but also costs and deters investment in small-hydro projects. A more definable framework is therefore necessary to streamline regulations, with a view to rendering hydropower a more attractive source of renewable energy as part of the energy mix. To address these regulatory challenges, governments must be focused on creating policies that reduce bureaucratic inefficiencies while not overlooking responsible stewardship of the environment. International cooperation is an important strategy here, through which learning and best practices across regions inform more effective regulatory frameworks. Neuhoff et al. (2006) emphasizes that improving access to capital and encouraging research and development (R&D) investment is critical in developing a robust emerging energy technology market, including hydropower. By learning from foreign models and energizing domestic governance regimes that are supportive of hydropower development, policymakers can make the permitting process easier. This not only encourages innovation but also encourages the bias of the stakeholders towards initiating projects, thereby enhancing a larger-scale application of renewable energy technologies. Furthermore, ownership patterns at the local scale, as explained by Mazza (2008), can play an important role in avoiding regulatory and permitting challenges associated with hydropower development. By putting local projects in the hands of local communities, such models often build more public acceptance and participation, which can make the regulation process easier. Local ownership could also enhance the economic benefit from small-hydro projects, which provides local communities with an incentive to ensure efficient implementation and operation of these systems. Such projects would still need to have facilitative state and federal policy frameworks, however. Such frameworks need to have the aim of overcoming some of the challenges that revolve around local projects, such as the availability of finance and technology, thereby setting an enabling environment for small-hydro integration into the overall energy sector.

6.4.3. Technological Limitations and Maintenance Issues

Technological limitations are significant challenges to the use of hydropower and small-hydro systems. While these are renewable sources that are gaining popularity for their value in reducing the consumption of fossil fuels, they also have limitations in terms of energy density and scalability. The site-bound nature of hydropower and uncertainty in water flow present difficulties for exploitation on a large scale. For instance, renewable energy generation has limited power density that restricts ability to generate enough energy on a regular basis, which can complicate infrastructural development necessary in order to cater to global demand. Furthermore, the technology employed in effective hydropower production has a tendency to demand colossal initial capital and extended developmental periods, and it is difficult to expand operation size quickly or easily, thus limiting its flexibility in an emerging energy market (Abolhosseini et al., 2014). Factors concerning maintenance are at the core of keeping the durability and operational efficiency of hydropower stations. Unlike the rest of the energy sources, hydropower plants require continuous maintenance and monitoring to counteract the effects of continuous exposure to water and fluctuating environmental conditions. Component failures have the potential to lead to prolonged downtimes, compromising production and affecting the consistency of energy supply. Human factors should also be taken into account in the maintenance; breakdowns in process procedures can add to these challenges, and accidents can be catastrophic for the ecology. Additionally, the intricate synergy of machine components necessitates a skilled workforce to undertake routine maintenance inspections and ad-hoc repairs, underlining training and investment in providing peaceful working conditions and minimizing risks related to technology gaps (David et al., 2025). With increasing needs for renewable energy, bridging technological disparities and solving maintenance issues in hydroelectric facilities is imperative for promoting sustainable development. The subtle interplay of power generation, environment management, and regulation showcases the need for adaptive solutions that ensure optimal operation reliability. Initiatives such as the adoption of advanced monitoring systems can provide instant data, which enables pre-emptive maintenance and power generation optimization. Furthermore, immense investment in research and development can provide the impetus to innovate, which enables transcending limitations inherent in hydropower plants. By focusing on optimization of technology efficiency and improvement of maintenance procedures, the hydropower sector can more seamlessly integrate small-hydro facilities, enabling a diversified and healthy energy portfolio to complement global sustainability aims (Abolhosseini et al., 2014).

6.5. SMALL-HYDRO INTEGRATION

The integration of small-hydro power with the overall energy situation presents a chance to raise the sustainability and resilience of power production. Small-hydro plants, widely recognized as plants of up to 10 megawatts in size, have been preferred with their environmentally friendly footprint and ability to adapt to the local surroundings. These systems can be utilized strategically to supplement other renewable energy sources, providing the supply of energy with a firm and solid foundation. The fluctuating characteristics of wind and solar power have the tendency to cause mismatches between consumer demand and energy supply. On this front, the contribution of small-hydro cannot be overemphasized since it can generate a consistent level of power, particularly during periods of low output from such variable power sources, again highlighting the need for their integration into diversified energy portfolios (Beluco et al., 2019). The hybridization process, or blending of different renewable power sources, also enhances the efficiency and reliability of the energy system. Through the integration of small-hydro with other technologies, such as solar and wind, complementary nature of these resources can be harnessed. For instance, peak generation from solar occurs on sunny afternoons, while small-hydro generation can be made available during evenings and periods of poor weather. Such time-based complementarity can be quantified well and harnessed for efficient energy dispatch and grid stability improvement. The ongoing research into the synergy between variable renewable sources brings to the fore the value of integrating small-hydro systems, cementing their position in making possible a cleaner, stronger energy future (Beluco et al., 2019). Besides the performance advantages, integrating small-hydro systems speeds up economic development at the local level. Communities stand to gain from the construction and maintenance of these plants, generating jobs and increased energy independence. The small nature of these projects renders them extremely adaptable in complex geographic and socioeconomic contexts, allowing the generation of energy even in remote or far-flung areas. Secondly, as nations embark on ambitious renewable energy projects, small-hydro provides a crucial pathway for achieving local and national energy aspirations. By satisfying both energy needs and local economic powers, small-hydro integration appears not only as an environmental necessity but also as a socio-economic pusher which may flourish sustainably (Beluco et al., 2019).

6.5.1. Definition and Characteristics of Small-Hydro

Small-hydro is the name given to hydropower projects that generate electricity from the kinetic energy or potential energy of descending or flowing water, typically with an installed capacity of less than 10 megawatts (MW). This categorization differentiates small-hydro projects from large hydroelectric power plants that have massive environmental and social impacts. Small-hydro projects are less invasive to the environment and host communities, making it possible to install them in different geographical locations. Furthermore, these installations can be designed for run-of-the-river operations, where water is channeled without fundamental alteration to the environment, and hence minimizing environmental impact. As a new component of renewable energy options, small-hydro has also been of interest due to its ability to provide local and sustainable sources of power, which will facilitate transition to a low-carbon energy system. One of the most characteristic aspects of small-hydro is its ease of integration with other renewable sources of energy, enhancing overall efficiency and reliability. The natural variability in climate and weather-driven sources of energy has a propensity to lead to mismatch between energy demand and supply. Small-hydro can be harnessed as a remedy for this by running in synergy with solar or wind power schemes. This hybridization exploits the complementary nature of different energy sources so that it is more stable and more predictable in energy production. For instance, during low levels of solar energy generation, small-hydro plants can provide energy support to fulfill energy requirements, providing a balanced mix that optimizes resource utilization. Such integration not only promotes energy resilience but also concurs with the quest to reduce fossil fuel consumption in a bid for sustainable alternative energies (Beluco et al., 2019). The economic and operational advantages of small-hydro only add to its merit as a power source. Small-hydro plants have fewer capital expenditures compared to large hydroelectric stations, which makes it possible for them to be viable to more stakeholders such as communities and local governments. Additionally, most small-hydro plants can be built without large reservoir construction or revolutionary land use alteration, implying fewer operational complexities and regulatory problems. This reduced environmental footprint is complemented by the ability to produce jobs, where small-hydro plants are able to contribute to local economies through creation of construction and operational jobs. Therefore, small-hydro's multifaceted benefits of being environmentally compatible, integrable, and economically viable render it a critical component of sustainable renewable energy strategies, confirming its significance in the global initiative toward sustainable energy infrastructure (Cai et al., 2025) (Figure 6.13).

Figure 6.13. *Small-hydropower.*

Source: https://cdn.prod.website-files.com/64f9d0036cb97160cc26feba/66bb1738 9bc1aa844694734b_2c4b7245.png.

6.5.2. Advantages of Integrating Small-Hydro into the Energy Grid

Incorporation of small-hydroelectric power into the grid offers numerous reasons, primarily through enhancing the reliability and resilience of the supply of power.

With still growing energy demand, incorporating different sources of renewable energy is essential in countering the volatile nature of traditional sources of renewable energy such as wind and solar. Small-hydro plants can provide a stable and uninterrupted supply of generation to supplement other renewable energy sources, thus offering an enhanced energy supply. Hybridization of this nature, as much as it combines the benefit of various renewable sources, mitigates variability characteristic of weather-based electricity systems, thus making small-hydro an apt reservoir for smoothing peaking in power supply and power generation, thus ensuring grid stability (Beluco et al., 2019). Furthermore, small-hydro power plants provide significant environmental benefits through their lower ecological footprint compared to large hydroelectric projects. They tend to require less invasive damming, resulting in minimized local effects on biodiversity and water quality. Small-hydro can thus help achieve sustainable development objectives while reducing greenhouse gas emissions associated with fossil fuel production. Moreover, the local character of small-hydro schemes guarantees production of energy close to the points of consumption, thereby minimizing transmission losses and optimizing overall energy efficiency. Through the localized approach, energy independence is also fostered as well as increased community-based investment in renewable energy technologies (Beluco et al., 2019). Additionally, small-hydro projects can become pivotal in generating economic development, particularly in rural and disadvantaged communities. By diversifying energy portfolios and offering employment opportunities in hydro facility construction, maintenance, and operation, small-hydro stimulates local economies. Consumers are provided with short-term investment in small-hydro projects that bring long-term savings on the energy cost, and reduced reliance on imported fossil fuels, contributing to more energy security. Thereafter, introducing small-hydro into the energy mix not only adds to the efficiency of the overall electricity system but also helps to address socio-economic problems, paving the way towards a sustainable future of energy that is good for the environment and for the communities it will serve (Beluco et al., 2019).

6.5.3. Case Studies of Successful Small-Hydro Projects

Small-hydro schemes have been extremely effective in large regions of the globe, demonstrating their feasibility as a source of clean energy and helping local communities. For instance, schemes in nations like Nepal have delivered desperately needed electricity to remote villages, advancing rural development and enhancing resilience through economic avenues. Integration of such schemes into the domestic energy mix has allowed communities to utilize sources of clean energy while supporting agriculture and small industry. These schemes indicate the decentralizing power of small-hydro to provide energy and decrease dependence on larger, often less sustainable energy production. This approach not only minimizes environmental effects but also enables local stakeholders to be empowered with shared ownership and management, with increased community engagement being critical to long-term sustainability. In contrast with more significant hydroelectric schemes, small-hydro schemes provide some unique advantages, especially in regions with infrastructure shortages. Mozambican experience indicates that small-scale, community-level hydroelectric schemes can be able to address local energy deficits while ensuring environmental sustainability. The Mozambican problems up to now relying on large dams like Cahora Bassa highlight the need for a shift towards diversified renewable power. Centralizing project implementation at the local level can lead to increased, more equitable benefits allocation to the population to support socioeconomic growth. Through coordinated development of small-hydro, Mozambique can enhance its energy security and set up mechanisms that adapt to the climate risks of larger schemes, as per reports on energy planning in Mozambique (Hankins, 2009). Moreover, experience from Nepal suggests that environmental hazards posed by hydropower development need to be acknowledged. While the country grapples with earthquakes, Nepal's Indigenous hydropower community has been prone to 'socially organized denial' of such natural hazards to favor immediate energy requirements at the expense of long-term sustainability (Butler et al., 2017). Successful Nepalese small-hydro projects have shown resilience in the face of adversity by incorporating robust risk analysis and adaptive management principles. By borrowing a lesson from the experience of large hydro projects, small-hydro developments are better placed to avoid both geopolitical and environmental hazards. Not only does this improve the technical viability of renewable energy schemes but also highlights the imperative of interdisciplinary cooperation by stakeholders in reconciling Indigenous knowledge with scientific expertise to ensure provision for resource use by future generations.

6.6. FUTURE TRENDS IN HYDROPOWER

The future of hydropower is poised for revolution with the creation of technology and policy frameworks to address new energy needs. As countries in all parts of the world shift their focus to renewable energy sources, the construction of small-hydropower plants can give a promising path toward building energy resilience and reducing greenhouse gas emissions. With the global hydropower potential yet to be largely developed, investment in this field can give a big economic return along with environmental gains. In nations such as Brazil, decision-making on investments in energy infrastructure can have the capacity to slow down or speed up progress toward low-carbon energy goals. The choices will shape the energy makeup of the country for decades to come, necessitating visionary planning balancing immediate energy production with longer-term climate targets (Lucon et al., 2015). Climate change adds a new dimension to directions in the future of hydropower. The uncertainty of hydropower potential, which is mainly caused by great-scale climatic events, necessitates a more adaptable management policy. Research indicates that it is crucial to understand hydropower availability and future tendencies around 2050 using related factors. Particularly, this is more significant in regions susceptible to climatic changes such as El Niño. The combination of these climatic factors with hydropower generation influences not only energy production but also investment policy. For instance, the research conducted on Chile's central watersheds brings to the fore the importance of factoring in climate variability in hydropower planning because not doing so could lead to overinvestment in projects that cannot produce expected returns (Dieppois et al., 2019). In the future, good execution of hydropower strategy will require collaborative effort that includes technological innovation, environmental sustainability, and participation from local communities. Existing trends are moving towards more decentralized energy supply systems, where small-hydro plants can meet local needs and reduce dependence on conventional sources of power. Coupled with technological innovation in the field of energy storage, small-hydropower schemes can be a valuable complement to energy security in addition to reducing the impacts of climate change. While governments and businesses focus their sights on low-carbon energy systems, small-hydro projects could be a critical component of a diversified mix of energy, ultimately helping to transition to a more sustainable low-carbon future (Lucon et al., 2015) (Figure 6.14).

Figure 6.14. *Future of hydropower.*

Source: https://www.greenlivinganswers.com/blog/wp-content/uploads/2024/08/hydropower-future.webp.

6.6.1. Innovations in Hydropower Technology

The technology of hydropower has undergone revolutionary changes in its topography with a view to greater efficiency, reducing the environmental impact, and better integration with smaller-scale enterprises. Contemporary designs of turbines, such as variable-speed turbines, allow for more energy extraction in a range of water flow conditions. Such flexibility not only maximizes energy output but also minimizes ecological disturbance through an adaptation of the natural flow regime of water bodies. Furthermore, developments in computing modeling and simulation technology enable better understanding and forecasting of hydropower plants' effects on the environment. These technologies are critical considering the increasing oversight of the industry for its environmental footprint as well as the need for sustainable energy sources, particularly in small-hydro integration schemes. Integration of digital technologies is also a revolution in hydropower. Internet of Things (IoT)-powered remote monitoring and predictive maintenance solutions have the ability to greatly enhance operational efficiency. The systems are capable of real-time data, implying that operators can anticipate technical failures and optimize performance without unnecessary downtime. Furthermore, the use of advanced data analytics enables smarter water management, particularly where the water supply is intermittent. As smaller hydropower plants have been operating with limited budgets, taking advantage of these innovations will better equip them to compete and become sustainable, addressing growing demands for the generation of green power. Policy frameworks are important in encouraging innovations in hydropower technology. Studies suggest that countries with pro-environment policies and deregulated energy sectors will witness spectacular advancements in renewable technologies, such as hydropower. By minimizing entry barriers and offering a conducive environment for research and development, policymakers can help stimulate innovation and adoption in the sector. Moreover, government R&D investment, particularly in small-hydro systems, complements these efforts and stimulates patent production. The policy support complementarity effect coupled with market structures can ultimately lead to more technological developments needed to support decentralized and sustainable energy solutions, further vindicating the importance of the strategic interventions of policies towards progress (Gelil et al., 2012; Nicolli et al., 2025).

6.6.2. Policy Developments and Funding Opportunities

Policy development holds vital implications for hydropower integration and small-hydro, particularly in the wake of the move to more sustainable energy networks as exerted by governments. The transition is underscored by recent strategic outlays of approximately 0.5 trillion United States dollars projected for Brazil's infrastructure as a whole in the coming decade. Such monetary outlays represent a vital mechanism that supports either continuation of carbon-based infrastructure or facilitates development towards low-carbon-oriented economies. These developments point towards the imperative of a uniform policy regime which harmonizes energy production with climate goals so that investments make way for sustainable operations. As Brazil synchronizes its pollution profiles and targets, the imperative of policies enhancing dependency on renewable energy increases progressively, serving as the core of hydropower growth in accordance with environmental sustainability as well as energy security (Lucon et al., 2015). Furthermore, to enhance the viability of small-hydro projects, funding opportunities must be simultaneously accessible and strategically targeted. Countries like Mozambique exemplify this necessity, as they currently face significant disparities in electricity access. With most energy production concentrated on large-scale projects that primarily serve wealthier nations, there is a crucial need for policies that advocate the decentralization of energy infrastructure. The finding of renewable energy resources in Mozambique not only offers a path toward universal energy access but also promotes climate resilience through the country's electricity grid diversification. The approach not only empowers native communities but also minimizes ecological footprints, allowing for the sustainability of the Zambezi River, a vital resource (Hankins, 2009). Financing supporting small-hydro schemes is instrumental in bridging these gaps, encouraging movement toward a more inclusive energy industry. Altogether, the intersection of policy development and financing options creates an active ground for hydropower promotion, particularly that of small-hydro integration. Consolidated strong policies with targeted financial assistance can drive development in renewable energy projects to the benefit of both environmental and social justice interests. Such concerted approaches yield multiple dividends in the form of increased local energy production, reduced reliance on foreign energy importation, and improved resiliency to climate change effects. Sustained funding for strong policy frameworks and diversified finance will be key in supporting future hydropower ventures that serve society and the environment. By such targeting, countries will be able to maximize the use of their natural resources while promoting socio-economic advancement in a more carbon-conscious world.

6.6.3. Potential for Hybrid Systems with Other Renewable Sources

With the evolving energy scene globally as it shifts towards sustainability, the hybridization of hydropower with other forms of renewable energy is a highly promising avenue for improving grid resilience and reliability. Hybrid systems that combine hydropower with solar or wind power can harness temporal and spatial complementarity between them to counterbalance the inherent variability of weather-dependent electricity generation. For instance, hydroelectric power provides baseload generation reliably, while solar and wind energy provide peak generation on sunny or windy days. This dynamic interaction minimizes supply-demand mismatch, thereby ensuring a more stable power supply. Heightened interest in hybrid systems as the answer to the energy integration challenge has been reflected in several studies of their potential, leading to the focus on the need for further investigation in this area (Beluco et al., 2019). Quantification of temporal and spatial complementarity between renewable sources is essential for hybrid energy systems optimization. Emerging techniques, such as methods based on correlation coefficients and compromise programming, allow one to analyze extensively how different energy sources can function in combination effectively. A recent study, which had taken into account complementarity between wind, solar, and hydro resources, showed that selection of temporal scales significantly affects system-wide performance. This indicates the necessity to embrace holistic approaches in hybrid system design, as different combinations can lead to different levels of efficiency when considering local environmental conditions and demands. By scientific exploration of such correlations, researchers can engineer hybrid systems to achieve the optimal output and fossil fuel minimization, thereby increasing sustainability goals (Beluco et al., 2019). The ability of hybrid systems to increase energy security extends beyond supply increase; it also encourages resilience against climatic uncertainty. With more frequent and intense extreme climate events due to climate change, diversification of the energy mix with hybrid systems mitigates individual energy source risks. For instance, even though hydropower is traditionally dependable, it may be severely diminished with droughts. Solar power, however, may thrive during periods of dryness, providing much-needed electricity when hydropower generation is diminished. Through the integration of the blend of various renewables strategically, electricity suppliers can ensure that the system becomes more adaptable and well-suited to meet changing conditions. Eventually, this adaptability paves the way to a cleaner energy future that validates the pivotal role of hybrid systems in modern power generation (Beluco et al., 2019).

SUMMARY

Hydropower and small-hydro system unification is an important strategy for the transition towards a low-carbon future. As global energy demands continue to increase, so do climate change pressures, compelling the imperative of innovative solutions to decrease greenhouse gas emissions. Investment in hydropower infrastructure not only satisfies energy demands but also can be integrated into broader climate ambitions. As indicated in recent analyses, Brazil has crucial policy choices that will shape its energy future and influence its economic competitiveness and global climate commitments (Lucon et al., 2015). Stakeholders need to be aware of the hydropower and small-hydro integration potential for decoupling from carbon-hungry fuel sources and making considerable strides toward achieving the required decarbonization target in the energy sector. Financial sustainability of integrating hydropower with other sources of renewable energy, particularly wind power, also underscores the benefits of looking into hydropower options. Experiences prove that tapping available hydroelectric capacity can be effective in stabilizing the grid as well as facilitating the integration of variable renewable sources such as wind (Cornelis, 2025). Such a dynamic interplay can result in significant cost savings in terms of reductions in carbon emissions, especially when wind output is supported with large-scale fossil fuel facilities. By creating a diversified portfolio of energy that leverages the strengths of multiple renewable sources, governments and utilities can decrease overall greenhouse gas emissions as they improve the dependability of energy. This not only highlights the worth of hydropower but also indicates how strategic integration can lead to more durable, sustainable infrastructure. In short, embarking on hydropower and small-hydro integration demands concerted effort by policymakers, utilities, and communities across the globe. As climate issues deepen, adopting environment-friendly energy solutions becomes more and more important. Nations such as Brazil are at crossroads, and choices made today could influence coming generations (Lucon et al., 2015). With a view to hydropower and integration with other renewable energy resources, one can build a robust energy system that meets the needs of today while responding to climate necessities. Ultimately, collective collaboration in energy planning, investment, and technology development will determine the success of integrating hydropower into the world energy system as a step towards a more sustainable and resilient future for all.

When talking about integrating hydropower systems such as small-hydro projects, one should consider their ability to provide sustainable energy solutions to global climate change initiatives. Small-hydroelectric power plants provide a sustainable option compared to conventional energy production, enabling the generation of energy at the local level with reduced transmission losses and enhanced energy security. Besides, their connection to national grids has the ability to generate enormous savings in terms of greenhouse gas emissions, as stated by the urgent need to move away from carbon-based systems to cleaner energy sources. Investments in small-hydro are not only capable of facilitating Brazil's ambitious climate targets but are also going to drive the country toward a low-carbon economy, enhancing its global competitiveness in an increasingly environmentally conscious market (Lucon et al., 2015). Also, equitable distribution of energy resources is one of the main issues when hydropower integration in developing nations is being discussed. Mozambique's energy status is the very representation of the imbalances inherent in megaproject hydropower, where most of its power produced is serving neighboring South Africa at the expense of the local population. This case presents the urgent need for a more consolidated approach to energy development, where renewable sources of energy are utilized to benefit local communities and provide diversification of energy supplies. Through the utilization of small-hydro schemes, Mozambique can establish a clean energy sector that not only addresses short-term energy deficit but also enhances mitigation against exposure to the impacts of climate change. This strategic shift can promote socio-economic development by ensuring that hydropower gains are equitably shared and improve national resilience (Hankins, 2009). Last but not least, it is important to achieve the broader implications of hydropower and small-hydro integration into the energy infrastructure sector. The need for a one-stop strategy to harmonize energy generation, transmission, and environmental sustainability intensifies daily. Policy makers must consider the ability of small-hydro to complement large grid systems and join with other renewable sources, such as solar and wind power. Such multi-strategy is advantageous in the optimization of grid resilience and flexibility in meeting fluctuating energy demands while maximizing the utilization of local resources. Furthermore, promoting innovation and investment in small-hydro has the potential to spur economic growth, create employment, and strengthen energy independence, paving the way for a wide-ranging energy future that is reconcilable with climate goals and social justice (Hankins, 2009; Lucon et al., 2015).

REFERENCES

1. Abdel Gelil, I., Aguayo, F., Boyle, G., Breukers, S., et al. (2012). Policies for capacity development. https://core.ac.uk/download/9836421.pdf (accessed on 30th September 2025).
2. Abolhosseini, S., Altmann, J., & Heshmati, A. (2014). A review of renewable energy supply and energy efficiency technologies. https://www.econstor.eu/dspace/bitstream/10419/96759/1/dp8145.pdf (accessed on 30th September 2025).
3. Ballesteros, A. R., & Nakhooda, S. (2010). Investing in sustainable energy futures: Multilateral development banks' investments in energy policy. https://core.ac.uk/download/71352408.pdf (accessed on 30th September 2025).
4. Beilfuss, R. (2012). A risky climate for Southern African hydro. https://core.ac.uk/download/71361025.pdf (accessed on 30th September 2025).
5. Beluco, A., Canales, F. A., Guezgouz, M., Jurasz, J., et al. (2019). A review on the complementarity of renewable energy sources: Concept, metrics, application and future research directions. https://core.ac.uk/download/286642706.pdf (accessed on 30th September 2025).
6. Beluco, A., Canales, F. A., Jurasz, J., Kies, A., et al. (2019). Assessing temporal complementarity between three variable energy sources by means of correlation and compromise programming. http://arxiv.org/abs/1905.00117 (accessed on 30th September 2025).
7. Boero, D., Chiaberge, M., Girardello Detoni, J., Tonoli, A., et al. (2012). Trade-off analysis and design of a hydraulic energy scavenger. https://core.ac.uk/download/pdf/11430131.pdf (accessed on 30th September 2025).
8. Butler, C., & Rest, M. (2017). Calculating risk, denying uncertainty: Seismicity and hydropower development in Nepal. https://core.ac.uk/download/141737712.pdf (accessed on 30th September 2025).
9. Cai, X., McKinney, D. C., Ringler, C., Rosegrant, M., et al. (2025). Modelling water resources management at the basin level: Review and future directions. https://core.ac.uk/download/pdf/6405323.pdf (accessed on 30th September 2025).
10. Deeb, A., Leino, I., Toba, N., Cramton, P., & Vergara, W. (2025). Wind energy in Colombia: A framework for market entry. https://core.ac.uk/download/pdf/6362847.pdf (accessed on 30th September 2025).
11. Dieppois, B., Killingtveit, Å., Moriasi, D., Sidibe, M., Link, O., Arriagada, P., et al. (2019). Impacts of climate change and climate variability on hydropower potential in data-scarce regions subjected to multi-decadal

variability. https://core.ac.uk/download/304335299.pdf (accessed on 30th September 2025).

12. Elverhøi, M., Fleten, S. E., Fuss, S., Heggedal, T., et al. (2025). Evaluation of hydropower upgrade projects–A real options approach. https://core.ac.uk/download/pdf/6507487.pdf (accessed on 30th September 2025).
13. Hampl, N., Stoiciu, A., Szabo, E., Totev, V., et al. (2014). Assessing the disruptiveness of new energy technologies–An ex-ante perspective. https://core.ac.uk/download/35453965.pdf (accessed on 30th September 2025).
14. Hankins, M. (2009). A renewable energy plan for Mozambique. https://core.ac.uk/download/71361024.pdf (accessed on 30th September 2025).
15. Inyang, B., Ingwe, R., & Ering, S. (2025). Sustainable energy implementation in urban Nigeria. https://core.ac.uk/download/pdf/6276590.pdf (accessed on 30th September 2025).
16. Joseph, P. (2006). The potential significance of micro hydro systems. https://core.ac.uk/download/56703214.pdf (accessed on 30th September 2025).
17. Li, J. (2025). Mitigation country study–China. https://core.ac.uk/download/pdf/6248883.pdf (accessed on 30th September 2025).
18. Lucon, O., Fransen, T., & Romeiro, V. (2015). Bridging the gap between energy and climate policies in Brazil: Policy options to reduce energy-related GHG emissions. https://core.ac.uk/download/75782972.pdf (accessed on 30th September 2025).
19. Mazza, P. (2008). Community wind 101: A primer for policymakers. https://core.ac.uk/download/71350352.pdf (accessed on 30th September 2025).
20. Neuhoff, K., & Sellers, R. (2006). Mainstreaming new renewable energy technologies. https://core.ac.uk/download/1248990.pdf (accessed on 30th September 2025).
21. Nicolli, F., Vona, F., & Nesta, L. (2025). Determinants of renewable energy innovation: Environmental policies vs. market regulation. https://core.ac.uk/download/pdf/6441011.pdf (accessed on 30th September 2025).
22. Pittock, J., Meng, J., & Geiger, M. (2009). Interbasin water transfers and water scarcity in a changing world: A solution or a pipedream? https://core.ac.uk/download/75783296.pdf (accessed on 30th September 2025).
23. Van Kooten, G. C. (2025). Wind power: The economic impact of intermittency. https://core.ac.uk/download/pdf/6922535.pdf (accessed on 30th September 2025).

24. Wade, D. C., & Kessides, I. N. (2025). Towards a sustainable global energy supply infrastructure: Net energy balance and density considerations. https://core.ac.uk/download/pdf/6667875.pdf (accessed on 30th September 2025).

CHAPTER 7

BIOMASS AND OTHER RENEWABLES

Contents

INTRODUCTION

The quest for biomass and other alternative sources of energy is an essential element of contemporary discourse for sustainable energy options. As the world grapples with the urgent need to seek alternatives to fossil fuels, the integration of renewable resources into energy strategies becomes more imperative. Renewable energy, particularly biomass, is not only a way of reducing greenhouse gas emissions but also a way of energy diversification and energy security enhancement. The character of the regenerative systems, as introduced in a series of papers, emphasizes the need to understand the differences between fossil-fuel systems and renewable systems. This comparative analysis reveals the fundamental features that direct energy applications planning and spatial distribution in the built environment and make a more sustainable energy future possible (Koh et al., 2009). Agriculture is important in this movement towards renewable energy through the generation of biomass. The integration of renewable energy source development and sustainable farming is more and more recognized, considering the contribution of farming to greenhouse gas emissions. The reduction of these emissions by the European Union depicts the dual role of the industry: minimizing its environmental impact and, at the same time, working towards the exploitation of renewable energy opportunities. With incorporation of sustainable methods, agriculture is not only addressing non-CO_2 greenhouse gas emissions such as methane and nitrous oxide but also emerging as a new source of renewable energy (Wysokińska, 2012). This collaboration strengthens the position of innovation in agricultural methods to pave the way for a greener energy future. Impacts of utilizing biomass in the overall renewable resources' context extend far beyond energy production. The establishment of a clear vision for the role of biomass involves considering its environmental, economic, and social dimensions. Biomass system use depends on adequate research into prevailing resources, technological change, and policy frameworks conducive to sustainable practice. In pursuing practical solutions, the integration of biomass as part of a multi-pronged renewables strategy has potential for energy self-reliance but also for rural economic development. Proactive involvement of all parties is called for to fully unleash the potential of biomass and other renewables so that their deployment supports the larger goals of environmental conservation and sustainable development (Figure 7.1).

Figure 7.1. *Biomass: A renewable energy resource.*

Source: https://www.energy.gov/sites/default/files/styles/full_article_width/public/Biomass%20Resources.png.

Definition of Biomass and Renewable Energy

Biomass as a renewable fuel source contains organic plant and animal matter that is harvested and processed into energy utilizing various technologies. Biomass contains everything from wood and agricultural products to waste byproducts. Biomass is special because it captures the carbon when it grows and has the theoretically carbon-neutral process if it is harvested and processed appropriately. The addition of biomass in energy systems can minimize reliance on fossil fuels and thus improve both energy security and sustainability. Moreover, wood-based biomass is recognized to be an essential resource by the European Union in its energy policy, encouraging it to be put into use as a primary element of combined heat and power plants, with special reference to efficiency and diversity in renewable energy applications (N/A, 2008). Renewable sources of energy, including biomass, are imperative to global effort to arrest climate change and limit the release of greenhouse gases. Together with other renewables, renewable energy is any energy derived from sources that can naturally be replenished, such as solar, wind, hydro power, and geothermal. Biomass is commonly categorized with the other renewables because it too is regenerable, and it has intrinsic value in an energy blend. However, the possible integration of high amounts of renewable energy sources, including biomass, is accompanied by considerable technical challenges, primarily in energy storage technologies.

Figure 7.2. *Biomass.*

Source: https://sites.lafayette.edu/coynee/files/2019/05/biomass-energy-source.jpg.

According to findings in current research, new storage technologies are required for managing the day-to-day variation in energy demand, further justifying the contention that the effective harnessing of renewable energy sources is contingent on the improvement in energy storage technologies (Villanueva et al.,

2019). The pairing of biomass with other renewable energy systems illustrates one of the most compelling paths forward for sustainable energy systems. The focus increasingly shifts toward hybrid systems that optimize energy output and minimize environmental impacts. Combinations can involve anything from pairing biomass energy with solar thermal storage systems to looking toward new breakthroughs in carbon dioxide-based energy cycles with maximum efficiency and minimum cost. This cross-fertilization of the technologies is instrumental in creating a more sustainable energy infrastructure that can cope with the fluctuations of energy demand and resource supplies. As the argument for renewable energy plays out, it is evident that the definition and use of biomass will be instrumental in shaping future energy profiles, facilitating a transition towards enhanced sustainability and energy autarky (N/A, 2008; Villanueva et al., 2019) (Figure 7.2).

Importance of Renewable Energy in Combating Climate Change

As the rate of climate change accelerates, it has become a priority of utmost importance for governments and green activists alike to transit towards renewable sources from fossil fuel. Renewable energy, exemplified by biomass and wind power, is of immense potential in reducing greenhouse gas emissions significantly. Biomass, which is composed of organic waste products, not only generates energy but also facilitates sustainable farming through the utilization of waste products. This twofold benefit works to cancel out the carbon footprint of traditional power sources. The switch to biomass and other resources signifies a mission for sustainability in energy systems that is imperative to the resolution of the pressing matters presented by climate change. By integrating biomass into the energy mix, countries can decrease their fossil fuel dependence and advance towards a carbon-free world, inherently emphasizing the significance of renewable energy in collective action against climate change (Bricas et al., 2019). The role of renewable energy in climate change mitigation extends beyond emissions reductions; it also encompasses socio-economic drivers affecting resilience at the community level. Where nations are trying to minimize their carbon footprint, laws that promote renewable energy development become a key component of this process. In the majority of nations, including the UK, onshore wind turbines are in a good position to contribute significantly towards electricity generation from low-carbon sources. Opposition at the local level to specific projects, though, is likely to complicate the deployment of renewable energy technologies. Public skepticism research in communities shows that public suspicion may be founded on misapprehension instead of outright opposition, presenting the need for evidence-based decision-making processes. Closing this gap and building support for renewable initiatives depends on communication and public engagement (Ackermann et al., 2009). The quest for renewable energy, particularly in terms of biomass and wind energy, fundamentally changes energy prospects while addressing the requirements of sustainable development. Transition to renewables not only mitigates greenhouse emissions but also guarantees energy security and fosters economic growth. From the development and use of local biomass resources, societies can create jobs, minimize waste of resources, and stimulate rural economies. Renewable energy projects also bring dual environmental benefits, improving the air quality and reducing dependence on imported fuels. Integrating renewables into national policy is therefore a commendable response to climate change. As countries unite to tackle this world challenge, the need for biomass and other renewables becomes central to ensuring energy security in the future, accentuating their position in a low-carbon economy (Bricas et al., 2019) (Figure 7.3).

Figure 7.3. *The importance of renewable energy in combating climate change.*

Source: https://www.hulkapps.com/cdn/shop/articles/scaled_IMAGE-FILE-UPLOADS-the-importance-of-renewable-energy-in-combating-climate-change-101020_1024x.jpg?v=1722251459.

7.1. UNDERSTANDING BIOMASS

The idea of biomass involves not only the realization of its potential as a renewable energy source but also a study of how its production and utilization are complex. Biomass, derived from organic matter such as plant refuse and residues, can be transformed using processes such as anaerobic digestion and gasification into usable forms of energy. Biomass as a renewable source of energy holds a lot of promise as a means of reducing greenhouse gas emissions and decreasing dependency on fossil fuels. But achieving its highest potential is still plagued with challenges. For instance, the efficiency of biomass utilization can be compromised by poor harvesting conditions and low-grade feedstock. Thus, achieving localized renewable energy production goals, such as required in European Union policy, means having intimate knowledge of the feedstock quality and processing technologies applied in biomass energy production (Bekkering et al., 2016).

Figure 7.4. *Understanding biomass energy.*

Source: https://www.greenlivinganswers.com/blog/wp-content/uploads/2024/08/biomass-energy.webp.

The Fischer–Tropsch (FT) synthesis process is a critical use of biomass, reducing the latter to liquid fuel. The conversion process is a carbon-neutral process relative to conventional fossil fuels and even more so in view of global warming issues as well as dwindling petroleum reserves. However, conventional gasification processes for biomass have low rates of carbon utilization ranging from 25% to 40% of the biomass feedstock. This limitation negatively impacts the overall economic feasibility of biomass-to-liquid (BTL) fuel. For higher carbon efficiency and reduced emissions, innovation in hybrid technologies, such as integration of concentrated solar power (CSP) with the BTL cycle, has emerged as a winning choice. This hybrid approach not only enhances carbon use but also addresses the general concerns of energy storage and sustainability in the changing context of renewable energy (James et al., 2013). Moreover, the environmental effects of biomass energy must be examined carefully, keeping sustainability and life cycle analysis in view. For example, some biomass pathways, like green gas production, hold significant potential for producing substantial energy output but also run the risk of undermining the integrity of the overall environment if managed in a non-sustainable way. Other pathways, like waste management strategies, can be demonstrated to have reduced emissions and smaller environmental footprints, thus emphasizing the requirement for balance in biomass utilization. The findings from the examination of different biomass pathways underscore the importance of comprehensive energy policies that, alongside promoting renewable supply of energy, can further stimulate ecological sustainability. As discussions regarding energy transition continue to progress, it is imperative to develop well-directed approaches for sustainable uses of biomass resources with limited adverse environmental impacts (Bekkering et al., 2016) (Figure 7.4).

7.1.1. Types of Biomass Sources (e.g., Wood, Agricultural Residues, Waste)

To understand the diversity of biomass sources, wood must be taken as among the most significant material employed in energy production. Wood biomass encompasses varied forms such as forest and timber industry logs, chips, as well as sawdust. Its applicability is evident in urban and rural applications of energy, ranging from residential heating to industrial power generation. Biomass use of wood is also friendly to the energy profile as a renewable alternative to fossil fuels. Additionally, wood's carbon-neutral position, when harvested sustainably, also reduces greenhouse gas emissions. However, there should be intelligent management of forest resources so as not to over-harvest, which would have negative ecological impacts and therefore render the very sustainability that wood biomass seeks to provide useless (Zolin, 2025). Crop wastes are also a source of vital biomass, such as crop production residues such as straw, husks, and other organic wastes. These have great potential for energy production, commonly referred to as second-generation biofuels. The conversion of energy of agricultural residues not only adds economic value to what would otherwise be a waste material but also aids in managing farm residues that could otherwise end up in soil depletion or pest problems. The utilization of these residues as an addition to the energy mix also aids in a circular economy with waste products being transformed into useful materials. By utilizing agricultural residues, farmers and landowners are able to economically benefit while reducing dependence on non-renewable energy (English et al., 2025). Agricultural residues thus hold the key to making energy production systems sustainable. Waste biomass, finally, is comprised of biomass harvested from industrial and municipal sources, which is often an underutilized but valuable renewable energy production source. Waste materials like food waste, garden trimmings, and industrial residues may be reused through methods like anaerobic digestion or burning to generate energy. This process not only minimizes the amount of waste destined for landfills but also helps abate environmental concerns with respect to waste disposal. By incorporating waste biomass into energy systems, urban areas are able to overcome problems related to landfill overcrowding and greenhouse gas emissions while at the same time generating an important energy resource. The workability of this strategy reflects increasing realization of the potential for waste as an energy source, radically changing our conceptualization of waste management in the renewable energy context (Zolin, 2025) (Figure 7.5).

Figure 7.5. *Types of biomass sources.*

Source: https://www.greenlivinganswers.com/blog/wp-content/uploads/2024/08/biomass-energy-sources.webp.

7.1.2. The Process of Converting Biomass into Energy

Biomass energy conversion entails a series of processes that leverage organic matter to yield renewable energy alternatives. Biomass, which comes from plant and animal materials, can be utilized as an immediate source of energy through combustion or converted into more sophisticated forms of bioenergy. This process develops in two principal avenues: thermochemical and biochemical conversions. Thermochemical conversion involves the process of heating biomass to high temperatures to produce gases, liquids, and char, whereas biochemical conversion utilizes microorganisms to break down organic materials to produce biofuels like ethanol and biodiesel. The mode of conversion is based on the category of biomass utilized and the output form of energy desired. Through improvements in these conversion technologies, biomass energy can make a significant contribution to reducing dependency on fossil fuels and reducing greenhouse gas emissions, highlighting its position in the greater scheme of renewable energy development. The integration of biomass conversion processes with current energy infrastructure optimizes overall efficiency and sustainability of energy production. Combining waste heat recovery with biomass conversion is a new approach that is crucial to obtaining the best efficiency from energy systems.

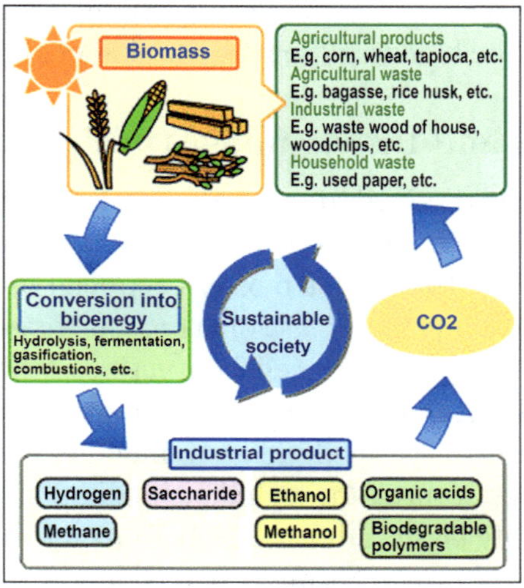

Figure 7.6. *Thermochemical processes to convert biomass into energy.*

Source: https://blog.upsbatterycenter.com/wp-content/uploads/2014/07/biomass.gif.

Through recovering heat generated during the processing of biomass, plants can direct such energy into the production process, conserving resources and promoting energy recovery. The application of a Virtual Power Plant (VPP) facilitates the integration of biomass excess power into conventional energy sources such as gas or coal. The method not only optimizes the generation of energy but also provides grid services since it smooths out the fluctuating output of renewable energy. This combined strategy hints at the potential that biomass can serve both as an energy source and as a means of enhancing the resilience of energy infrastructure to aid in constructing sustainable energy systems (Eetvelde et al., 2013). The benefit of biomass lies in its ability to be transformed into various forms of energy, with wide-ranging environmental and economic consequences. Advanced conversion technologies make possible the conversion of biomass into solids, liquids, and gases, such as biochar, biodiesel, and biogas, respectively. They all have their applications, contributing to various industries, ranging from heating and the generation of electricity to transportation fuel. For instance, biomass-derived biofuels can readily replace conventional fuels, lower carbon prints while promoting energy autonomy. Furthermore, biomass not only provides renewable energy but also value-added chemicals as a by-product, adding more to its utilization. With the continuous advances in the technology of biomass conversion, it is of the utmost significance to create a sustainable energy system that not just reduces the production of greenhouse gases but also stimulates economic growth by creating more jobs in the bioenergy sector (Bhaskar et al., 2020) (Figure 7.6).

7.1.3. Advantages and Disadvantages of Biomass Energy

Biomass energy, which is derived from organic matter such as plant biomass and animal wastes, benefits in many ways, with the prime characteristics being the diversification of energy. Biomass is mostly beneficial in terms of substituting fossil fuels, thereby reducing the release of greenhouse gases and environmental pollution. Biomass can be transformed into various types of energy such as biofuels, biogas, and direct combustion, thus offering variety in energy production processes. This versatility is particularly convenient for those communities who wish to enhance energy security through the exploitation of local resources, as biomass is usually easily produced sustainably from agricultural waste products or residues, all within the principles of circular economy thinking. Furthermore, the installation of biomass energy systems can stimulate rural economies by creating jobs in agricultural, processing, and energy-producing sectors, encouraging economic stability in rural areas as discussed in debate over the sustainability of these steps (José et al., 2025). However, use of biomass energy is not free from challenges. One of the main disadvantages is the possibility of conflict in land use, especially where biomass crops invade food crops or natural ecosystems. As demand grows for biomass material, there is always the danger of overutilization that can lead to deforestation or adverse impacts on diversity, canceling out the very environmental benefit being sought in the first place. Secondly, energy conversion from biomass does have a process with emissions that can negate some of its advantages over fossil fuels. For instance, burning biomass can release pollutants, and that raises concern where the air quality and health effects for neighboring communities are being questioned. Evaluating such trade-offs is essential so that biomass energy would not inadvertently be a contributor to environmental degradation, so it makes its role in the renewable energy landscape problematic (Agnolucci et al., 2007). In considering biomass energy, one is able to see that its efficacy relies on proper management and technological innovation. The production of more effective biomass conversion methods can reduce emissions and increase efficiency, allowing cleaner energy to be derived from organic materials. Further, integrating biomass into a broader scheme of renewable energy—wind, solar, and hydroelectric—can offset its disadvantages while generating the greatest overall environmental benefits. Policymaking towards sustainable biomass operations, say, promoting agroforestry or waste energy systems, can also establish a balance between food crop production needs and energy generation. Ultimately, although biomass energy provides a potential solution to carbon footprint reduction and energy resilience, its utilization needs to consider ecological and socioeconomic settings to avoid potential drawbacks that would offset its role towards a transition to renewable energy (José et al., 2025) (Figure 7.7).

Figure 7.7. *Challenges in biomass energy.*

Source: https://www.greenlivinganswers.com/blog/wp-content/uploads/2025/01/biomass-energy-challenges.webp.

7.2. THE ROLE OF BIOMASS IN THE ENERGY MIX

Incorporating biomass into the energy mix is a crucial approach to meeting energy demands at the global level while minimizing environmental issues simultaneously. Biomass as a renewable energy source can be derived from organic wastes like agricultural waste, wood, and residuals, thus posing as a sustainable energy source instead of fossil fuels. Wood-based biomass use, in particular, is promoted as a key component of energy policy in the European Union. This approach identifies the potential of biomass in cogeneration power plants capable of efficiently converting organic matter to energy while delivering significant reductions in greenhouse gases. Such projects focus on the double benefit of biomass: enhancing energy security and ensuring environmental sustainability, ultimately setting it as the solution to transitioning to a low-carbon economy (N/A, 2008). Furthermore, the utilization of biomass in energy generation is associated with a range of problems that must be carefully managed. They include competition for resources, as an increase in demand for biomass feedstocks could lead to conflict with agriculture and land use. In order to manage these issues, policymakers must accord top priority to the creation of sustainable biomass supply chains that address both the environmental and social aspects. For instance, adopting stringent data collection methods can enhance biomass availability and its probable environmental effects. Addressing these challenges, thus, ensures maximum use of biomass in order for it to still be an effective energy source without breaking the integrity of ecosystems as well as surrounding populations. Thus, resolving biomass production and resource preservation is critical towards maximizing its contribution to the energy mix (N/A, 2008). Educational and research efforts also play an essential role in shaping the future position of biomass in the energy mix. The introduction of multidisciplinary postgraduate programs, for instance, the JELARE project carried out in most countries, reflects the necessity to teach students who possess the necessary capabilities to handle the intricacies of renewable energy. With the incorporation of sustainability and energy economics courses, such programs prepare graduates to address real problems, fostering creativity and problem-solving in biomass management. Not only does this added academic framework aim to modernize the curriculum but also to create an educated workforce with the capacity to advance renewable energy technologies. Biomass, in this case, is not only used as energy but also acts as a driver for broader social and economic development according to sustainable development goals.

7.2.1. Current Global Biomass Energy Production Statistics

The global situation of biomass energy production has undergone significant progress in the past decade, which reflects a turn towards a greener energy future. Biomass, comprising organic material such as plant life and animal residues, is one of the largest contributors to the pool of renewable energy as countries look to reduce greenhouse emissions and reliance on fossil fuels. As of recent time, biomass comprises about 60% of the global consumption of renewable energy, attesting to its pivotal contribution to energy transition policies globally. Of special mention are countries such as Brazil that have become forerunners in biofuel production, utilizing their agricultural resources optimally to drive cleaner energy options. The joint effort of countries in designing renewable energy promotion policies has initiated a global movement, which has allowed biomass technologies to compete with traditional fossil fuel markets for electricity and transport fuels, necessitating ongoing investment in this area (N/A, 2005). With the increase in biofuel production, available biomass energy statistics reflect growth patterns in geographically different contexts. For instance, sophisticated biomass energy technology has become increasingly well-known in developing countries where access to energy remains a painful concern, particularly rural regions. Few regions, including East Africa and South Asia, exhibit phenomenal potential in developing biomass resources in a sustainable manner, upholding regional energy security and economic development. Additionally, as urbanization increases, the demand for clean and efficient energy sources validates the case for the utilization of biomass in urban and peri-urban zones. These dynamics reveal how biomass production of energy is not merely a figure but part of the greater social and economic growth, which acts to enhance energy resilience toward shifting energy trends in addition to climate change pressures (Feng et al., 2019). While the path of biomass energy production is bright, it remains to be fully optimized as a resource. There is need for management interventions and technology to prevent threats to alternative land use, environmental degradation, and social equity concerns. Production of biomass needs to be well managed to ensure that procurement does not compromise food security or lead to deforestation, thereby maintaining key ecosystems and biodiversity. Researchers and policymakers must emphasize research and development that will enhance biomass conversion processes to be more efficient and sustainable. Shifting towards improved biofuels and optimal utilization of biomass in energy portfolios will require inclusive regimes ensuring sustainable practices and recognizing various purposes different forms of biomass can serve. Finally, energy needs and conservation of the environment go hand in hand as a balance is more crucial in utilizing biomass energy to fullest as one of the pillars of renewable resources (N/A, 2005).

7.2.2. Comparison of Biomass with Other Renewable Sources

In the field of renewable energy, biomass is a special source that deserves close comparison to alternatives like solar, wind, and hydroelectric power. While solar and wind energy utilize natural forces, biomass is produced from organic matter, such as agricultural wastes and special energy crops. This trait gives biomass clear advantages, including being able to contribute to energy supply while directly raising agricultural productivity and land use efficiency. Yet, in contrast to actually inexhaustible sources, the production of biomass is limited by the availability of arable land and thus risks competing with food and feed production. Thus, as the study results reflect, biomass energy and farm markets are interdependent in a highly intricate way and need to consider socioeconomic and environmental objectives (Becker et al., 2025). In contrast to the comparatively passive nature of wind and sun energy, production of biomass feedstock is characterized by advanced agricultural management with profound implications for land utilization, nutrients, and ecological structure. As hinted, increased biomass yield of energy can add nutrient loads and chemical use, which presents potential environmental issues (Becker et al.,

2025). Furthermore, although biomass energy can contribute to stabilization of the supply of energy with ongoing availability, it also presents the issue of greenhouse gas emission as well as land-use change. The dual role of biomass in energy supply and contributing to food and feed safety complicates policy frameworks promoting the sustainable production of energy. Policy makers therefore have to strike a balance between these myriad factors as they come up with regulations that will allow the production of biomass as a viable alternative energy source compared to its counterpoints. In determining the full spectrum of renewable sources, there are some considerations to mention: each source of energy possesses both advantages and disadvantages. Solar power, for instance, while abundant and capable of providing electricity with no emissions during application, is geographically and climatically dependent. Similarly, wind power also suffers from concerns regarding intermittency as well as impacts on wildlife. Biomass, when positively controlled, contains a unique synergy between farm practices and energy production, with the ability to lower the output of carbon dioxide through sustainable means. However, as bringing biomass into broader energy policy evolves, it must consider how it gains from cross-links with other renewable types and how it aligns with legislative agendas, such as those legislatively driven by statutes dictating renewable energy production. This multi-dimensional comparison shows that there is need for a holistic outlook where biomass finds its position in the overall context of renewable energy plans (Figure 7.8).

Figure 7.8. *Biomass energy versus other renewable energy sources.*

Source: https://www.greenlivinganswers.com/blog/wp-content/uploads/2024/08/biomass-comparison.webp.

7.2.3. Biomass's Potential for Future Energy Needs

The potential of biomass to serve as a renewable energy source is being recognized more and more while looking for greener energy solutions. Biomass, which is derived from organic feedstocks such as agricultural residues, forestry, and municipal waste, provides an opportunity for energy production while solving waste disposal problems. Evidence indicates that the biomass energy industry in certain regions, such as Mexico, is most favorable; the biomass energy potential for the country lies between 2635 and 3771 PJ/year, making it a strong candidate among other renewable options (Alemán-Nava et al., 2014). Its inherent potential coupled with advancements in conversion technology makes biomass a key player in future energy scenarios, particularly with policy interventions shifting towards cleaner energy initiatives. Further, introduction of biomass energy into the overall renewable portfolio aligns with national objectives of reducing carbon footprints and ensuring energy security. For instance, Mexico's ambition to produce 35% of its electricity from renewable energy by 2024 demonstrates the strategic importance of biomass in achieving such ambitions (Alemán-Nava et al., 2014). In economic terms, the consequences of large-scale development of biomass energy are important, with the sector holding promise in job creation and driving rural economies. As development of bioenergy continues, it is essential to have policies in place to facilitate investment in research, infrastructure, and human capacity development to ensure biomass will assist in the realization of a sustainable energy future. There are, nevertheless, obstacles to unlocking the complete potential of biomass energy. Despite enormous resources and research invested, there remains limited availability, technology adoption, and market development. Building a secure supply chain for biomass feedstocks and sidestepping regulatory hindrances are crucial to overcoming these challenges. Furthermore, measures need to be taken to stimulate public acceptance and awareness of biomass technologies, which are imperative to stimulating investment and innovation in this sector. By encouraging a shared environment among the stakeholders like government, private companies, and educational institutions, the biomass sector can thrive, ultimately playing an important role in meeting future energy demands backed by protection of the environment and economic stability.

7.3. OTHER RENEWABLE ENERGY SOURCES

Within the renewable energy category, there are several sources besides biomass worth mentioning, specifically solar, wind, and hydropower. All three of these sources are valuable in diversifying the energy mix, ensuring energy security, and limiting dependence on fossil fuels. Solar power, harnessed by photovoltaic cells or solar thermal units, has seen spectacular technological advancements, making it increasingly competitive in terms of cost. Wind energy, similarly, has witnessed unprecedented expansion, particularly in the windier regions, thus contributing significantly to the provision of energy to most countries.

Hydropower, long known for providing mass energy solutions, continues to be a reliable source and also provokes concerns regarding environmental impacts and habitat disruption. In examining the possibility of other renewable sources of energy, their strengths and limitations must be considered in their application. Solar and wind power, for instance, enjoy environmental benefits such as zero greenhouse gas emissions; they are, however, marred by intermittent supply, which necessitates complex energy storage facilities. In contrast, hydropower can potentially be used to add more reliably to the energy pool but typically involves significant environmental and social trade-offs, such as effects on aquatic ecosystems and community relocation. The integration of such renewable sources into the energy mix therefore requires careful management, balancing clean energy needs against ecological sustainability and social equity. Moreover, strategic decisions in electricity generation are made upon careful analysis and consideration of external costs. With the use of a life cycle assessment of different renewable technologies, it is more convenient to determine the long-term impacts associated with electricity generation processes. For instance, there is a rising trend in the utilization of renewable technologies, including biomass and wind power, as the energy mix evolves to minimize external costs. Notably, the recognition of these expenses will make renewable technologies like wind and hydro significantly more competitive, even replace biomass-generated energy in some situations (Georgakellos et al., 2014). This intricate dynamic of cost, sustainability, and impact on the environment highlights the necessity of multidisciplinary policy for energy, which is centered on the integration of various renewable sources into one and the same sustainable energy landscape.

7.3.1. Overview of Solar Energy and Its Applications

Solar energy is one of the most promising substitutes to conventional forms of energy, harnessing the suns abundant light to generate heat and electricity. Photovoltaic (PV) panels, which convert sunlight directly into electricity, have made significant strides in domestic, business, as well as industrial applications. This technology is particularly appealing in that it is scalable; solar panels can be installed on rooftops or in massive solar farms, satisfying different energy needs and geographic locations. Additionally, including solar power in the energy mix can reduce greenhouse gas emissions significantly while contributing to energy independence. While the world grapples with how to adjust to climate change, solar energy not only offers a clean source of energy but also responds to the world's movement away from the use of fossil fuels. Solar energy is used for many different applications, not just electricity generation, but also heating, such as in solar water heaters. Solar heaters utilize solar collectors to absorb the sun's rays, which are then utilized to heat household water or water for industrial use.

Solar thermal technology is applicable across various settings, such as domestic dwellings, swimming pools, and large-scale industrial heating processes. Schemes and incentives by the government have assisted in the application of the technology. Furthermore, advances in solar energy storage systems, such as battery technology, enhance the dependability of solar heating systems, making it possible to provide consistent hot water even during unfavorable weather. Such flexibility says a lot about the tremendous potential of solar energy in the renewable energy market. As per the Global Status Report, the increasing demand for renewable energy has triggered dramatic expansion in the adoption of solar power, with a boost from commendable policies from various governments of the world. Several countries are beginning to add a range of different renewable sources to their energy mix, with mandates driving biofuels blending and growth in solar installations. This diversity within renewables like solar, wind, and biomass aids the multi-faceted approach needed in meeting sustainable energy goals. With ongoing expansion of solar markets, particularly in regions with abundance of sunshine, economic opportunities and employment in this sector are considerable. As a result, solar power not only serves to meet energy needs but also works towards a paradigm shift towards a sustainable and resilient energy future, which goes a long way towards proving its importance in the entire debate about Biomass and Other Renewables (Gomberg et al., 2009; N/A, 2005) (Figure 7.9).

Figure 7.9. *Overview of solar energy.*

Source: https://www.greenlivinganswers.com/blog/wp-content/uploads/2024/08/solar-energy-sustainability.webp.

7.3.2. Wind Energy: Technology and Impact

Wind energy technology is a critical component of the world shift towards the sustainable power solutions. With its ability to harness the kinetic energy from the wind through turbines, wind energy never releases greenhouse gases when it is in operation, and thus it is clean as compared to fossil fuel technology. The application of wind power to feed energy systems illustrates how renewable technologies can downscale the impact of climate change by a considerable margin. For example, the development of hybrid renewable energy systems combining wind, solar, and biomass technologies emphasizes the versatility of wind energy. The systems are designed to work in synergy, ensuring energy resilience and sustainability, particularly in off-grid societies with anxieties towards energy insecurity and environmental degradation (Orozco et al., 2025). Industries are benefited by this technology not just in terms of reduced operating expenditure but also via job creation linked with its development and maintenance.

Figure 7.10. *Understanding wind energy.*

Source: https://www.greenlivinganswers.com/blog/wp-content/uploads/2024/08/wind-energy-1.webp.

The socio-economic benefits from wind power are massive, triggering local employment and economic stability throughout communities. As highlighted by recent studies, deployment of wind energy solutions has the capacity to create full-time equivalent (FTE) jobs, alongside triggering skills enhancement within the local populace. Moreover, the integration of wind energy into a diversified renewable energy strategy will play a critical role in regional energy independence, reducing the demand for imported fossil fuels. This is particularly the case in circumstances like the one observed in Cuba, where the implementation of a hybrid renewable system incorporating wind and other renewals resulted in considerable greenhouses gas emission and fossil fuel use reductions (Orozco et al., 2025). This transition positions wind energy not just as an environmental solution but as an economic force, capable of changing local economies and supporting broader climate goals. The transition to wind power is not without its set of challenges, however. Technological deployment challenges, resource competition, and the need for infrastructure upgrades must be addressed in order to capitalize on wind energy's benefits. For instance, ongoing innovation in hydrogen production from renewables further muddies the waters, triggering potential changes in energy dynamics. The interaction between wind power and hydrogen production is supplemented by the evolution of technology in electrolysis, which can significantly reduce carbon emissions from hydrogen production up to 2045. By transforming policies aimed at decommissioning fossil fuels, more imperative is the consideration of how wind power can be included in an integrated renewable system including biomass and other green technology to construct a future of sustainable energy (Figure 7.10).

7.3.3. Hydropower: Benefits and Environmental Concerns

Hydropower is amongst the globe's most significant renewable energy sources that has played a major role in reducing the utilization of fossil fuels. By tapping the kinetic energy of flowing water, hydropower plants can generate electricity with negligible greenhouse gas emissions. This transition not only helps combat climate change but also helps improve air quality, which is an essential issue regarding public health. The recent technological advances, such as smart grid integration, have increased the cost-effectiveness and efficiency of hydroelectric systems to make them more appealing than previously. These developments resonate with the growing importance to shift into cleaner energy technologies, as required in recent renewable energy reviews (Udele et al., 2025). Therefore, while hydropower has several benefits, there is a requirement for thorough knowledge of its environmental impacts in order to guarantee sustainable use. While blessed with its advantages, hydropower infrastructure construction has a heavy price tag attached to it for the environment. Building dams and reservoirs could result in people displacement, ecosystem transformation, and disruption of aquatic life. Even more so, these changes are likely to introduce issues like sedimentation and water quality degradation, which can be a danger to the surrounding flora and fauna. Dams also create extreme fluctuation in the water level, which impacts fish migration as well as breeding behavior. These environmental effects are usually quoted by the detractors as the principal drawback, challenging the sustainability of the hydropower projects. Therefore, it is very important to have stringent environmental studies and stakeholder consultations while planning and executing hydropower projects to avoid adverse effects while still making them profitable (Dwiputra et al., 2024). Apart from environmental factors, sustainability for the long term for hydropower must consider socio-economic factors as well. For instance, resettlement of people due to new dams can cause social unrest and lead to complex legal litigation. Furthermore, reliance on hydropower in desert regions is problematic in terms of energy security. Prolonged droughts can significantly lower water levels, thus limiting power generation and jeopardizing energy supply. This disparity emphasizes the need for a varied portfolio of clean energy that includes other types like biomass, solar, and wind. With the use of various renewable technologies and policies that promote sustainable utilization, communities can more effectively secure their power future while eliminating the environmental concerns linked with hydropower (Udele et al., 2025). Planning and innovative approaches are essential in ensuring a balanced approach in energy generation (Figure 7.11).

Figure 7.11. *Understanding hydropower.*

Source: https://www.greenlivinganswers.com/blog/wp-content/uploads/2024/08/hydropower.webp.

7.4. ENVIRONMENTAL IMPACTS OF BIOMASS AND RENEWABLES

The environmental impacts of renewable energy and biomass are complex and diverse. While the production of biomass holds significant potential for offsetting greenhouse gas emissions against fossil fuels and a lower climate change impact, this comes with some significant provisos. Biomass is intensive in land use and is capable of leading to competition between energy requirements and other required resources such as food and animal feed. This competition will strain agricultural systems and alter regimes of nutrient loading in soil, with consequences for ecosystem health overall. The sustainability of biomass then as a renewable energy source then depends on careful management choices to balance ecological health with energy needs for production, demonstrating the necessity for a strong framework to evaluate these interactions and their long-term impacts on agriculture and the environment (Becker et al., 2025). In addition, the use of biomass energy is also closely linked to land-use decision-making, which has several environmental effects. Land-use choices regarding which biomass feedstocks to utilize have important consequences for ecosystems and biodiversity locally, as well as social processes of land management. Land utilized for biomass can result in habitat loss, fragmentation, and land-use change endangering existing biodiversity. x

Figure 7.12. *Environmental benefits of biomass energy.*

Source: https://www.greenlivinganswers.com/blog/wp-content/uploads/2025/01/biomass-energy-environmenta-benefits.webp.

The socio-economic benefits from wind power are massive, triggering local employment and economic stability throughout communities. As highlighted by recent studies, deployment of wind energy solutions has the capacity to create full-time equivalent (FTE) jobs, alongside triggering skills enhancement within the local populace. Moreover, the integration of wind energy into a diversified renewable energy strategy will play a critical role in regional energy independence, reducing the demand for imported fossil fuels. This is particularly the case in circumstances like the one observed in Cuba, where the implementation of a hybrid renewable system incorporating wind and other renewals resulted in considerable greenhouses gas emission and fossil fuel use reductions (Orozco et al., 2025). This transition positions wind energy not just as an environmental solution but as an economic force, capable of changing local economies and supporting broader climate goals. The transition to wind power is not without its set of challenges, however. Technological deployment challenges, resource competition, and the need for infrastructure upgrades must be addressed in order to capitalize on wind energy's benefits. For instance, ongoing innovation in hydrogen production from renewables further muddies the waters, triggering potential changes in energy dynamics. The interaction between wind power and hydrogen production is supplemented by the evolution of technology in electrolysis, which can significantly reduce carbon emissions from hydrogen production up to 2045. By transforming policies aimed at decommissioning fossil fuels, more imperative is the consideration of how wind power can be included in an integrated renewable system including biomass and other green technology to construct a future of sustainable energy (Figure 7.10).

7.4.1. Carbon Footprint of Biomass Energy Production

The carbon fingerprint of biomass energy production is complex, both reflecting the material itself and the method of generating energy. Biomass, derived from organic materials such as wood, agricultural waste, and residues, is often perceived as a renewable energy. But it is not that simple. The carbon released through the production of biomass will be significantly varied depending on land use, type of biomass, and transportation. For instance, when forests are being cut down in an unsustainable way, the carbon emitted through combustion is greater than the carbon captured during the biomass's growth, increasing overall global carbon emissions. It is therefore necessary to consider the whole lifecycle of biomass energy production so that its carbon footprint can be accurately estimated and its role in climate abatement measures can be determined in the context of renewable energy sources as a whole. Use of carbon emissions measurements with consumption accounting principles can unveil the actual environmental implication of biomass energy production. The consumption accounting principle (CAP) highlights the carbon footprint of the whole supply chain, starting from raw material extraction and culminating in ultimate energy use. This method brings out the complexities and volatilities of biomass production and shows how trade and consumer demand can drive carbon outputs. When analyzing emissions this way, policymakers can better understand the full environmental impact of biomass energy and thus make more informed decisions regarding energy policy and environmental management. By integrating this understanding into national and regional energy policy, production techniques can be brought into alignment with carbon-reducing initiatives and, in the end, a clean energy future. Finally, the implementation of sustainable biomass techniques is essential in reducing the carbon footprint of producing biomass energy. Policy and technological advancement will go a long way toward decreasing the negative impacts of biomass on the environment. For example, the utilization of agricultural waste or sustainably forested forests' waste not only promotes carbon sequestration but also improves the general efficiency of biomass as an energy source. Moreover, stimulating local supply chains can minimize transport emissions and thus align closer with CAP-oriented policies. In the wider environmental context, these actions highlight not only the potential for biomass to be a renewable resource but also emphasize the need for strategic planning and local involvement in promoting sustainable energy practice. The discourse on carbon footprints, aligning with the values of sustainability, can help produce an energy landscape that is most in line with climate action and environmental protection.

7.4.2. Land Use and Biodiversity Considerations

Land use and biodiversity interaction is a critical aspect where biomass and other renewable fuels are concerned. With increasing populations and energy demands, land use towards the production of biofuel and biomass has immense environmental effects. This underlines the merit of an Integrated Environmental Assessment (IEA) approach, one that evaluates environmental effects along a series of dimensions from primary biomass production to final energy services delivery. Using this tool, stakeholders can better gauge the effect different land management practices are having on biodiversity and on greenhouse gas emissions. The historical research on agricultural practices, particularly in France, shows how effective land use policy can promote yield without compromising huge areas of forests, thus mitigating climate change, and maintaining biodiversity (Riedacker, 2025). The results are critical in developing sustainable energy policies that are friendly to the environment at a global perspective. In addition, the growing biofuels industry also raises pressing issues of food security and land use, especially in nations where agricultural potential is already threatened. Competition between food crops and biofuel crops for land can lead to food shortages, as evidenced by studies that blame the 2007–2008 global food crisis on biofuels. These evaluations show variations in the economic implications of biofuel policies, noting that biofuel-induced reallocation of productive land to energy crops can have adverse implications on food prices and supply (Shrestha et al., 2025). Policymakers thus need to tread a fine line; whereby increased development of biofuels does not occur at the cost of critical food production or biodiversity conservation. Sustainable alternatives, like second-generation biofuels derived from non-food biomass, represent new opportunities, but they must also be done with an eye to their effects on the environment. In the end, the challenge of balancing land use for growing biomass versus the necessity of preserving biodiversity and food security is complex but not insurmountable. Solutions must be multi-faceted, embracing agricultural technology, government policy, and local outreach to encourage sustainable solutions. For instance, increasing yield per hectare via improved farming practices can restrict the demand for additional land conversion, enhancing food security without expanding deforestation pressure. In addition, restoration initiatives on biodiversity conservation can supplement bioenergy activities so that there can be an integrated approach to land management. Such collaborative efforts not only bring value to climate change mitigation but also align with global sustainability goals, and the attainment of an energy, food, and environmental well-being future becomes a possibility.

7.4.3. Pollution and Waste Management Issues

Pollution control through effective waste management systems is of paramount significance in the pursuit of sustainable energy commodities, particularly in the context of biomass. Biomass, derived from organic matter, can reduce waste while generating energy simultaneously. The practice identifies biodegradable waste as a useful material that can significantly reduce greenhouse gas emissions. For instance, a study on Myanmar waste management found that enhanced use of biodegradable waste led to significant reductions in greenhouse gas emissions and fossil fuel utilization, representing a double positive impact of recycling organic waste (Juchelková et al., 2018). By waste-to-energy conversion, societies are not only reducing the pollution problems but also generating a cleaner environment, thus demonstrating that waste management is interconnected with renewable energy sources. The transition to biomass and other renewable energies also necessitates a clear understanding of the environmental impacts of waste dumping practices. Urban areas, which are known to have high volumes of waste generation, are also troubled by serious pollution challenges due to ineffective waste management. Traditional waste disposal practices, such as open dumping, contribute to increased emissions of greenhouse gases, thus degrading air quality and public well-being. The above-mentioned study emphasized the fact that as much as 1800 Gigagrams of CO_2-eq emissions were caused by inappropriate waste disposal in Myanmar, which emphasizes the urgent need for efficient management practices (Juchelková et al., 2018). The transition to biomass as a source of energy not only offers emission reduction but also encourages economic growth and employment, which ought to be of interest to policymakers' keen on developing green energy activities. On top of that, economic advantages of enhancing waste management methods through promotion of biomass are imperative. Through establishing a robust biomass industry, states have the ability to offer employment and support the local economies while reducing pollution concerns. The valuation of the renewable energy sector, in this instance, biomass, registers potential for major economic value as well as strategic development potential. Policymaking favoring sustainable strategies can lead to the use of technologies that convert waste into an asset. This lessens the use of fossil fuels to a large extent while boosting environmental sustainability, rendering biomass a front-runner among renewable energy sources. Thus, incorporating sound waste management practices into renewable energy development is necessary for us to attain a greener and more sustainable future.

7.5. ECONOMIC ASPECTS OF BIOMASS AND RENEWABLE ENERGY

The economic viability of biomass and other alternative energy is increasingly considered a solution to the transition to sustainable energy systems. Biomass fuels offer an attractive alternative to fossil fuels, reducing greenhouse gas emissions and increasing energy independence. The European Union's Directive 2009/28/CE sets ambitious objectives for member states to produce and consume renewable energy, including biomass, requiring a model that is sustainable based on economic, environmental, and social criteria. This standard emphasizes effective management practices to avoid land-use conflicts and harmonize energy production with nature protection, thereby forming a crucial nexus between economic development and environmental sustainability (Gissi et al., 2025). By nations embracing these standards, it is imperative to balance economic objectives with sustainable land management strategies to maximize long-term sustainability. In Poland, the cultivation of specific energy crops, like Virginia mallow and silver-grass, enables us to gain greater insight into the economic characteristics of biomass production. A detailed analysis indicates that they are one of the most important sources of agrobiomass that can enter the Polish national renewable energy system. The economic analysis considers not only the initial establishment and sustainable profitability but also projects that investment in biomass plantations would yield massive returns, provided logistical planning for biomass supply is optimized (Grzesik et al., 2012). The approach involves a dual consideration of economic viability and environmental responsibility, considering the cost of establishment and maintaining such energy crops. These analyses are crucial for policy formulation, making sure that new energy plans work in tandem with both market forces and environmental considerations. Additionally, the dynamics between economic factors and the production of biomass go beyond profit margins; they also include overall socio-economic considerations. Proper utilization of biomass energy can boost local economies through the provision of employment in agricultural activities, transportation, and the energy sector, as well as rural development. However, the resulting economic potential must be crossed in harmony with ecological considerations, particularly regarding biodiversity and landscape integrity, as highlighted by the EU guidelines (Gissi et al., 2025). Integration of biomass into the renewable energy industry through policy must involve local communities to ensure social considerations are integrated. These issues can be tackled to enhance public endorsement and support, thus facilitating a smoother transition to a more sustainable energy future with optimum economic returns of biomass production (Figure 7.13).

Figure 7.13. *Economic impact of biomass energy.*

Source: https://www.greenlivinganswers.com/blog/wp-content/uploads/2025/01/biomass-energy-economic-impact.webp.

7.5.1. Cost Analysis of Biomass versus Fossil Fuels

The economic comparison of biomass and fossil fuel is essential in analyzing the viability of renewable energy resources. Biomass from organic materials presents an appealing alternative to fossil fuels because it has the potential for greenhouse gas mitigation along with resource sustainability. Biomass technology can be more expensive compared to traditional fossil fuels in most cases, primarily infrastructure and purchasing feedstock cost. However, the eventual financial implication to biomass may be significantly lower, given that it is able to access renewable energy resources without the volatility associated with the fossil fuel industry. Further, with the development of technologies, converting biomass also has its own costs becoming increasingly competitive, thus there is a necessity to check these economic factors whenever one is advocating for the transition to alternative energy. A vital consideration in the cost accounting is the externalities of fossil fuel use. The impacts of climate change and environmental degradation on the economy are at times not reflected in the market price of fossil fuels, therefore the underestimation of their true cost. For instance, fossil fuel emissions have negative health effects and cause environmental damages at a cost to public health systems and the economy as a whole. The fact that these externalities are included in the price of fossil fuels makes a hugely dramatic impact on the energy economics scenario, possibly to balance the situation in favor of biomass and other alternative energy sources. Here, the smart policy intervention, such as carbon pricing, can render cleaner technology more attractive, making biomass economical compared to fossil fuels in the long term. Moreover, the role of government regulation and subsidies cannot be excluded when it comes to pricing biomass versus fossil fuel. Different nations have used different types of energy policy models, whereby some have encouraged renewable resources through economic incentives and support. For instance, the integrated Energy/Climate package of the European Union's significantly works in the direction of limiting the use of fossil fuels as well as enabling the growth of renewable resources (Devogelaer et al., 2025). Alongside this, however, are global efforts to reduce climate change itself, which further make energy transition to sustainable energy sources more imperative. With energy security being achievable through biomass, its economic implications, when appropriately balanced against its benefits, become economically viable but also necessary in their role towards reduced reliance on fossil fuels (Elliott et al., 2003). Hence, there is a need for an integrated strategy for fostering an appreciation of biomass cost dynamics among policymakers and renewable energy stakeholders.

7.5.2. Job Creation in the Renewable Energy Sector

The renewable energy sector is the leading sector in addressing both environmental sustainability and economic growth, primarily through the provision of jobs. As nations move towards cleaner sources of power, sectors such as biomass, wind, and solar are turning out to be the powerhouses of the job market. Beyond emphasizing renewable resources being a means of addressing climate change, it is also an economic policy for stimulation of local economies. In Iowa alone, for instance, significant investments have been undertaken through programs like the Iowa Power Fund that finances projects in wind, solar, and biofuels. The initiatives not only work toward enhanced energy independence but also stand to provide substantial jobs. The blending of environmental goals with economic development illustrates the possibility for the renewable energy sector to bring new opportunities while paving the way for a sustainable future (N/A, 2008). Transitioning to green economy entails more than technological innovation; it calls for alterations to the labor structure. Green jobs are characterized by their focus on sustainability, seeking a balance of environmental results and economic viability. This new form of employment requires highly trained employment in various industries, including manufacturing, research, and installation of renewable technology. The establishment of these jobs can be viewed as evolution in accordance with the demand for clean energy alternatives, which is crucial on economic, environmental, and social levels. Countries around the world, particularly in Europe, are looking for best practices in green job integration in their economies. Successful adoption of such practices yields long-term advantages, making nations not only renewable energy giants but also economic rebirth precursors (Kryk et al., 2015). Renewable energy investment is, therefore, investment in the future workforce. Continual development of biomasses, as well as other renewable resources, represents a soil fertile with possibilities for employment creation. Besides production and installation direct employment, jobs related to renewable energy management, policymaking, and education are also increasing. This multi-faceted strategy for job development enhances regional economies' resilience by introducing varied skill sets into the marketplace and reducing reliance on fossil fuels. As states and nations increasingly make renewable energy projects a priority, further investment in training and education initiatives will be necessary. Lastly, having an ability for the human resource to succeed in the renewable sector is essential for long-term sustainability and economic growth (N/A, 2008).

7.5.3. Government Policies and Incentives for Renewable Energy

Effective government policies and incentives occupy a key position in accelerating the process of transition to renewable energy, including biomass and other renewable resources. Policies are a crucial factor in addressing the interrelated challenges of climate change, energy security, and economic growth. By providing an enabling environment for innovation and industry growth through policy interventions promoting investment in renewable energy, governments can facilitate such transitions. For instance, targeted subsidies and tax credits can significantly reduce the cost of adopting renewable technology so that it becomes more attractive to both businesses and consumers. Furthermore, effective renewable energy laws and supportive regulation can facilitate market entry and expansion for renewable energy suppliers, ultimately triggering competition and reducing cost. Hence, renewable energy policy guidelines are required not just for curbing greenhouse gas emissions but also to frame the future energy situation in a sustainable framework (N/A, 2010). Countries around the world, like China, illustrate how much impact government policies can contribute to the renewable energy sector. The Chinese government has established a strong regulatory and legal platform, for instance, the Renewable Energy Law of 2005, which was aimed at promoting the utilization of renewable energy and energy diversification. The law includes incentives in the form of subsidy for renewable electricity and strategic investment in capacities for world-class domestic manufacturing that will be able to compete in future world markets. Solar and wind energy investment has picked up pace, with impressive capacity growth that shows Chinas commitment to renewable energy. Through recent years, China is aiming for 15% of the total energy consumption to be fulfilled by non-fossil fuels by 2020, showing policy objectives being coordinated with ambitious renewable energy targets (Campbell et al., 2010). On the other hand, the United States faces difficulties with a disunited national policy on encouraging renewable energy. The disunited one, it is argued, hampers market uptake of renewable technologies on the primary reasons that they are more expensive to generate electricity from renewable sources. But local incentives and state schemes are the significant driving forces in the growth of the renewable energy sector, with most areas having their own special policies toward the encouragement of biomass and solar energy take-up. Furthermore, increased public concern with the environment and climate change has gradually begun to put the issue on the political agenda, with sustainable energy as part of it. Since both nations are trying to meet the challenge of climate issues, the development of strong and harmonized policies for renewable energy remains important to initiating a revolutionary transition towards cleaner and adaptive energy systems globally (N/A, 2010).

7.6. TECHNOLOGICAL INNOVATIONS IN BIOMASS AND RENEWABLES

The transition to renewable energy sources is ever more driven by technological advances that optimize the efficiency and sustainability of biomass and other renewables. Underpinning this innovation is the concept of cascading utilization, which emphasizes the several material applications of renewables before their conversion into energy. This approach maximizes raw material usage while minimizing waste, emulating principles of sustainability and efficiency. Emerging technologies in bio-refineries exemplify this innovation, where biomass undergoes processes that yield both energy and high-value products, thereby increasing overall productivity. The transdisciplinary nature of research being undertaken today, as revealed in research on increased resource efficiency, calls out the realm where biomass can not only be a source of energy but also a fundamental material input in various industrial processes, ushering in new possibilities towards future sustainability in energy and production (Geldermann et al., 2015). In discussing the role of technological innovation in biomass consumption, there is a need to examine the disruptions these innovations bring to traditional energy markets. Traditional firms are likely to respond to these disruptions in disbelief due to the inherent nature of being able to make correct predictions in evolving environments.

Figure 7.14. *The future of biomass energy – advancements and potential.*

Source: https://www.greenlivinganswers.com/blog/wp-content/uploads/2024/08/biomass-energy-future.webp.

Disruptive technologies require evaluation frameworks, particularly when it comes to assessing their feasibility across various industries. For instance, the valuation of five technologies that can be used to generate renewable energy, including biomass, offers insights into how market forces are being modified as new technology takes center stage. The process is reflective of the pivotal necessity for organizations to be flexible and open to adjusting to disrupting technologies, which can redefine their operational landscape and enhance their competitiveness in the energy sector and thereby resolve market demand and environmental challenges (Hampl et al., 2014). Besides, IT innovations are at the heart of integrating biomass and other renewable energy sources into modern energy supply systems. Development of the smart grid, facilitated by IoT and analytics, allows for optimum management of energy resources, including biomass. The technologies facilitate real-time tracking and control of energy production and consumption, making the grid more reliable and resilient. In addition, co-design platforms that facilitate collaboration among producers and consumers allow for a supply chain that is more responsive to utilize the strengths of biomass as a renewable input. By using practices that have consumer attitudes aligned with sustainability at their core, the sector can influence technological innovations to make a transition towards cleaner energy. This shift not only addresses environmental demand but also enhances public involvement in renewable energy initiatives, demonstrating the entwining of technology, sustainability, and community engagement (Geldermann et al., 2015) (Figure 7.14).

7.6.1. Advances in biomass conversion technologies

The field of biomass conversion technologies has witnessed considerable advancements, which are crucial for the sustainable production of chemicals and fuels. The main processes are separated into thermochemical, chemical, and biochemical methods, each possessing its strengths. Thermochemical methods such as gasification and fast pyrolysis enable direct lignocellulosic biomass conversion into useful energy forms such as syngas and bio-oil. These technologies not only enhance the efficiency of biomass conversion but also facilitate the production of a variety of biofuels necessary for reducing dependence on fossil fuels. Interconnection of these processes into biorefineries fosters a circular economy system in which waste material is recycled, encouraging sustainable energy options and reducing greenhouse gas emissions (Maity et al., 2015). Chemical conversion technologies are also a promising area of innovation in biomass utilization. Triglyceride conversion by transesterification is a successful process that produces biodiesel, a renewable fuel alternative to conventional petrochemical fuels. The exploration of microalgae as a rich source of triglycerides also further enhances the potential for the production of biodiesel with high productivity and minimum land use. These technological advancements not only improve the sustainable manufacture of biodiesel but also pave the way for diversification to other biomass-derived fuels. As technologies evolve, they are poised to meet rising energy demands while improving environmental sustainability and energy security (Maity et al., 2015). In addition to advancements in thermochemical and chemical technologies, biochemical conversion technologies offer an alternative pathway for biomass application. Technologies that focus on carbohydrate conversion to alcoholic biofuels, such as bioethanol and bio-butanol, are quintessential examples of advances. Such biochemistry-driven conversions utilize natures large reservoirs, in particular lignin, to produce value-added fuels and chemicals in an efficient manner. Furthermore, new methods such as hydrodeoxygenation and biosynthetic routes not only render production more efficient but also increase the economic feasibility of biomass conversion. By improving the efficiency and reducing the costs of such processes, it is more likely that these processes will be accepted universally in various sectors, thereby making energy accessible and facilitating sustainable development goals globally.

7.6.2. Integration of Smart Grids with Renewable Energy Sources

The integration of smart grids with renewable energy sources is a fundamental step toward achieving a sustainable future for energy. Through making use of advanced technology, smart grids allow for the efficient management of intermittent renewable energy sources, e.g., solar and wind energy, which are being harnessed in huge quantities on an ever-growing scale. Smart grids enable a two-way communication system between producers and consumers of energy that enables real-time data exchange to enhance efficiency in the use and distribution of energy.

With the prices of renewable energy technologies coming down and their availability rising, the installation of smart grids seems like a necessary infrastructure that can meet enhanced energy demands without relying primarily on fossil fuels. This technologic synergy slots into bigger objectives aimed towards greenhouse gas reduction, and therefore it is an essential component of contemporary energy policy and market change programs (Flavin, 2008). In addition to applying towards operations efficiency, smart grids enable renewable integration by overcoming supply and demand variability. Weather-dependent intermittent sources of power are needed with accurate management systems to maintain grid reliability and stability. Smart grids utilize sophisticated information technology systems to balance supply volatility with consumer demand, reducing reliance on traditional baseload power generation facilities.

Furthermore, technology innovations in the guise of energy storage facilities have emerged as key components of smart grid infrastructure, enabling release of stored renewable energy during periods of peak demand. Since this integration provides for a more responsive and dynamic energy system, it also opens new economic and job opportunities in the renewable sector, highlighting the massive socio-economic benefits of transitioning towards smart grid systems (Axon et al., 2014). Smart grids do have promising potential, but their proper integration with renewable sources depends on comprehensive public policies and regulatory systems.

Political and financial support is crucial in driving smart grid development, as it not only stimulates innovation but also tackles the multiple challenges of energy transition comprehensively. However, public acceptability of smart technology, with challenges in respect of data privacy, security, and costs, has to be addressed to ensure smooth transition. Decision-makers need to include various stakeholders, e.g., customers, regulators, and the energy industry, in order to come up with a shared vision for undertaking smart grid. Through cooperation and the coming together of varied viewpoints, we can create a resilient and responsive energy system that not only meets current demand but also lays the foundation for a green energy future with greenhouse gas emissions significantly cut down (Flavin, 2008).

7.6.3. Future Trends in Renewable Energy Technology

As the world's energy map is transforming fundamentally, future trends in renewable energy technology will redefine energy use and generation. The integration of biomass and other renewables into grid power generation is becoming increasingly important in solving both environmental and economic objectives. Governments around the world are accelerating the support for renewable energy, creating policy frameworks that promote investment in such technologies. For instance, increasingly more countries are enacting blending mandates on biofuels, which is a true paradigm shift in energy policy in favor of sustainability goals. Aside from enhancing the energy security of countries, this also offers economic opportunities in rural communities, which emphasizes the numerous benefits of converting to renewable fuels (N/A, 2005). In addition, innovation within the biomass sector will remain a key area of attention with technology advancements driving increased efficiency in energy conversion. Emerging technologies such as new biogas plants and second-generation biofuels will ensure that the full potential of biomass resources is maximized while addressing food security and land use concerns. These technologies are dependent on the utilization of renewable energy sources and are gaining momentum in various markets, varying from off-grid rural power supplies to supplying energy for cities.

Additionally, since the recognition of global warming has grown, countries are placing more emphasis on the generation of cleaner energy sources as a foundation of their ecological policies. It has been reported that by 2025, the majority of the regions should have a significant percentage of their energy mix constituted by biomass and other renewable sources, which will be cost-competitive in the global energy market (N/A, 2010). Finally, the integration of scattered renewable technologies will unlock the gate to unified energy systems that opt for maximum resilience and sustainability.

For instance, biomasses' hybridization with wind and solar power technologies has the potential to offer a more stable and secure energy source, particularly in emerging economies. Furthermore, cross-border alliances, like those facilitated by institutions such as REN21, reinforce the need to share knowledge and strategy in the development of stronger implementation of renewable energy technology across the world. Through collaboration and innovation, those trends will decrease costs and increase availability, resulting in a diversified and sustainable energy future. The growing market dynamism opens up the opportunities for investment and growth, further establishing renewables position as a central component of international energy policy (N/A, 2005, 2010).

SUMMARY

In short, the role played by biomass in the wider renewable energy complex is large and in motion. As sustainable sources of power become increasingly necessary worldwide, biomass stands out due to its potential for scaling up and interoperability with existing agriculture and forestry sectors. According to the Community Strategy and Action Plan on Renewable Energy Sources, double the share of Renewable Energy Sources (RES) is the target that unequivocally identifies biomass as a significant key to energy independence as well as environmental sustainability. This approach emphasizes not just the necessity for technical innovation in bioenergy and energy crops but also the economic value that accrues for farmers as well as for European energy producers. The sophisticated development of biomass technology, therefore, has the dual benefit of being economically viable and environmentally friendly, and this is why innovative policy is necessary to promote growth in this sector (Latorre et al., 2001). Apart from this, the use and generation of wood biomass is also an imperative aspect of the energy policy of the European Union, particularly with the context of attainment of both climate and energy goals. As the EU wants to further enhance its energy security while advancing towards cleaner sources of energy, biomass then is a valuable commodity for the generation of combined heat and power. However, such an approach needs special attention to resource competition, data gathering, and environmental problems inherent in biomass production. Resolution of these problems is of maximum significance to ensure that the benefits of biomass are accessed without compromising ecological balance. Ultimately, the development of biomass should not only be concerned with the quantity aspect of energy generation but also focus on environmentally friendly practices that assure natural systems and promote biodiversity (N/A, 2008). In conclusion, the future of biomass and other forms of renewability relies on our commitment to creative solutions, green approaches, and comprehensive energy strategies. The future of biomass making a valuable contribution to renewable energy goals as well as climate action is clear, yet it requires an integrated strategy that spans technological innovation, regulatory support, and public sensitivity. As research for biomass continues to advance and its impacts on the environment, stakeholders must get together and create mechanisms to allow us to utilize resources wisely and make renewable energy accessible to everyone equitably. The path forward is not as much as creating more energy but getting the most out of resources so that sustainability and resilience are established within our energy systems. The integration of this vision for biomass and other renewables is essential in achieving a sustainable future for energy that benefits society as well as the environment.

The finding of biomass and other alternative forms of energy has gained a tremendous amount of momentum as the globe turns to sustainable sources for the replacement of fossil fuels. When considering the move towards clean energy, it is essential to note how cities such as Boston are adopting strategies that are centered around energy efficiency and decarbonization. Carbon Free Boston plan heavily suggests that the city's electricity demand will grow and reliance on petroleum fuels and natural gas will fall. This shift signals a commitment to meet the targets of carbon reduction using clean energy sources, which must also be implemented in line with the principles of social equity and economic development (Cleveland et al., 2019). Such strategies mention the importance of comprehensive energy policies that take into account not only environmental concerns but also optimize the quality of life in societies. A critical part of making the transition towards renewable energy is electrifying building heating and cooling, which was identified in a series of strategic reports. Successful implementation of economical energy efficiency will reduce dramatically the consumption of fossil fuels, and the urban landscape will become more sustainable. The focus on technologies such as roof-fixed photovoltaic installations shows the rise in in-boundary clean energy generation. Moreover, with cities aiming for carbon neutrality by 2050, it is necessary to expand electricity distribution networks and acquire renewable fuels in a sustainable way, which requires energy policy to be addressed through an interdisciplinary framework (Cleveland et al., 2019). This intersection of energy efficiency, technology, and urban policy is key to attaining wider climate aspirations, setting an example for effective sustainability projects. Under the lens of international renewable energy strategy, case studies in countries like China yield valuable insights into the efficacy of aggressive electrification programs. Ongoing technological change implies that the optimization of renewable technology adoption may be able to plateau CO_2 emissions. Recent studies show that by leveraging non-conventional electric and renewable energy sources, China can peak its nationwide CO_2 emissions as soon as 2025. This proactive move not only addresses near-term climate change objectives but also represents the need for new policies encouraging technology adoption in various industries. Ultimately, combining these strategies holds great potential for significant CO_2 savings, highlighting the essential balance between electrification and sustainable energy practices in an effort to shift towards a less carbon-intensive future.

REFERENCES

1. Ackermann, T., Barrow, C., Bell, D., Benson, J., Bornstein, M., Breukers, S., et al. (2009). Identifying predictors of attitudes towards local onshore wind development with reference to an English case study. https://core.ac.uk/download/50223.pdf (accessed on 30th September 2025).
2. Agnolucci, P., Arnstein, S., Bäckstrand, K., Bana e Costa, C. A., Brans, J. P., Cavallaro, F., et al. (2007). Is small beautiful? A multicriteria assessment of small-scale energy technology applications in local governments. https://core.ac.uk/download/51100.pdf (accessed on 30th September 2025).
3. Alemán-Nava, G. S., Casiano-Flores, V. H., Cárdenas-Chávez, D. L., Dallemand, J. F., et al. (2014). Renewable energy research progress in Mexico: A review. https://core.ac.uk/download/77019055.pdf (accessed on 30th September 2025).
4. Allan, G., McGregor, P., & Swales, J. (2011). The electricity generation mix in Scotland: The long and windy road? https://core.ac.uk/download/19609893.pdf (accessed on 30th September 2025).
5. Axon, C., Balta-Ozkan, N., Cipcigan, L., Connor, P., et al. (2014). Scenarios for the development of smart grids in the UK: Literature review. https://core.ac.uk/download/155776124.pdf (accessed on 30th September 2025).
6. Becerra Villanueva, J. A., Calderón, B., Carvalho, M., Chacartegui, R., et al. (2019). Trans critical carbon dioxide charge-discharge energy storage with integration of solar energy. https://core.ac.uk/download/322845447.pdf (accessed on 30th September 2025).
7. Becker, A. (2025). Biomass for energy production in the context of selected European and international policy objectives. https://core.ac.uk/download/pdf/6397072.pdf (accessed on 30th September 2025).
8. Bekkering, J., Benders, R. M. J., Moll, H. C., Pierie, F., et al. (2016). Spatial and environmental assessment of energy potentials for anaerobic digestion production systems applied to the Netherlands. https://core.ac.uk/download/159430117.pdf (accessed on 30th September 2025).
9. Bhaskar, T., Chen, W. H., & Ong, H. (2020). Biomass processing for biofuels, bioenergy and chemicals. https://core.ac.uk/download/478135262.pdf (accessed on 30th September 2025).
10. Bricas, N. (2019). The scope of the analysis: Food systems. https://core.ac.uk/download/232200252.pdf (accessed on 30th September 2025).
11. Campbell, R. J. (2010). China and the United States—A comparison of green energy programs and policies. https://core.ac.uk/download/5130065.pdf (accessed on 30th September 2025).

12. Cleveland, C. J., Fox-Penner, P., Pollack, A., Walsh, M., et al. (2019). Carbon free Boston: Energy technical report. https://open.bu.edu/bitstream/2144/39216/2/CFB_Energy_Technical_Report_190514.pdf (accessed on 30th September 2025).

13. De Fence, J., Jensen, C., McIntyre, S., Munday, M., et al. (2011). Reconsidering the calculation and role of environmental footprints. https://core.ac.uk/download/19767453.pdf (accessed on 30th September 2025).

14. Devogelaer, D., & Gusbin, D. (2025). Working paper 16-09–EU energy/climate package and energy supply security in Belgium. https://core.ac.uk/download/pdf/6537844.pdf (accessed on 30th September 2025).

15. Dwiputra, F., & Saputri, F. (2024). Analysis of renewable energy potential in South Sulawesi, Indonesia. *G-Tech: Jurnal Teknologi Terapan*. https://www.semanticscholar.org/paper/4dbf35a887320c167382e55f90a42db22bfec244 (accessed on 30th September 2025).

16. Elliott, D. (2003). *Sustainable Energy: Choices, Problems and opportunities*. https://core.ac.uk/download/479.pdf (accessed on 30th September 2025).

17. English, B. C., Jakus, P. M., Jensen, K. L., Menard, J., et al. (2025). Consumer preferences for electricity from bioenergy and other renewables. https://core.ac.uk/download/pdf/6408000.pdf (accessed on 30th September 2025).

18. Feng, W., Fridley, D., Karali, N., Khanna, M., et al. (2019). Energy and CO_2 implications of decarbonization strategies for China beyond efficiency: Modelling 2050 maximum renewable resources and accelerated electrification impacts. https://core.ac.uk/download/323308754.pdf (accessed on 30th September 2025).

19. Flavin, C. (2008). Low-carbon energy: A roadmap. https://core.ac.uk/download/71340916.pdf (accessed on 30th September 2025).

20. Geldermann, J., Kolbe, L. M., Krause, A., Mai, M., et al. (2015). Improved resource efficiency and cascading utilisation of renewable materials. https://core.ac.uk/download/151157330.pdf (accessed on 30th September 2025).

21. Georgakellos, D., & Rentizelas, A. (2014). Incorporating life cycle external cost in optimization of the electricity generation mix. https://core.ac.uk/download/19332248.pdf (accessed on 30th September 2025).

22. Gissi, E., & Siciliano, G. (2025). Biomass production and land use management in the Italian context: Regulations, conflicts, and impacts. https://core.ac.uk/download/pdf/6580176.pdf (accessed on 30th September 2025).

23. Gomberg, A., & Kruse, W. (2009). Growing Ohio's green energy economy. https://core.ac.uk/download/71351635.pdf (accessed on 30th September 2025).

24. Grzesik, M., Pszczółkowska, A., Pszczółkowski, W., Romanowska-Duda, Z., et al. (2012). *Biomass Production of Selected Energy Plants: Economic Analysis and Logistic Strategies*. https://core.ac.uk/download/71974358.pdf (accessed on 30th September 2025).

25. Guerra, J. B., Buch, L. F., & Amaro, N. (2025). *E-Learning: Sustainability, Environment and Renewable Energy, A Multinational Training Pilot Module at The Postgraduate Level*. https://core.ac.uk/download/pdf/9306808.pdf (accessed on 30th September 2025).

26. Hampl, N., Stoiciu, A., Szabo, E., Totev, S., et al. (2014). Assessing the disruptiveness of new energy technologies–An ex-ante perspective. https://core.ac.uk/download/35453965.pdf (accessed on 30th September 2025).

27. Ishido, H. (2025). *Financing Rural Energy Services in the Philippines: Global Environment and People-Centred Development as Public Goods*. https://core.ac.uk/download/pdf/6506213.pdf (accessed on 30th September 2025).

28. Israel Herrera Orozco, S., Banacloche, Y., Lechón, Y., & Domínguez, J. (2025). Assessing the socioeconomic and environmental impact of hybrid renewable energy systems for sustainable power in remote Cuba. *Sustainability*. https://www.semanticscholar.org/paper/7d34f8411eaabdfede777d21fbb889f45b92b97f (accessed on 30th September 2025).

29. James, O. O. (2013). Improving carbon and process utilization for biomass-to-liquid fuel operations. https://core.ac.uk/download/16436077.pdf (accessed on 30th September 2025).

30. Juchelková, D., Raclavská, H., Sassmanová, V., Tun, Z. M., et al. (2018). Utilization of biodegradable wastes as a clean energy source in the developing countries: A case study in Myanmar. https://core.ac.uk/download/161968078.pdf (accessed on 30th September 2025).

31. Kamenský, P. (2008). Renewable energy sources in CR. https://core.ac.uk/download/30290482.pdf (accessed on 30th September 2025).

32. Koh, J., & Stremke, S. (2009). Sustainable energy transition: Properties and constraints of regenerative energy systems with respect to spatial planning and design. https://core.ac.uk/download/pdf/29249814.pdf (accessed on 30th September 2025).

33. Koj, J. C., Zapp, P., Wieland, C., Görner, K., & Kuckshinrichs, W. (2024). Life cycle environmental impacts and costs of water electrolysis

technologies for green hydrogen production in the future. *Energy, Sustainability and Society*. https://www.semanticscholar.org/paper/3ec668338b0cde9ba2404229d414a1508a70ca81 (accessed on 30th September 2025).

34. Kryk, B. (2015). Green jobs–good practices. https://core.ac.uk/download/71980475.pdf (accessed on 30th September 2025).
35. Latorre, M. P. (2001). The contribution of bio-energy to meeting renewable energy sources: European Union objectives. https://core.ac.uk/download/5074141.pdf (accessed on 30th September 2025).
36. Maity, S. K. (2015). Opportunities, recent trends and challenges of integrated biorefinery: Part II. https://core.ac.uk/download/38677882.pdf (accessed on 30th September 2025).
37. N/A. (2005). *Renewables 2005: Global Status Report*. https://core.ac.uk/download/71338835.pdf (accessed on 30th September 2025).
38. N/A. (2008). *Charting Our Own Course: Today's Challenges, Tomorrow's Opportunities, December 2008*. https://core.ac.uk/download/11350328.pdf (accessed on 30th September 2025).
39. N/A. (2008). *Powering Europe Sustainably: A Discussion Paper on the Potential Role of Biomass in the EU*. https://core.ac.uk/download/71353894.pdf (accessed on 30th September 2025).
40. N/A. (2010). *Who's Winning the Clean Energy Race? Growth, Competition and Opportunity in the World's Largest Economies: G-20 Clean Energy Factbook*. https://core.ac.uk/download/71355815.pdf (accessed on 30th September 2025).
41. Riedacker, A. (2025). A global land use and biomass approach to reduce greenhouse gas emissions, fossil fuel use and to preserve biodiversity. https://core.ac.uk/download/pdf/6461240.pdf (accessed on 30th September 2025).
42. Shrestha, A., & Timilsina, G. R. (2025). *Biofuels: Markets, Targets and Impacts*. https://core.ac.uk/download/pdf/6280431.pdf (accessed on 30th September 2025).
43. Udele, N. O., Oladeinde, J. O., Smart, E. E., Goni, A. A., Idoko, D., Musa, J. M., & Agboola, P. O. (2025). Advancements in renewable energy technologies and their impact on environmental sustainability. *World Journal of Advanced Engineering Technology and Sciences*. https://www.semanticscholar.org/paper/b682c7d38b86eea9c6c129e3b31ec21cd92a897d (accessed on 30th September 2025).

44. Van Eetvelde, G., Vandevelde, L., Vandoorn, T., Vansteenbrugge, L., et al. (2013). Renewable energy balancing with thermal grid support. https://core.ac.uk/download/55732711.pdf (accessed on 30th September 2025).
45. Wysokińska, Z. (2012). Mutual dependence between sustainable energy and sustainable agriculture policies–From the global and European perspective. https://core.ac.uk/download/71974354.pdf (accessed on 30th September 2025).
46. Zolin, M. B. (2025). *Biomass Energy: New Opportunity for Agriculture and Marginal Areas?* https://core.ac.uk/download/pdf/6234092.pdf (accessed on 30th September 2025).

CHAPTER 8

POWER ELECTRONICS FOR RENEWABLE INTEGRATION

Contents

INTRODUCTION

The integration of renewable energy sources (RES) into the current grid has become an energy planning priority in today's world. As global energy needs continue to rise in parallel with environmental concerns, the transition towards cleaner energy sources becomes increasingly unavoidable. RES is one among these solutions that can prove sustainable and environmentally stable. However, these sources of energy possess certain characteristics, the primary reason being their intermittent nature, which makes integration into the power grid challenging. The challenge of reliable power availability is one such issue, and it calls for effective means of power management and integration of different power sources. Power electronics is critical in power management of such challenges by providing technological leverage to convert, control, and deal with electrical power. This field of research is essential to creating a stable energy infrastructure that integrates traditional systems with innovative, innovative renewable sources in a way that enhances the stability and resilience of the energy supply.

Figure 8.1. *Power electronics in renewable energy.*

Source: https://www.electropages.com/storage/temp/public/3e6/60f/dc9/power-electronics-in-renewable-energy__750.jpg.

Power electronics advancements play a critical role in high-efficiency and reliable operation of renewable energy systems. Recent research has led to the development of highly complex converter topologies, which are specifically aimed at the unique operating conditions of RES. Such multi-port converters enable the integrated utilization of various forms of energy to allow for concurrent transition from generation to usage. That these systems are efficient is also significant, with increased losses threatening to undermine the environmental benefit that comes with renewable generation. Research is thus underway to further improve converter design in an effort to curb inefficiencies while maximizing capability to handle variable loads. Moreover, the design of these converters must consider not only component count and size but also overall system performance over a range of operating conditions. The increased complexity of energy demand points to the necessity for innovative power electronic solutions that provide reliable, high-performance systems to cater to renewable energy integration (Rehman et al., 2017). Universities are also making an effort to study power electronics and renewable energy systems in preparation for future engineers. For instance, the *Universitat Politècnica de Catalunya* has introduced tailor-made programs that focus on imparting a strong understanding in critical areas such as the Bachelor's Degree in Energy Engineering. This is an education that instructs particularly on the sizing and design of renewable energy systems and focuses on interdisciplinary learning for optimal future energy solutions. This structure of learning is important as it produces a new generation of engineers that are capable of developing innovative power electronics solutions for connecting the renewables. By means of an understanding of theoretical and practical fundamentals, students can contribute significantly to the creation of advancing technologies that enhance energy system sustainability and resilience. Such a convergence of education with industry demand is a testament to the active reaction to global energy challenges, leading to more efficient and resilient power infrastructure (Martínez et al., 2017) (Figure 8.1).

Definition of Power Electronics

The power electronics definition is the technology to supply power conversion and control of electrical energy efficiently. Power electronics, in short, is the science and application of semiconductor devices to control high-voltage and high-current electricity power. Power electronics is important to enhance the integration and performance of renewable energy systems, which tend to use advanced methods of power conversion such as inverters, converters, and rectifiers. These devices help in the regulation of voltage levels, AC-to-DC conversion, and ensuring efficient utilization of renewable energy sources like solar and wind within the existing power grid infrastructure. By providing proper control over electrical energy, power electronics plays an important role in the optimization of energy usage and overall reliability and sustainability of renewable energy projects. In the utilization of renewable energy, power electronics enables free interfacing of alternative forms of energy with traditional energy supply systems. Microgrids, for instance, which highlight local energy production and use, benefit a great deal from power electronics innovation. The technology enables such systems to be transferred between standalone mode and grid-connected mode, hence enabling energy resilience and stability. With supporting federal and local laws driving such undertakings in the US, microgrid-based projects are gaining traction. Power electronics' openness to advanced control methods enables techniques capable of dynamically interacting with the electricity market, thus optimizing energy distribution and load management when integrating solar and wind renewable resources in microgrid integration (Bao et al., 2018). Besides, as the field of power electronics has developed, intelligent materials and systems are increasingly emphasized with the aim of enhancing functionality in energy applications. It is mirrored in the design of responsive, adaptive smart materials that significantly influence the efficiency of energy conversion technology. Through the integration of organic thin film electronic devices (OFEDs) and bio-responsive materials, possibilities expand for new applications in renewable energy. These technologies not only revolutionize the efficiency of photovoltaics but also lay the groundwork for visionary designs in systems of energy that are centered on sustainability and human interaction. By depicting these materials, the integration of power electronics in renewable energy technology has the capability to power emerging technologies that respond to current environmental and consumer needs (Oliver et al., 2009) (Figure 8.2).

Figure 8.2. *Power electronics.*

Source: https://media.licdn.com/dms/image/v2/D5612AQFsLAYRf-lydA/article-cover_image-shrink_600_2000/article-cover_image-shrink_600_2000/0/1701445952768?e=2147483647&v=beta&t=j45jt-txXfpp3EUufy2_OCBPlPHuTJnHUIZ-rGFzWZlY.

Importance of Renewable Energy Integration

Integration of renewable sources into the conventional power grid is most critical to sustainability and energy security. As world energy demand rises, fossil fuel reliance is beset with immense environmental risk and ambiguity of energy supply. Adding renewables, particularly solar and wind energy, bridges the gap between supply and demand, thereby enhancing grid stability. Power electronics plays an integral role in this integration by way of efficient energy conversion and management systems. Advanced converters and inverters developed by power electronics technology provide a seamless integration of the renewable sources, with an assurance that generated power can be effectively utilized and stored. The developments in such systems are critical in overcoming the constraints around variable energy generation, paving the way to a robust and more resilient energy network. Besides, the economic viability of renewable energy systems has improved substantially, making them good alternatives for numerous applications. According to recent research, solar energy systems, particularly photovoltaic (PV) technologies, have emerged as viable options for residential, business, and industrial uses. The decline in the price of PV modules and installation, along with power electronics used for energy transformation, has significantly contributed to market penetration. Technologies that enable grid-connected applications to have also made it easier to integrate solar power into installed systems, cut down on operational expenses, and assuage climate change. With emphasis in current research, understanding the dynamic interactions of PV technology and power electronics will further enhance system efficiency and optimize future installations, thereby ushering in an energy future that is sustainable (Abu-Rub et al., 2017). With the advancement of the energy sector, one can never lay too much stress on the importance of innovative materials and technology in the utilization of renewable energy. The production of intelligent materials such as organic photovoltaics and smart textiles offers promising avenues for increasing functionality and performance in energy uses. The advanced materials can be embedded in everyday products, providing additional avenues for energy generation and utilization. The union of electro-optical and bio-responsive materials, in addition to enhancing user interaction, also adheres to principles of ambient living where technology becomes a seamless part of living at the individual and urban levels without intrusion. Focusing on these state-of-the-art designs, the range of renewable energy systems is extended beyond traditional applications to an even greater domain. This systems-oriented thinking of design and integration will create broader deployment of renewables, bringing in much-needed change to innovative solutions that harness the complication of energy demands today (Oliver et al., 2009).

8.1. FUNDAMENTALS OF POWER ELECTRONICS

The function of power electronics in the integration of renewable energy systems cannot be overstated. Power electronics refer to the technology of converting and controlling electric power with maximum efficiency by employing switching electronic devices. Power electronics have a crucial role in enhancing the performance as well as reliability of renewable energy technologies, such as wind generators and solar panels. Through management of output current and voltage, power electronic converters enable renewable energy systems to connect to electricity grids reliably. More significantly, concepts such as converter topologies and pulse width modulation are critical towards achieving the maximum output of energy with minimum losses. Integration appropriately through power electronics not only improves the efficiency of a system but also enables grid stability, enabling there to be a smoother transition to reliance on renewable energy. With the growing demand for sustainable energy options, the power electronics education system will have to reform. Recent shifts in curriculum at universities such as the Technical University of Catalonia lead the way in such a reform. The inclusion of specialized courses on renewable energy systems in engineering curricula is aimed at equipping the future generation of engineers with power electronics expertise.

Figure 8.3. *Power electronics fundamentals.*

Source: https://kajabi-storefronts-production.kajabi-cdn.com/kajabi-storefronts-production/file-uploads/themes/2156580793/settings_images/678d5d0-c835-4187-62-4237751ee776_vBYXOU9xSI6c1PCOLCqi_file.jpg.

These training programs are vital as they address the complexities involved in renewable energy technologies, particularly in terms of sizing and designing various systems. Blending theoretical principles with practical uses, students can acquire a sophisticated understanding of how power electronics functions in renewable energy systems, therefore enhancing innovation in the sector. This academic approach ultimately culminates in the commercialization and utilization of renewable energy technologies in real-world applications, alleviating both energy requirements and environmental issues. The dynamic context of solar technology evolution highlights the increasing role of power electronics in renewable energy integration. Advances in photovoltaic (PV) systems have established them as a feasible energy solution for various applications, necessitating efficient power conversion technologies. The power electronics converters are required to optimize the performance of the solar installations so that they are capable of efficiently converting and managing the variable output of the solar panels. Major trends in the PV industry and technology of these systems require power electronics research and development to manage specific challenges like installation and grid compatibility. As demonstrated by current studies, the destiny of solar energy lies in the ongoing improvement of connected conversion systems, a domain in which power electronics is of key importance, underlining the need for industries and universities to work in concert more effectively in meeting rising energy demands sustainably (Abu-Rub et al., 2017; Martínez et al., 2017) (Figure 8.3).

8.1.1. Basic Components of Power Electronics

The field of power electronics is an important interface between the renewable sources of energy and their efficient usage in electrical systems. Power electronics has at its core several basic elements, which include power semiconductor devices, passive components, and control systems. Power semiconductors, such as diodes, transistors, and thyristors, are at the core of converting, controlling, and delivering electrical power. They facilitate the conversion of direct current (DC) delivered by renewable energy sources like solar panels and wind turbines into alternating current (AC), which is necessary for grid supply. This feature is significant as renewable energy sources possess the potential to deliver power in an intermittent manner, at times necessitating reliable conversion procedures in order to ensure stable and efficient power delivery (Rehman et al., 2017). In addition to power semiconductor devices, passive elements like inductors, capacitors, and transformers are essential to power electronic systems. Inductors and capacitors are employed for filtering and smoothing voltage and current waveforms and hence for enhancing the quality of supplied power. Transformers are employed for voltage stepping-up or stepping-down as per requirement to enable smooth interface between the renewable energy systems and the local grid system. These passive elements have a crucial function in making the power conversion operations stable and reliable, as their performance directly affects the operation of the interconnected system. Correct design and utilization of these elements will enhance the general efficiency of power electronic interfaces utilized in renewable energy systems (Rehman et al., 2017). Finally, control systems round out the power electronics building blocks by ensuring the system operates in its maximum state regardless of shifting conditions. Advanced control strategies, particularly those employing digital signal processing, enable very precise voltage and current regulation required when handling intermittent renewable resources that vary with the surroundings. An example here is the design of synchronization structures that determine harmonic frequencies in three-phase grid-connected systems, which is essential in maintaining power quality and energy efficiency. These cutting-edge control techniques are imperatively needed, given the complexities in integrating multiple renewable sources into a single energy system. As research continues to enhance these underlying building blocks, the ability to create more efficient and robust power electronics systems will increase, thus advancing the utilization of renewable energy (Figure 8.4).

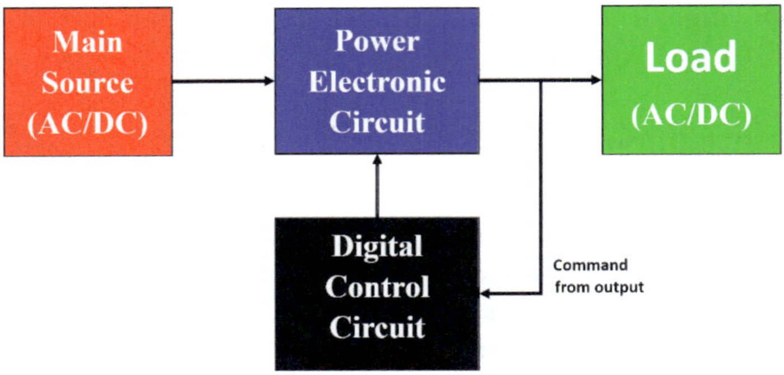

Figure 8.4. *General block diagram of power electronic circuit.*

Source: https://instrumentationtools.com/wp-content/uploads/2023/08/Basic-components-of-power-electronics.png.

8.1.2. Types of Power Converters

Power electronics is a key discipline in the effective incorporation of renewable energy resources, particularly through power converter operations. Power converters are commonly classified into two general types: DC-DC converters and DC-AC converters. DC-DC converters, such as buck converters and boost converters, play a pivotal role in bridging the voltage levels to adapt to the operational requirements of various energy systems in order to realize optimal output from renewable energy sources like photovoltaic cells and batteries. With effective conversion of output from photovoltaic cells, these converters enhance the overall energy output and stabilize the electric grid. Conversely, DC-AC converters, which are also referred to as inverters, are used for interfacing renewable energy systems and the current AC grid. This interfacing is critical to allow consumers to consume renewable energy simultaneously while ensuring grid stability and reliability. The increasing role of high-power converters has made it ever clearer in recent years with the increased interconnecting of renewable energy resources to the grid. The rapid growth in installed photovoltaic (PV) and wind generation plants reflects the need for robust power electronics able to efficiently handle and control such energy inputs. The intricacies of renewable energy systems, such as fluctuating power generation and dynamic load profiles, necessitate sophisticated control schemes in these converters. Special caution needs to be taken in design and control system tuning to cater to less predictable energy outputs. As expounded in the current research, the optimal tuning of these converters has been the priority to ensure their operation under dynamic conditions, hence they are efficient and reliable in energy integration as seen from the recent advancements in AC-DC power converters (Alloza et al., 2016). Additionally, power converter technology has also provided several solutions to renewable energy integration challenges. Specifically with regard to new developments in photovoltaic technology and adjacent conversion systems, the continuing evolution in power electronics has been highlighted in this context (Abu-Rub et al., 2017). Innovations in converter design now facilitate intelligent control approaches with plug-in enabled integration of generation sources into today's grid. These technologies are especially relevant since energy storage solutions rely greatly on the efficiency of power converters to store delivered energy for future use. The application of such sophisticated power electronics not only solves current installation problems but also makes renewable energy systems capable of more effectively meeting future energy demands, further securing the crucial role of power converters towards sustainable energy solutions (Figure 8.5).

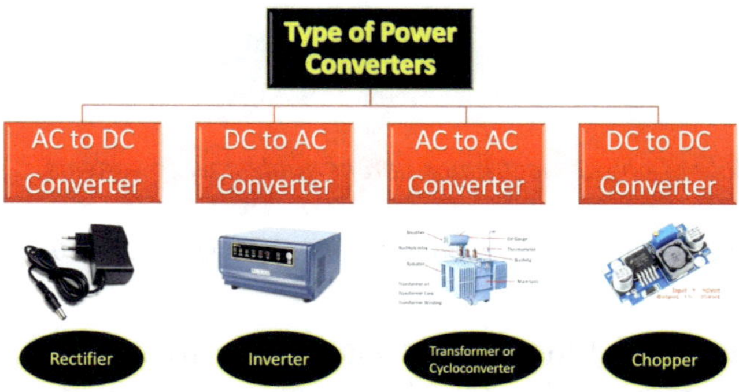

Figure 8.5. *Types of power converters.*

Source: https://i.ytimg.com/vi/ydNU7hTmS2s/maxresdefault.jpg.

8.1.3. Role of Power Electronics in Energy Systems

Power electronics are of critical significance in maximizing the performance and efficiency of energy systems, particularly in the integration of renewable energy. With renewable means of energy, such as solar and wind power, on the ascendancy, the requirement for effective power conversion and management has thus increased. Power electronics technology enables variable direct current (DC) from renewable sources to be converted to alternating current (AC), which is beneficial for grid interfacing. The conversion is important since traditional power systems operate primarily on AC generation and therefore it becomes difficult to incorporate multiple renewable sources into the system. By implementing advanced control algorithms, power electronic devices can further optimize the output from renewable sources, obtaining the maximum power in variable environmental conditions, demonstrating their fundamental importance to the stability and dynamics of an energy system (Xiao et al., 2024). Power electronics have a further role in efficiency and reliability in wind energy generation.

Figure 8.6. *Role of power electronics in renewable energy systems.*

Source:https://cdn.everythingpe.com/community/resW_638700973102649299.png.

Utilizing variable speed drives allows wind turbines to operate at optimal levels, responding to varying wind speeds. This not only improves turbine energy output but also facilitates grid connection through sophisticated power regulation and fault condition protection. Furthermore, the ability of power electronics to regulate the reactive supply of power is also a major contributor to grid stability. Consequently, this technology not only enhances the performance of independent wind power systems but also contributes to the overall robustness of the electrical grid towards the reception of renewable sources (Blaabjerg et al., 2023). The more complicated the acceptance of huge renewable energy systems becomes, the more power electronics need to be developed to counteract obstacles presented by conventional grids. Solar power systems also identify the essential role of power electronics in integrating renewable energy. Photovoltaic inverters and Maximum Power Point Tracking (MPPT) technologies assist in maximizing the energy extraction from solar panels and ensuring credible grid synchronization. As solar energy generation gains popularity, the reliability and efficiency provided through power electronics are essential to push solar systems to the next utility level. Not only do such inverters enable energy conversion but also the integration of the storage batteries, hence load balancing and better overall energy management. Such a dual role of the power electronics in facilitating the simultaneous utilization of the solar energy and storing excess for the future enables them to be the central devices in countering the energy system complexities in the face of increased reliance on renewable energy sources (Xiao et al., 2024) (Figure 8.6).

8.2. RENEWABLE ENERGY SOURCES

The growing significance of renewable energy cannot be overstated, particularly in terms of addressing global energy demands and reducing the effect of climate change. A good case in point is solar energy, which has demonstrated vast potential for large-scale deployment due to its scalability and decreasing prices of photovoltaic (PV) technologies. With the increase in power electronics, solar systems are now a potential substitute for various applications—ranging from home rooftops to industrial-scale commercial operations. This trend not only contributes towards energy self-sufficiency but also towards reducing the level of greenhouse gas emissions. The present work discusses the intricacy of the solar energy conversion systems, highlighting the necessity of efficient power electronics to tap, convert, and feed solar energy into the grids and ultimately gear towards a sustainable energy future (Abu-Rub et al., 2017). In parallel, in tandem with solar technology, the proper utilization of renewable energy resources is reliant on sophisticated control mechanisms that provide maximum energy supply to the grid. Since renewable generation has the potential to be intermittent, having strong systems controlling both active and reactive power flow is very important. Advanced control techniques, such as using a virtual flux approach, boost this ability, with the remote control of power injection at the Point of Common Connection (PCC) being possible. This technology enables one to accurately control energy delivery, managing energy variability issues, and maintaining grid stability. Through overcoming traditional limitations of distance and infrastructure, such power electronics solutions facilitate seamless integration of diverse renewable energy sources into the energy mix, thereby enhancing the reliability of modern electrical grids (Garc et al., 2018). The constantly evolving power electronics is thus at the core of renewable energy sources' integration into an effective energy future. Technological advancement in converters improves the efficiency of energy conversion and allows for integration of various sources to provide optimized power. Integration of diverse renewable technologies, such as wind, hydro, and solar, provides an opportunity to diversify energy portfolios and lower the risk of reliance on one source. With ongoing research, intelligent systems will continue to increase operational reliability and flexibility, which will lead to a more robust energy grid. In the end, successful collaboration between researchers, engineers, and policymakers will be essential to pushing forward these technologies while ensuring compatibility with environmental objectives and long-term energy security (Abu-Rub et al., 2017) (Figure 8.7).

Figure 8.7. *Renewable energy sources.*

Source: https://www.ecoredux.com/wp-content/uploads/environmental-sustainability.jpg.

8.2.1. Overview of Renewable Energy Types

The speedy transition to renewable energy is imperative in an attempt to address global climate change and reduce the consumption of fossil fuels. Among the numerous types of renewable, solar power is the most notable based on its accessibility and technological advancements. Solar power is primarily delivered via photovoltaic (PV) systems, which are used to directly convert sunlight into electricity. The price competitiveness of solar installations has rendered this technology a competitive option for several applications, ranging from residential to commercial and industrial uses. As highlighted in recent studies, the improvement of PV module efficiency and cost is of higher priority for large-scale application. In addition, the incorporation of sophisticated power electronics converter systems is essential to optimize the performance of solar installations. These systems allow for control of electrical output to grid specifications, offering a stable and secure source of power, and thus underscore the importance of these advancements in this technology within the broader context of bringing renewable energy on board (Abu-Rub et al., 2017). Wind power, another prominent renewable energy, harnesses the kinetic energy generated as a result of wind currents for the generation of electricity in turbines. This form of power constitutes a significant portion of the energy supply in the world, particularly where consistent wind patterns prevail. The production of energy has been significantly improved by the technological advancements in turbine technology, particularly for offshore wind farms. However, in order to utilize this capacity to the maximum, proper integration into existing power systems requires the application of power electronics. They assist in managing variability and intermittency associated with wind power generation so that more stable delivery of power is ensured and the grid becomes more stable.

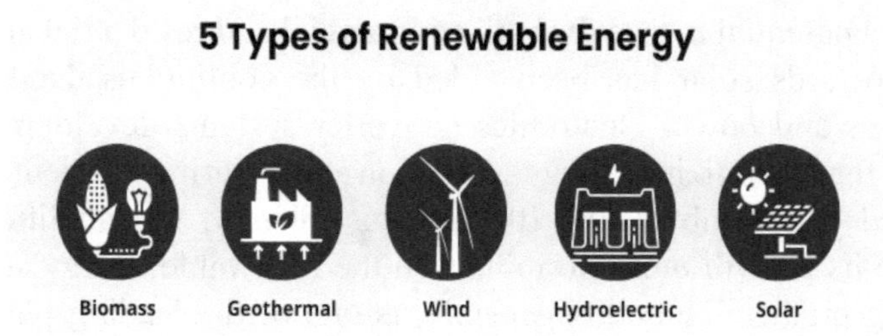

Figure 8.8. *Five different types of renewable energy.*

Source:https://safetyculture.com/_next/image/?url=https%3A%2F%2Fwp-website.safetyculture.com%2Fwp-content%2Fuploads%2Fsites%2F3%2F2024%2F05%2F5-Types-of-Renewable-Energy.png&w=1920&q=75.

Therefore, investment in smart grid technology becomes crucial because it enables real-time control and monitoring of energy flow. Through the utilization of advanced systems that allow for the co-integration of wind energy with other renewals, we are able to build a more sustainable and secure energy future (Copeland et al., 2019). Hydropower is perhaps one of the oldest and most established types of renewable energy and utilizes the gravitational power of flowing water to generate electricity. As a renewable source of power, hydropower is particularly desirable for its potential to provide base-load power with consistent and dependable supply of energy. However, the ecological price of big dam projects has led to the re-appraisal of hydropower methods, bringing attention to small and run-of-river plants. These cause less damage environmentally without being outside the sphere of the renewable energy input. For power electronics, connectivity between hydropower and the modern grid network requires the application of sophisticated technologies that allow for dependable energy supply. Consequently, power electronic converters become central in energy generation control as it varies with water flow, ultimately enhancing hydropower reliability and efficiency towards achieving an integrated system for renewable energy (Abu-Rub et al., 2017) (Figure 8.8).

8.2.2. Characteristics of Solar and Wind Energy

The characteristics of solar energy are primarily established by its availability, technology, and environmental considerations. Solar energy has significantly advanced as it is capable of harnessing the energy from the sun to create usable electricity through photovoltaic (PV) cells. This characteristic not only serves the world's sustainability needs but also reduces the utilization of fossil fuels. The advancements in solar technology, particularly in PV systems, have made solar power a potential means for both commercial and residential applications. The trend towards solar has been aided by the continuous decrease in PV module prices and power electronics converter systems development, which are required for maximizing energy extraction and ensuring efficient integration into the grid (Abu-Rub et al., 2017). Consequently, the qualities of solar power make its growth more prominent in the renewable energy scene. Wind power, being predictable and expansible, is one of the leading players in the renewable energy industry. Relying on wind turbines converting wind kinetic energy into electrical energy, this renewable energy is expansible in installation levels, ranging from small turbines for sole use to large wind farms that can provide electricity to communities at large. One of the distinguishing strengths of wind power is that its operating cost after installation is low, and therefore it is economically sound everywhere except in very few locations. Wind power is also, however, subject to prevailing local wind conditions, which can introduce uncertainty in the supply of energy. To counteract this intermittency, efficient power electronic interfaces are essential, allowing for the integration of electricity produced by the wind into smart grid applications and ensuring an availability of energy supply (Guarnieri et al., 2016). The attributes of wind energy, therefore, highlight its promise as a reliable and scalable source of renewable energy. These wind and solar energies are tied into existing power grids through advancements in power electronics, which serve as the translator between consumers and renewable resources. The diversity of the energy sources presents unique challenges and opportunities for integration. Solar energy, for instance, is available in the daytime but solar radiation and sunlight do not always match demand. Wind energy, in contrast, can provide energy during other periods and therefore establish a complementarity link between the two sources. Application of high-power electronics not only stabilizes the grid but also optimizes the supply of energy from such renewables, riding over intermittency and load balancing issues. Further, with the development of the technology for energy storage, the role of power electronics in managing energy supply and enabling smart grid attributes becomes increasingly critical. Together, solar and wind energy, supported by innovative power electronics, pave the way towards a clean energy future (Figure 8.9).

Figure 8.9. *Regional disparities and features of solar and wind energy potential in Bulgaria.*

Source: https://researchleap.com/wp-content/uploads/2021/10/renewable-enegy.jpeg.

8.2.3. Challenges in Harnessing Renewable Energy

The integration of renewable resources into the power grid raises numerous issues that have to be addressed to optimize their share in the overall energy supply. Natural variability of renewable resources, particularly solar and wind resources, is a primary problem. They are naturally intermittent in nature, resulting in fluctuations in energy output that can destabilize the grid. Therefore, power electronics systems will have to be designed to handle these fluctuations efficiently, facilitating smooth integration into conventional power systems. In addition, the need for a high-strength communication infrastructure for these power electronics is essential. There is a requirement for an advanced communication network to facilitate real-time monitoring and control, whereby power from renewable sources can be effectively utilized and matched with demand in industry and domestic households, as highlighted by Qian et al. (2013). The second crucial challenge is with regard to the technologies that would be needed to facilitate a shift towards renewable energy. Conventional innovation activity generally focuses on incremental advances to established technologies that, in most instances, are not nearly as large as the needs of a mass-scale switch to renewable energy systems. To achieve quantitative reductions in greenhouse gas emissions, it is important to become more revolutionary in new energy carrier technologies. The limitations of current strategies imply that enhancing the performance of renewable alternatives will not suffice; what the case requires are mission-oriented programs aimed at fundamentally transforming energy production, supply, and consumption systems. Such an identification underscores the need for new policy steps and strategies that positively impel transformative power electronics advancements in the entire integrating process, underscored by Smith (2025). Finally, the extensive use of renewable energy is typically discouraged by economic and infrastructural limitations. The high upfront capital for renewable technology serves as a fundamental discouragement for both consumers and producers. Moreover, the prevailing infrastructure may not be sufficiently adapted to handle the special requirements posed by an energy mix diversified to encompass renewables. The deployment of a smart grid with intelligent power electronics to maximize energy transfer from any source is instrumental to overcoming these barriers. With millions of devices and consumers connected through a smart grid, reliability and security in the energy system become ever more critical. Overcoming all these demands requires an overarching approach that weaves together technological innovation and regulatory frameworks to establish a supportive environment for sustained growth in the renewable energy industry, guided by suggestions from Qian et al. (2013) (Figure 8.10).

Figure 8.10. *Renewable energy transition challenges: Harnessing innovation for a sustainable future.*

Source: https://prismecs.com/storage/blogsImages/1707743694.png.

8.3. POWER ELECTRONICS IN SOLAR ENERGY SYSTEMS

Use of power electronics in solar energy systems is a critical area to ensure higher efficiency and reliability. Due to growing needs for renewable energy worldwide, photovoltaic (PV) systems are now a viable technology for use in residential rooftops to industrial installation. The process of converting sunlight into usable electric energy is largely a function of highly developed conversion mechanisms that stabilize the output voltage and current of the solar panels. Advances in power electronics, particularly inverter technologies and energy storage devices, have helped improve integration of solar energy into the grid. This advancement not only facilitates seamless operation within intra-connected systems but is also compatible with regulatory standards and utility requirements, which reveals the potential of power electronics to propel solar energy technologies towards more comprehensive adoption (Abu-Rub et al., 2017). A critical aspect of power electronics used in solar systems is the implementation of Maximum Power Point Tracking (MPPT) technology. This technology makes it possible for systems to constantly optimize the solar panel energy output by adjusting the electrical operating point. Because solar irradiance conditions change due to environmental conditions such as shading or cloud cover, MPPT algorithms allow solar power systems to operate at maximum power. This is significant particularly because standard PV systems that lack MPPT can lead to high energy losses. The engineers have been knowing about this problem and designing more stable designs that incorporate MPPT under inverter systems for maximum performance even in suboptimal conditions (Mensah et al., 2004).

All these technological advancements contribute considerably to making the whole solar installation more stable, further opening the possibilities for greater acceptance and faith in the use of solar power. Projected into the future, power electronics will be driving innovation in solar energy systems as efficiency and integration challenges are being overcome. Engineers are developing new materials for PV cells and next-generation inverter technologies that promise to increase performance while lowering the cost. Furthermore, streamlining system integration with modular architecture and smart grid capabilities will allow for more versatile, scalable solar power solutions. With evolving energy requirements, the ability of power electronics to suit and optimize solar systems will be pivotal. The rising focus on sustainability and lowering carbon footprints highlights the importance of continuous research and development in the field, so power electronics can remain superior in terms of renewable energy integration and progress towards a cleaner energy future (Abu-Rub et al., 2017).

8.3.1. Function of Inverters in Solar Applications

Inverters are crucial to solar uses, as they serve as the link between solar photovoltaic (PV) systems and the electric grid. By converting solar panel-generating direct current (DC) into alternating current (AC), inverters make it possible for the integration of solar power into the current power grid. This is so because most of the electrical devices and appliances in the commercial and residential areas operate using AC power. Besides, the inverter optimizes the output by using maximum power point tracking (MPPT) methods in order to ensure that the solar panels operate in their most efficient levels throughout varying environmental conditions. Such functionality is at the core of optimizing overall energy harvesting from solar installations and hence providing renewable energy input to the grid and towards a sustainable energy market (Arghandeh et al., 2013).

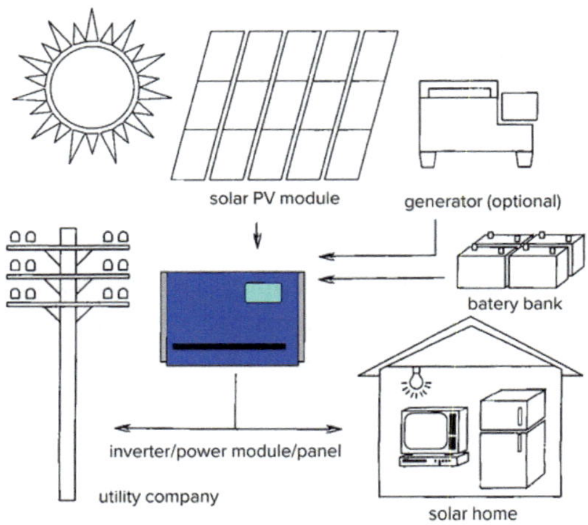

Figure 8.11. *The role of an inverter in a solar electric system.*

Source: https://cdn.shopify.com/s/files/1/0714/3365/0395/files/role-of-an-inverter.jpg?v=1724799662.

In solar applications, inverters also contribute to grid stability and reliability. As the penetration level of renewable energy resources like solar increases, the management of voltage profile in the grid becomes more complex. Inverters assist in controlling the voltage, which is significant for the purpose of maintaining power quality delivered to the consumers. Inverters today are typically provided with advanced features that include grid support functionalities, i.e., reactive power supply and frequency control. Both these features are of utmost importance since more levels of distributed energy resources are being connected to the grid. Coping with the challenge of high penetration of renewables and electric vehicles integration is imperative to prevent voltage variation and ensure uninterrupted supply (Arghandeh et al., 2013). Along with this, the design of inverters is evolving with the enhancement of technology, resulting in enhanced performance for solar uses. Designs lately involve the two-stage inverter system, where a boost converter first increases voltage from PV modules and then a single-phase full-bridge inverter supplies AC output. This technique not only ensures effective energy conversion but also allows for stand-alone operation, free of the need for complex grid connections. The transformer-less inverter types have additional benefits of reduced costs and sizes, although they do require careful attention in safety protection against leakage currents. With continued research in the area of power electronics, the application of inverters will increasingly find their place in the transition toward a sustainable energy future (Figure 8.11).

8.3.2. Maximum Power Point Tracking (MPPT)

The integration of renewable sources into the grid needs sophisticated solutions that optimize their efficiency, and Maximum Power Point Tracking (MPPT) is one such technology. MPPT is employed particularly in photovoltaic (PV) systems to optimize energy harvesting under varying conditions of solar illumination and temperature. Through constant readjustment of the modules' electric operating point, MPPT algorithms ensure harvesting of the maximum available power. This role is rudimentary especially in environments where solar irradiation could vary throughout the day, impacting the energy output of the PV system considerably. Therefore, the use of MPPT not only improves the reliability of solar systems but also their economic viability, resulting in improved integration of renewable sources into the grid (Jayasankar et al., 2023). The development of MPPT methods has seen the implementation of various optimization approaches to further enhance power harvesting from solar panels. Some of them involve the application of Genetic Algorithm with Partial Swarm Optimization (GA-PSO), which is special in its versatility to enhance the tracking process. This algorithm features the greatest advantages of both genetic algorithms and particle swarm optimization to end up with more stable performance in variable environmental conditions. The use of these new control techniques assists PV systems in maintaining their optimal power output despite the occurrence of unbalanced voltage levels or sudden irradiance changes. In essence, clever utilization of MPPT not only guarantees maximum energy harvesting but also enhances the overall resilience of renewable energy systems and makes them position themselves as stable players in modern-day energy supplies (Jayasankar et al., 2023). Additionally, as solar energy systems gain commonality in residential and commercial applications, it becomes more imperative to comprehend the interaction of MPPT and power electronics. The DC-to-AC conversion technology also has an effect on the efficiency of MPPT. High-performance power electronic converters allow solar power to be integrated efficiently into the grid while addressing challenges that are intrinsic to issues of power quality such as harmonics and voltage fluctuations. As recent literature has compiled, the ongoing research in PV technology and associated power electronics highlights the necessity to develop advanced conversion systems that will be capable of keeping pace with a growing solar market. Through constant innovation in the field of MPPT and power electronics, the stakeholders can enhance grid resilience and enable the energy transition towards a cleaner energy industry (Figure 8.12).

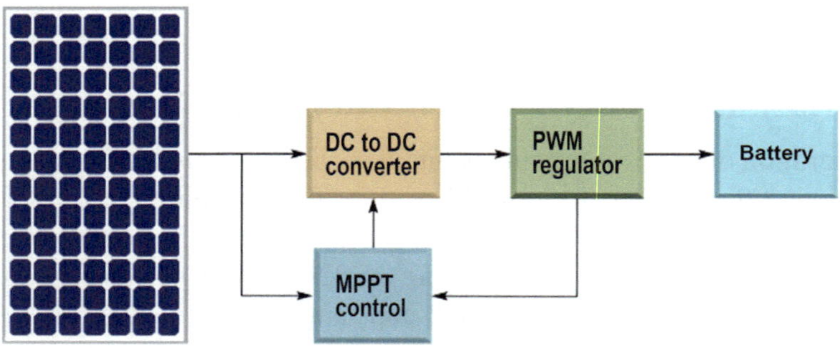

Figure 8.12. *Maximum power point tracking (MPPT) charge controller circuit diagram.*

Source: https://electricalacademia.com/wp-content/uploads/2018/07/Figure-2-6.png.

8.3.3. Grid Integration of Solar Power

Integrating solar power into the grid must be done with a clear grasp of photovoltaic (PV) systems and the issues involved in their integration. As the need for clean energy sources is growing, engineers are turning to PV systems both for household and commercial purposes. Nevertheless, even after the technology advancements, the smooth integration into the grid is a complex matter. One of the main features is the implementation of Maximum Power Point Tracking (MPPT) technology, which ensures maximum energy output from solar arrays performing in variable conditions. Allowing systems to be at optimum efficiency, MPPT is one of the main features to reduce the loss through shading and other conditions. This effectiveness not only contributes to the stability of the system but also allows for a more stable energy supply to the grid, paving the way for commercial-scale applications of solar technology in many areas (Mensah et al., 2004). At the grid integration level, the power electronic converters are the integration infrastructure of solar generation to the overall electric network. Converter technologies have been central to overcoming the limitations of traditional power systems, which relied centrally on generation. With the growing use of distributed energy sources like solar power, the need for advanced inverter systems has never been greater. They facilitate effective energy transfer from the solar panels to the grid while making it possible for the systems to meet grid codes and enhance overall system flexibility. Moreover, they sustain system stability through voltage and frequency regulation based on solar output oscillations, consequently optimizing overall grid performance (Abu-Rub et al., 2017). As the solar market expands, power electronics innovation will be critical in closing the current challenges and in the mass adoption of solar energy. Lastly, understanding the economic side of integrating solar power onto the grid cannot be overlooked. The reduction in cost for solar PV technologies has rendered them suitable alternatives for energy supply in various industries like domestic, commercial, and industrial markets. However, this transition is beset with challenges associated with investment in the necessary infrastructure, e.g., better transmission infrastructure and sophisticated control systems. The solar market landscape of the present suggests an evolutionary transition, which calls for continuous research and development to address installation problems and optimize overall capital investment for grid integration. The future perspective suggests that with advancements in power electronics, they not only increase the efficiency but also reduce the operational cost, thus making it possible for a more sustainable and economically viable energy landscape (Abu-Rub et al., 2017) (Figure 8.13).

Figure 8.13. *Grid integration of solar power.*

Source: https://www.nrel.gov/images/libraries/news/program/2020/20200203-grid-integration-istock-1041657418.jpg?sfvrsn=85992434_0.

8.4. POWER ELECTRONICS IN WIND ENERGY SYSTEMS

The use of power electronics in wind energy systems is a major milestone towards a transition to renewable energy. Power electronics facilitate the connection between wind turbines and the electrical power grid, making it possible to efficiently convert mechanical power to electrical power. Power electronics maximize not only the generation process but also enhance power system stability and reliability that rely more on intermittent renewable energy sources. As wind power comes to play a significant role in global power capacity, the role of power electronics is more than mere conversion; it is the tactical manipulation of energy flow, which is crucial for the maintenance of grid stability. Further, there should be developments in power electronic interfaces to accommodate the varying nature of wind energy, where output variability should be effectively addressed to eliminate interference in the overall grid system (Bialasiewicz et al., 2006). The emergence of distributed energy resources demands a transformation in power system strategies, particularly as wind energy and other renewables are integrated. Power electronics are the central factor that allows this shift by making it possible to efficiently harness various sources of renewable energy and overcome the issues presented by their intermittence. Specifically, power electronics allow advanced algorithms to manage supply and demand in real-time. This is made necessary by the fact that traditional power generation systems that are high in inertia and stability yield to dynamic systems that are renewables dominant. The resultant change calls for state-of-the-art solutions with a focus on reliability and operational efficiency to facilitate the integration of wind energy into an expanded energy system (Fernández et al., 2020). With increasing application of power electronic converters in wind power systems, system stability and grid reliability concerns must be addressed in detail. The traditional reliance on rotating masses to offer grid inertia has diminished with the spread of wind power technologies, which calls for the invention of alternative methods of stabilizing systems. Synthetic inertia and virtual inertia have been suggested, which take account of the capability of power electronic converters in mimicking the stabilizing action of traditional generation. This shift is imperative since the dynamic behavior of modern power systems requires novel control strategies and reserve policies. With the assistance of next-generation power electronic technologies, wind energy can be utilized for frequency control and grid stability in general, building a sustainable energy future while following within the bounds by traditional models of electricity generation (Bialasiewicz et al., 2006; Fernández et al., 2020).

8.4.1. Role of Converters in Wind Turbines

The integration of power electronics, in this instance, converters, is crucial towards augmenting the performance as well as efficiency of wind turbines. Wind power generation is variable and intermittent in nature, and hence, there is a compelling need for systems that are robust and capable of handling these issues. Converters facilitate the conversion of the mechanical energy generated by the turbine into electrical energy to be injected into the grid. This has to be regulated very tightly in order to provide the power output that harmonizes with the needs of the grid, particularly in voltage and frequency stability. By applying the most modern converter topologies, the fluctuations can be regulated by engineers and guaranteed that the power quality is ensured, hence the overall reliability of wind energy systems to contribute effectively to the power supply (Allali et al., 2019). Secondly, developments in converter technology have offered better capability for wind turbines, particularly through application of multi-level converters. These systems are designed to minimize harmonic distortion and improve power quality, which is essential due to the natural fluctuation of wind that can burden the electrical system with harmonics. By employing configurations such as static AC/DC/AC multicell converters, scientists have developed improvements in the voltage waveforms at both the input and output segments of the wind energy conversion system. This transition not only optimizes the control of power but also incorporates energy storage technology that provides a stable power supply, thus improving the contribution of the converter in renewable integration (Allali et al., 2019). New wind energy conversion system (WECS) models have been put forth to account for the intricate dynamic aspects of renewable power generation. Various operating regimes and switching characteristics of power electronics that are instrumental in optimizing turbine operation are taken into consideration. A physical state-space model for permanent magnet synchronous generator-based WECSs has been suggested with the specific purpose of tracking the finesse of the power output under dynamic conditions. This model is intended to consider practical effects such as input saturation and integral windup, which can significantly influence the reliability of wind energy systems. By considering this close look at these control and operation management aspects, the application of converters as a fundamental aspect of guaranteeing efficiency and stability of energy systems based on renewable sources such as wind emerges.

8.4.2. Control Strategies for Wind Energy Conversion

The introduction of wind energy into the grid has sparked the development of advanced control techniques to offset the challenge posed by its natural variability and intermittency. With larger and more powerful turbines in large wind farms becoming prevalent, the quantity of good quality power introduced to the grid is highly dependent on it. The wind energy output oscillations can lead to voltage level and frequency changes, which are responsible for inter-harmonics and harmonics that degrade overall power quality. Recent novel approaches like a static AC/DC/AC multicell converter structure have been proposed for enhancing power quality in wind energy conversion systems. By utilizing serial switching cells, this solution not only allows for higher power output with fewer components but also significantly improves waveform integrity on both input and output points. Hence, having robust control strategies is essential for maximizing the efficiency and reliability of wind energy inputs to the grid (Allali et al., 2019). Another extremely critical field of wind energy conversion is the impact on power electronics driven system inertia and stability. Traditional methods of establishing power system inertia have predominantly entailed physical rotating masses directly connected with the grid. However, the new trend towards integrating renewable power supplies, namely wind energy, has introduced a notable difference in these dynamics. Because most wind power plants are connected to the grid through power electronic converters, grid inertia is really diminished, and stability and reliability are more difficult to achieve. In response to this, new concepts such as synthetic or virtual inertia have been proposed that represent synthetic inertia delivered through controlled converter devices. These control techniques are critical in replicating the missing rotational inertia, thus maintaining frequency control systems stable under high-penetration renewable conditions (Fernández et al., 2020). Moreover, effective power management strategies are critical in an effort to optimize wind energy conversion systems and address the operational challenges associated with their variability. Incorporating battery energy storage systems in wind power designs is a plausible solution to enhance supply continuity and reliability. The buffering ability in these systems facilitates the smoother integration of intermittent wind resources into the power grid. As demonstrated in recent simulations of systems involving a 149.2 kW wind turbine induction generator, both optimized control algorithms and energy storage can be utilized to significantly reduce harmonic distortion and improve transient response. By actively balancing demand and supply, sophisticated control measures and storage systems not only stabilize and enhance power quality but also lead to a more robust and flexible power grid as a response to increasing penetration of renewable energy (Allali et al., 2019).

8.4.3. Grid Connection Challenges for Wind Energy

Integrating wind power into the present grid network is fraught with countless challenges, mainly due to the fact that it is variable and uncertain in nature.

Wind power generation depends on environmental conditions that undergo constant changes during the day, thus affecting power generation greatly. This makes it challenging to guarantee grid stability and reliability as conventional power plants are likely to be more reliable in their generation patterns. Grid managers must invest in advanced power electronics and control systems to render wind power efficient by responding dynamically to the varying generation and consumption trends. These technologies are essential to facilitate the integration of renewable energy's intermittent nature with the constant nature of energy demand, ultimately enhancing the flexibility of the grids and capacity to adapt to renewable variability without compromising supply stability. It is critical to recognize these integration complexities for ongoing advancement in power electronics for the purpose of optimizing the utilization of renewable energy. Beyond the intermittent nature of wind energy, the location spread of wind farms also poses daunting technical challenges for grid integration. Wind resources are usually located at far-distant places where access to existing grid infrastructure is limited or economically uneconomical. This calls for the building of new lines for transmission, which can be costly and time-consuming. Advanced power electronics can help mitigate some of these challenges by providing means for long-distance low-loss transmission and facilitating distributed energy resources integration. Moreover, energy storage technology, such as Vanadium Redox Flow Batteries, is promising technology which could level out the intermittency of wind power so that it may be suitably integrated into the grid.

These technologies are essential in bridging the smooth transition from wind power generation and the existing grid infrastructure, thereby making the overall transition to a cleaner energy grid possible (Guarnieri et al., 2016). Ongoing evolution of mechatronic systems will play a pivotal role in addressing the grid connection problems in the case of wind energy. These systems, based on mechanical, electrical, and software components, can significantly enhance the management of the grid and the distribution of energy in the context of renewables. With the application of sophisticated control algorithms and sensors, the mechatronic systems can optimize energy production and consumption in real time. This approach not only increases the reliability of power supply from wind farms but also satisfies grid codes and standards, which are essential for system stability. Moreover, interdisciplinary design of mechatronic systems encourages collaborative research towards the energy ecosystem, including interactions between renewable energy generation, smart grids, and electric vehicles. As the technologies evolve, their use is a crucial solution to the problem of integrating wind power into our grid (Augsburg et al., 2017).

8.5. ENERGY STORAGE SOLUTIONS

Energy storage systems are a vital element in the smooth integration of renewable energy systems, primarily due to their ability to handle the natural intermittency of such resources. Due to the growing use of wind and solar power systems, energy storage technology such as batteries and flow batteries is becoming increasingly prominent. Not only do these systems provide excess energy generated during high production times but also distribute it when demand exceeds supply. Among numerous options, vanadium redox flow batteries are particularly notable for their scalability and long cycle life and are particularly favored for use at the level of the grid. Research has found that such systems are capable of managing supply and demand effectively with the objective of maximum economic efficiency, thus facilitating a seamless transition towards renewable energy sources in modern smart grid infrastructure (Guarnieri et al., 2016).

Figure 8.14. *Energy storage solutions.*

Source: https://be-cis.com/wp-content/uploads/2021/10/Energy-Storage-Solutions.jpg.

Along with this, the evolution of photovoltaic technologies has paralleled the need for efficient solutions for storing energy. With solar energy systems being cost-competitive for residential and commercial applications, corresponding advancements in advanced power electronics have also emerged. Advanced power electronics are critical to convert, regulate, and optimize solar energy for storage and utilization. The reduction in the price of photovoltaic modules has fueled research on developing both the efficiency of solar energy conversion and the efficiency of allied storage facilities. This combined approach—tying together the production of solar energy with energy storage devices by means of sophisticated power electronics—minimizes loss and maximizes usability, fostering a renewable energy culture (Abu-Rub et al., 2017). In addition to solar systems and batteries, several other energy storage devices offer complementary benefits that enhance both grid resilience and operational flexibility. For instance, pumped hydro storage, compressed air energy storage, and thermal energy storage are energy storage options. All of them possess distinct operating characteristics, strengths, and vulnerabilities that must be considered in the context of a decentralized energy landscape. The employment of energy storage systems not only makes electricity price stable but also assists in the integration of a greater mix of renewable energy resources. As the demand for clean, reliable energy continues to grow, a diversified approach based on a mix of energy storage technologies, combined with advances in power electronics, will be essential in the quest for a clean energy future (Figure 8.14).

8.5.1. Importance of Energy Storage in Renewable Integration

It is very difficult to integrate renewable power sources such as wind and solar into existing power grid systems, the most significant of these being the uncertainty of supply. Such sources are intermittent in nature, which creates a mismatch between demand and supply, potentially jeopardizing the stability of the grid. Energy storage systems (ESS) are a key component of the solution, as they serve as a buffer to absorb surplus energy when generation is high and release it when in shortage. Not only does this improve grid reliability, but it also minimizes dependence on fossil fuel-based peaker plants, providing a greener energy ecosystem. According to the latest study, proper use of energy storage, particularly together with upcoming power electronics, is paramount in facilitating large-scale renewable integration and ensuring a flexible energy infrastructure able to meet the needs of today (Blaabjerg et al., 2023). In addition, combining energy storage and power electronics emphasizes dependence on technology development for effective renewables integration. Power electronics allow easy interfacing of storage systems to the grid, which allows rapid response to intermittent energy supplies from the renewable sources. The technology is crucial in installing sophisticated control systems for optimizing charge and discharge cycles of energy storage for achieving higher overall system efficiency. Furthermore, dynamic energy management systems inform the best utilization of storage based on real-time data, which enhances a more agile and reactive power grid. Power electronics and energy storage synergies are thus essential in making renewable energy systems realize their full potential, not only raising integration, but also grid stability against disturbances and sudden demand surges (Tianyu, 2024). Thirdly, the future of renewable integration via energy storage is directly connected to ongoing research looking to accelerate not just the efficiency but the cost-effectiveness of storage technology. Developments in battery technology, for instance, in the case of solid-state batteries and enhanced lithium-ion systems, have the potential for increased energy density and longer life in storage systems. Additionally, emerging storage solutions such as pumped hydro, compressed air, and thermal storage provide diverse options for energy management dependent on application and geography. As the energy storage industry continues to grow, driven largely by the transition to renewable energy sources, it is also essential that policy structures similarly co-evolve in harmony with technological advancements. This will not only keep pace with renewable integration but also lead the way to a sustainable and resilient energy system as envisioned in the current debate regarding power electronics and renewable energy systems (Blaabjerg et al., 2023) (Figure 8.15).

Figure 8.15. *The growing importance of energy storage.*

Source: https://images.ctfassets.net/1u811bvgvthc/MABJBoHDM-rgiFDDYYiBLp/bda3be7c15cb33b507930456c84f196f/energy_storage.jpg?fm=jpg&fl=progressive.

8.5.2. Types of Energy Storage Technologies

The integration of clean power sources into the traditional power grids necessitates more efficient energy storage technologies, which are necessary components in the handling of intermittent power production. Various technologies have been engineered, each with intrinsic advantage and disadvantage. The most prevalent means is through battery energy storage, particularly lithium-ion batteries, favored for their efficiency and scalability of application from the residential level to grid-level storage. This technology enhances the storage of excess energy generated from photovoltaic (PV) installations, thus enabling access to energy during low generation times. The increasing rate of solar installations worldwide, as noted, demonstrates the increased competitiveness of PV installations that in turn attribute their increased reliability and performance to effective energy storage systems (Abu-Rub et al., 2017). Therefore, awareness of these different energy storage technologies has an important function in enhancing renewable integration into power electronics. In addition to batteries, pumped hydro storage is one of the oldest and most well-established types of energy storage which employs gravitational potential energy. It involves energy storage by lifting water to a higher elevation when there is surplus energy generation and its release to produce electricity when demand is high. While the initial investment in creating pumped hydro systems is substantial, they are otherwise extremely efficient and can level energy loads on a gigantic scale and also have a tendency to be the pillars of grid stability. With the evolution of traditional power grids into Smart Grids, the integration of such enormous energy storage systems becomes a necessity. The two-way energy transfer capability of Smart Grids is able to effectively utilize the flexibility offered by pumped hydro and other storage facilities in balancing energy distribution and utilization across sectors in a more sophisticated way (Copeland et al., 2019). Technologies such as flywheels and thermal energy storage also form part of the universe of energy storage solutions. Flywheels, which perform energy storage in the form of kinetic energy, have rapid responses and the ability to deliver high power outputs within short times and are consequently best utilized for frequency control and power system short-term fluctuation stabilizing. Thermal energy storage systems, which store heat to be used at another time, can help to deal with excess energy by converting it to thermal energy to use for heating applications. These technologies will complement pumped hydro plants and batteries by filling gaps during periods of high-speed energy demand oscillations. When power electronics continue to advance, the convergence of these various facilities for energy storage will definitely be a crucial element in bringing a more resilient and sustainable energy system to support the future transition to renewable energy.

8.5.3. Role of Power Electronics in Energy Storage Systems

The increasing integration of renewable energy sources in power grids has necessitated the enhancement in the functionality of energy storage systems, a need which is addressed by power electronics. Since renewable energy generation, such as solar and wind, is intermittent, energy storage systems play a critical role in balancing supply and demand. Power electronics provide this stability through effective bi-directional energy transfer between sources of energy, storage devices, and the power grid. Efficient high-performance power converters, the cornerstone of such systems, enable precise control of energy conversion processes, enhancing energy storage solution efficiency as well as reliability. Utilization of these converters not only optimizes the use potential of storage systems but also reduces grid integration issues such as intermittency in renewable supply and peaks in demand, thereby promising an enhanced energy infrastructure to meet modern grid requirements (Blaabjerg et al., 2023). Besides controlling mere energy flow, power electronics also entails the use of advanced algorithms and technologies to maximize system performance. Adaptive control systems incorporated in power electronics, for instance, allow real-time control over energy storage devices. The real-time function is required to maximize efficiency during periods of excess generation from renewable sources or rapid load changes. Moreover, power electronics provide essential functions of charging and discharging management, state of charge calculation, and protection against faults, that significantly contribute to operation longevity and safety of energy storage systems. As researchers explore new methodologies for integrating renewables, continuous advancement in power electronics is also essential in the management of the complexities elicited by multi-source energy systems and their dynamism (Tianyu, 2024). Finally, the integration of power electronics and energy storage systems is key in developing smart grid solutions to manage the growing phenomenon of distributed generation. Power electronics enable localized energy management, enhancing the capability of energy storage systems to respond instantly to grid loads and varying levels of renewable supplies. Through facilitating seamless interaction between generation, storage, and consumption, such technologies make the grid more resilient and promote the transition towards sustainable energy solutions. In addition, the new technology of vehicle-to-grid (V2G) systems shows the potential for power electronics to facilitate energy transfer from electric vehicles to the grid and further boost the importance of energy storage in renewable integration. This interconnected system underscores the significance of power electronics both as a technology necessity and as a change-making driver towards an efficient and sustainable energy infrastructure (Blaabjerg et al., 2023).

8.6. SMART GRIDS AND POWER ELECTRONICS

Smart grids have been at the center of transforming the generation, transmission, and use of power, achieving maximum reliability and efficiency in energy management. Compared to traditional power grids, which typically have uni-directional power and information flow, smart grids enable bi-directional power and information exchanges between power producers and consumers. The transformation is necessary to address contemporary needs like increased demand for energy, concern for reliability, and security threats of traditional systems. Integrating power electronics into smart grids facilitates this to some extent by enabling higher-level control and optimization of energy transfer, thereby fully integrating renewable energy sources into the grid. According to studies discussed in the recent past, newly emerging semiconductor technologies and innovative control architectures are instrumental in the development of integrated intelligent cyber-physical systems that enable heterogeneity of requirements for renewable energy integration in modern energy infrastructures (Balda et al., 2021). The core of smart grid operation is the application of power electronics for enabling dynamic interactions among renewable energy sources and the consumer grid. Power electronic devices, such as inverters and converters, are important in enabling renewable sources—wind turbines and solar panels—to connect to the installed grid infrastructure. They facilitate the transformation of the generated energy to a compatible form with grid specifications with stable and reliable supply. The delicate balance between generating and using energy is a matter of time-of-use management and tracking, made possible through advances in communication methods and device structure that are at the foundation of smart grids. Such research highlights today the importance of such innovation, not only as a means of accommodating renewables, but also as a way to enhance system resilience and operational effectiveness overall (Balda et al., 2021). Moreover, the successful integration of Internet of Things (IoT) technologies in smart grids is credited with strengthening the efficiency of power electronics. IoT technologies in the shape of sensors, actuators, and smart meters give complete monitoring and control over energy flows via various parts of the grid. IoT technologies enable real-time communication among energy producers and consumers to apply predictive energy management methods sensitive to real-time demand fluctuations. Coupled with high-performance power electronics, IoT integration provides enhanced connectivity and automation necessary for the evolving energy world. As examined in recent studies, the employment of IoT and smart grid technology synergy is a means of surmounting long-standing challenges of energy reliability and efficiency and paving the way for future innovation in power electronics for integration of renewables (Copeland et al., 2019).

8.6.1. Definition and Importance of Smart Grids

The concept of smart grids is a paradigm shift in the management of electricity; wherein traditional electric supply infrastructure is supported with advanced information technology and automation. Smart grids are designed to optimize the generation, transmission, and use of electric power while delivering reliability and quality. Through the use of today's power electronics, smart grids facilitate dynamic power management of energy resources, especially renewable energy sources like solar and wind that are gaining popularity day by day. Smart grids' integration of technology and renewable power is the key to managing consumers' growing demands and minimizing environmental impact, thus not only being an innovation but a necessity for sustainable energy infrastructure. Therefore, the transition to a smart grid model sets the foundation for a secure electricity network that can efficiently satisfy energy needs in the future (Denysiuk et al., 2016). Smart grids are not merely a technological advancement; they possess broader economic and environmental effects. Smart grids promote demand-side management, and the scope to distribute resources more efficiently by making use of real-time data on energy availability and usage. This utility enables grid operators to predict variations in supply and demand, thereby reducing the threat of blackouts or brownouts. Moreover, with nations focusing on decarbonization, smart grids become pivotal in facilitating the integration of renewable energy technologies that have fluctuating output. With the promise of decreased reliance on fossil fuels, smart grids are a critical move toward sustainable achievement as well as economic development through the establishment of new industries for renewable energy (Flavin, 2008). Besides promoting efficiency and sustainability, smart grids provide energy security and consumer power. By providing minute-by-minute data on their energy consumption to consumers, smart meters enable users to make intelligent decisions about their energy usage, promoting energy-saving behavior. In addition to this benefiting consumers through lower bills, it also assists the grid stability as a whole by alleviating peak demand pressures. With renewable energy installations becoming increasingly prevalent, smart grids facilitate integration by effective management of distributed energy resources. Hence, smart grids are the core components of building a flexible and resilient energy system that enables the vision for renewable integration and the imperative of climate resilience. Their role in building the future energy system makes them of paramount importance as the foundation for the alteration of today's electricity grids (Denysiuk et al., 2016; Flavin, 2008).

8.6.2. Integration of Power Electronics in Smart Grids

Smart grids, as smart grids are emerging, require the implementation of new technologies, particularly power electronics, to effectively coordinate the complex connection between the renewable energy sources. As traditional power grids have not proved good enough to address the challenges of overcoming issues such as wastage of energy and reliability, innovation in smart grids accompanies a two-way transmission of energy and information. This evolution is imperative in the integration of variable renewable sources of energy like solar and wind, which require sophisticated control systems to maintain grid stability. Power electronics play a vital role as they facilitate efficient conversion and control of power that allows for smooth interaction between the renewable generators and grid infrastructure, thus increasing overall system flexibility and efficiency. The ongoing innovation in new power-distribution models and protection equipment will complement further this integration to make it possible for the smart grid to possess a large-scale deployment of renewables without compromising reliability and security in power supply (Balda et al., 2021). Integration of power electronics into smart grids also continues to be at the forefront in enhancing communication and data management for existing energy systems. The use of Internet of Things (IoT) technologies increases the efficiency of smart grids by controlling and monitoring energy flows in real time. The connectivity facilitates efficient use of devices such as smart meters, sensors, and actuators, making it an integrated system for energy management. Power electronics are used by smart grids to facilitate these processes, enabling efficient communication between service providers and consumers. Consequently, this integration not only enhances the generation, transmission, and distribution efficiency but also empowers consumers who have more knowledge and control over their energy usage, hence an even more sustainable way of using energy for communities (Copeland et al., 2019). As regards the future of power electronics to be used in the smart grid, it is apparent that future research and development are key to unlocking present challenges. Improvements in semiconductor devices and control algorithms will significantly enhance power electronic devices' efficiency and reliability. Moreover, with increasing integration of renewable energy, the demand for advanced protection and communication schemes increases proportionally. Mitigating these challenges requires a multiparty effort from researchers, industry stakeholders, and policymakers. Further research in resource distribution, device resilience, and operational coordination will be essential towards meeting the future energy demands of tomorrow. Ultimately, the integration of power electronics not only ensures the sustainability of smart grids but also opens the door to sustainable energy future, indispensable in transitioning to a low-carbon economy (Balda et al., 2021).

8.6.3. Benefits of Smart Grids for Renewable Energy

The transition to Smart Grids represents a significant advancement in the integration of renewable sources of energy into the energy grid. Through enabling two-way transmission of energy, Smart Grids promote improved interaction between producers and consumers of energy, thereby enhancing the effectiveness of energy distribution and consumption. The transition is significant in its ability to support the erratic nature of renewable sources of power such as solar and wind power. Specifically, Smart Grids' ability to rapidly adjust to fluctuations in energy demand and supply minimizes waste and optimizes the utilization of cleaner energy resources. Thus, the incorporation of Smart Grids directly aligns with sustainability goals, making renewable energy more accessible and consistent for consumers and assisting in driving overall environmental missions. Smart Grids allow the creation of technologies that are the key to efficient energy management, the fore is the Internet of Things (IoT). It is a collection of devices, ranging from sensors and actuators to smart meters, all working in harmony to manage and monitor energy flow in real time across the grid. As a result, energy consumers can make more knowledgeable decisions concerning energy usage, thus guaranteeing energy savings and efficiency. Secondly, the data collected by IoT devices can be analyzed to improve efficiency in the utilization of renewable power sources, reducing operational costs as well as their dependability. These improvements not only make consumers more involved in energy management but also push the transition to a low-carbon energy system, resulting in a sustainable and resilient energy economy (Develder et al., 2010). The stability and security offered by Smart Grids also facilitate greater utilization of renewable energy sources. Traditional power grids are confronted with the problems of outages and linearity in electric flow, which might prove to be a limiting factor in the integration of decentralized energy systems. Smart Grids, however, facilitate a robust framework that can cope with varied energy supplies from renewable sources located in various geographic locations. Such facilities are in turn to form the foundation for guaranteeing reliability and security of electricity supply because Smart Grids can identify and respond to disturbances in real-time. Therefore, in further securing the synchronization of renewable generation and use, Smart Grids not only optimize the benefits of clean energy technology but also maximize increased energy independence for communities (Copeland et al., 2019).

SUMMARY

In concluding the discussion of power electronics for renewable integration, it is crucial to establish the over-riding need for effective and reliable interfaces to facilitate use of renewable energy resources. Integration of different systems of energy involves inherent challenges due to the fluctuating nature of renewable resources. For instance, variability of wind and solar power needs advanced solutions that can dynamically manage the variations without sacrificing stable and quality power delivery to customers. Efficient power electronic converters are crucial to develop since they are crucial to making renewable energy systems reliable and efficient. By bridging efficiency loss and advocating sustainable solutions, power electronics serve as a bridge between renewable energy production and use in the actual world in the electrical grid (Rehman et al., 2017). Moreover, with constant development of research in the area, new topologies and structures such as multiport converters have indicated promising potential in energy optimization integration. Such systems also allow for concurrent utilization of multiple energy sources, therefore promoting high resilience and minimizing single-source dependence. For example, the commissioning of a multiport power electronics interface not only allows for energy input variability but also enhances system stability and performance overall (Rehman et al., 2017). Furthermore, the potential presented by such innovations as indicated through successful simulation studies and prototypes validates the feasibility of integrating such technologies into already existing energy infrastructure. The ongoing research on more reliable, more efficient, and smaller converter designs is instrumental to meet future energy demands and push forward the renewable energy industry. The education sector's contribution to preparing the upcoming generation of engineers to combat these challenges cannot be overstated. By incorporating specialized courses focusing on the design, sizing, and integration of renewable power systems into engineering curricula, institutions can train students to innovate within this field. As the case at *Universitat Politècnica de Catalunya* illustrates, these programs are essential to developing a workforce equipped to handle the complexities of renewable energy and power electronics. Ultimately, the integration of highly advanced power electronic systems, creative engineering, and comprehensive education will be key to fully exploiting renewable energy incorporation, to a sustainable energy future.

The use of power electronics in renewable energy systems is crucial for maximizing efficiency and reliability in the process of energy conversion. As the demand for green sources of energy continues to grow, the application of power electronics of the advanced kind towards the maximum utilization of such systems takes an additional significance. These include the power electronics' ability to manage flow of energy, responsiveness to variations in renewable resources, and simplicity of integration into the grid. Further, advances in technology facilitate the establishment of wiser and more interactive systems whereby supply can be varied in real time to balance with demand. The impact of these advancements on the deployment of renewable energy technology is a brilliant example of the necessity of integrating strong power electronic solutions into existing systems as well as developing systems to ensure a safe and stable energy future. In training, learning institutions are increasing curricula to address the growing role of incorporating renewable energy through power electronics. For instance, the Technical University of Catalonia has adapted its bachelor's degree in Energy Engineering to include courses that are focused on the design and sizing of renewable energy systems, responding to industry demands for professionals with specialized training in this field (Martínez et al., 2017). The new curriculum outlines the necessity of equipping students with the right knowledge for tackling real problems in connection to renewable energy systems. By this combination of theoretical idea and practical application, students are equipped to innovate in system design, energy management, and technology optimization, ultimately resulting in the more effective use of renewable resources and enabling a sustainable energy economy. The overlap of state-of-the-art materials technology and power electronics also further enhances the potential of the use of renewable energy. Technologies such as electroactive polymers, bio-responsive hydrogels, and organic thin-film electronic devices are providing the gateway to designing intelligent materials that can enhance the performance of energy systems. These technologies are making products such as organic photovoltaics and solar textiles possible, which have the capability to significantly enhance energy capture and use while being variable in nature for different environments and social requirements (Oliver et al., 2009). By using these adaptive materials in power electronics, the sector can take advantage of the outstanding properties of emerging technologies to create adaptive systems that harmonize with consumer trends and environmental factors, making renewable energy systems not only practical but essential in resolving contemporary energy issues.

REFERENCES

1. Abu-Rub, A., León Galván, J. I., & Malinowski, M. (2017). *Solar Photovoltaic and Thermal Energy Systems: Current Technology and Future Trends*. https://core.ac.uk/download/286563376.pdf (accessed on 30th September 2025).
2. Allali, A., Denai, M., Kaddour, F., Khodja, M., et al. (2019). Power quality enhancement in electricity grids with wind energy using multicell converters and energy storage. https://core.ac.uk/download/287581569.pdf (accessed on 30th September 2025).
3. Arghandeh, R., Brown, M., Cibulka, L., Miller, R., et al. (2013). Distribution system voltage management and optimization for integration of renewables and electric vehicles: Research gap analysis. https://core.ac.uk/download/288432751.pdf (accessed on 30th September 2025).
4. Augsburg, K., Horn, M., Ivanov, V., Ruderman, A., et al. (2017). Interdisciplinary design methodology for systems of mechatronic systems: Focus on highly dynamic environmental applications. https://core.ac.uk/download/96766384.pdf (accessed on 30th September 2025).
5. Balda, J. C., Blaabjerg, F., Mazumder, S. K., Kulkarn, M., et al. (2021). A review of current research trends in power-electronic innovations in cyber-physical systems. https://core.ac.uk/download/480297724.pdf (accessed on 30th September 2025).
6. Bao, Y., Feng, W., Jin, M., Liu, Y., et al. (2018). A review of microgrid development in the United States–A decade of progress on policies, demonstrations, controls, and software tools. https://core.ac.uk/download/323310623.pdf (accessed on 30th September 2025).
7. Bialasiewicz, J. T., Carrasco Solís, J. M., Galván Díez, E., García Franquelo, L., et al. (2006). Power-electronic systems for the grid integration of renewable energy sources: A survey. https://core.ac.uk/download/51404452.pdf (accessed on 30th September 2025).
8. Blaabjerg, F., Kim, K. A., Rodriguez, J., Yang, Y., et al. (2023). Power electronics technology for large-scale renewable energy generation. https://core.ac.uk/download/588889893.pdf (accessed on 30th September 2025).
9. Candela García, J. I., Fazlin Roslan, N., Luna Alloza, Á., Rocabert Delgado, A., et al. (2018). Remote power control strategy based on virtual flux approach for the grid-tied power converters. https://core.ac.uk/download/187265192.pdf (accessed on 30th September 2025).

10. Cang, T. (2024). *Electrifying Innovation: The Transformative Impact of Power Electronics in Modern Infrastructure*. https://core.ac.uk/download/615429848.pdf (accessed on 30th September 2025).
11. Carrasco Solís, J. M., León Galván, J. I., Reyes Díaz, M. R., Sánchez Segura, M. I., et al. (2014). Adaptive vectorial filter for grid synchronization of power converters under unbalanced and/or distorted grid conditions. https://core.ac.uk/download/286564930.pdf (accessed on 30th September 2025).
12. Copeland, R., Crespi, N., Rehmani, M. H., Saleem, Y., et al. (2019). Internet of Things-aided smart grid: Technologies, architectures, applications, prototypes, and future research directions. http://arxiv.org/abs/1704.08977 (accessed on 30th September 2025).
13. Denysiuk, S., Opryshko, V., Strzelecki, R., Denysiuk, O., et al. (2016). Implementation of the Smart Grid concept through the use of demand-side management programs and modern power electronics systems. https://core.ac.uk/download/81630897.pdf (accessed on 30th September 2025).
14. Develder, C., Haerick, W., & Mets, K. (2010). A simulator for the control network of smart grid architectures. https://core.ac.uk/download/55823207.pdf (accessed on 30th September 2025).
15. Fernández Guillamón, A., Gómez Lázaro, E., Molina García, Á., Muljadi, E., et al. (2020). Power systems with high renewable energy sources: A review of inertia and frequency control strategies over time. https://core.ac.uk/download/304372333.pdf (accessed on 30th September 2025).
16. Flavin, C. (2008). *Low-Carbon Energy: A Roadmap*. https://core.ac.uk/download/71340916.pdf (accessed on 30th September 2025).
17. Guarnieri, M., Mattavelli, P., Petrone, G., Spagnuolo, G., et al. (2016). Vanadium redox flow batteries: Potentials and challenges of an emerging storage technology. https://core.ac.uk/download/80155277.pdf (accessed on 30th September 2025).
18. Hackl, C. M., Hohmann, S., Jané-Soneira, P., Pfeifer, M., et al. (2018). Full- & reduced-order state-space modelling of wind turbine systems with permanent-magnet synchronous generator. https://core.ac.uk/download/197491600.pdf (accessed on 30th September 2025).
19. Jayasankar, V., & Satyanarayana, V. (2023). Modelling and design of PV grid integration for GA-PSO based on fluctuating power quality. https://core.ac.uk/download/642650394.pdf (accessed on 30th September 2025).
20. Luna Alloza, Á., Mir Cantarellas, A., Remón Rodríguez, D., Rodríguez Cortés, J., et al. (2016). Dynamics estimation and generalized tuning of stationary frame current controller for grid-tied power converters. https://core.ac.uk/download/81579774.pdf (accessed on 30th September 2025).

21. Martínez García, H. (2017). *Renewable Energies Course for Energy Engineering Students at the Technical University of Catalonia (UPC).* https://core.ac.uk/download/157812010.pdf (accessed on 30th September 2025).
22. Mensah, A. (2004). Modelling and analysis of solar arrays for grid connected systems with maximum power point tracking. https://core.ac.uk/download/236254640.pdf (accessed on 30th September 2025).
23. Moghaddam, A. F., & Van den Bossche, A. (2020). Direct usage of photovoltaic solar panels to supply a freezer motor with variable DC input voltage. https://core.ac.uk/download/286548912.pdf (accessed on 30th September 2025).
24. Oliver, R., Tillotson, J., & Toomey, A. (2009). Bioengineered textiles and nonwovens – The convergence of bio-miniaturisation and electroactive conductive polymers for assistive healthcare, portable power and design-led wearable technology. https://core.ac.uk/download/9587153.pdf (accessed on 30th September 2025).
25. Qian, Y., Sharif, H., Tipper, D., Yan, Y., et al. (2013). A survey on smart grid communication infrastructures: Motivations, requirements and challenges. https://core.ac.uk/download/12211338.pdf (accessed on 30th September 2025).
26. Rehman, Z. (2017). *Multiport Power Electronics Circuitry for Integration of Renewable Energy Sources in Low Power Applications: A Thesis Presented in Partial Fulfilment of The Requirements for The Degree of Doctor of Philosophy in Electrical Engineering at Massey University, Palmerston North, New Zealand.* https://core.ac.uk/download/148645579.pdf (accessed on 30th September 2025).
27. Smith, K. (2025). *Climate Change and Radical Energy Innovation: The Policy Issues.* https://core.ac.uk/download/pdf/6484436.pdf (accessed on 30th September 2025).
28. Xiao, M. (2024). *Application of Power Electronics Technology in Renewable Energy Systems.* https://core.ac.uk/download/616937539.pdf (accessed on 30th September 2025).

CHAPTER 9

ENERGY STORAGE FOR RENEWABLE INTEGRATION

Contents

INTRODUCTION

The transition to the use of renewable sources of energy creates unprecedented challenges and opportunities for energy networks globally. As nations struggle to reduce their reliance on fossil fuels, the adoption of renewable energy in infrastructure of traditional grids necessitates the reformulation of energy management policies. Renewable sources of energy such as the sun and wind present natural variability and intermittence challenging grid managers who must maintain the equilibrium of supply versus demand. This challenge underlines the paramount significance of efficient energy storage technologies that have the capacity to hold excess power during peak hours and transfer power during off-hours. Therefore, the success of integrating renewable energy sources does not depend solely on technological advancements but also on the macro-structural changes of energy markets and policymaking that prioritize sustainability and reliability. In the context of the Association of Southeast Asian Nations (ASEAN), the goals of 13–23% penetration of renewable energy by 2025 highlight the criticality of grid flexibility. Flexibility is crucial in the capacity for adapting to changes in renewable generation without sacrificing electricity stability. A six quantitative indicators-based grid flexibility study reflects the startling disparity between ASEAN nations in their preparedness to adapt to higher penetrations of renewable energy.

Figure 9.1. *Renewable energy integration.*

Source: https://media.licdn.com/dms/image/v2/D5612AQGNWDBgA8pxcw/article-cover_image-shrink_600_2000/article-cover_image-shrink_600_2000/0/1707575803254?e=2147483647&v=beta&t=Jyn7AS8NJzUqNdHmvCmXvQu7HsODYMnah-uOIMvlFd4.

The lack of robust forecasting mechanisms and limited market access are specifically mentioned as key weak points that can be addressed through regional collaboration and technological innovation in energy storage. As the region continues to work toward renewable energy objectives, maximizing grid flexibility will be instrumental in lowering overall system costs as well as increasing member states' energy reliability (Huang et al., 2019). Future smart grid technology innovations will have a key role in developing a robust and efficient energy infrastructure. Smart grids can potentially enhance supply and demand management so that different processes of renewable energy sources can be aggregated and real-time communication between consumer and producer. Furthermore, they simplify the implementation of storage systems for energy to enable it to endure the inherent weaknesses in blending renewable sources. Though there remains disagreement on parameters and definitions, the ability of smart grids to spur the creation of jobs, reduce greenhouse gases, and enhance the general stability of the grid is generally accepted. As political and financial backing for smart grid initiatives increases in the EU and globally, regulatory concerns and public opinions need to be addressed appropriately in ensuring widespread uptake and allowing for maximum utility from energy storage systems (Axon et al., 2014) (Figure 9.1).

Definition of Energy Storage

Energy storage is basically the process of harvesting and retaining energy to be used in the future. This becomes really important while adding renewable energy sources, whose variable and intermittent nature usually characterizes them. With the growing spread of renewable energy technologies, the ability to store energy is crucial in achieving supply-demand equivalence, thus rendering the energy system reliable. Energy storage facilities, ranging from batteries to thermal systems, are employed to fill in the daily fluctuation of use. The systems offer a chance for energy to be stored during periods of peak generation, for example, windy or sunny days, in order to be consumed subsequently, smoothing out the gap between production and consumption as well as availing a stable supply of energy according to grid load. Among the several energy storage technologies, technological progress has been highly promising, particularly in grid-scale installations. Current methods, like compressed air and pumped hydro, have proved to be good, but new technologies, like transcritical carbon dioxide systems, are gaining huge popularity. Not only are the alternatives as efficient as the existing ones, but they also show flexibility and scalability, which make them economically viable options in modern energy infrastructures. With an estimated round-trip efficiency of over 60% when paired with solar power, these systems indicate the convergence of energy storage and renewable generation. The deployment of such advanced technologies is at the heart of addressing the issue of the non-dispatchable character of renewable energies and is driving the push towards a sustainable energy future (Villanueva et al., 2019). Other than enhancing energy reliability, energy storage takes on an instrumental role in peaking deployment of renewable sources by enabling hybrid energy systems. Synergy in integrating different renewable energy sources has the impact of offsetting the natural variation in energy generation brought about by climatic factors. A close examination of hybridization between energy sources, such as solar-wind or hydro-solar, illustrates that the integration of two or more systems is capable of effectively mitigating supply-demand disparities by utilizing the complementary nature of such sources. Integration of energy storage in these hybrid schemes also increases their reliability and efficiency, with a higher scope for utilization of renewable energy. It can be observed that energy storage not only enhances the performance but also helps to reach broader objectives of large-scale deployment of renewable energy into the grid and opening up pathways to a sustainable energy future (Beluco et al., 2019) (Figure 9.2).

Figure 9.2. *Energy storage technologies.*

Source: https://www.azonano.com/image-handler/ts/20230719120057/ri/950/src/images/Article_Images/ImageForArticle_6513_16897824560828570.jpg.

Importance of Renewable Energy Integration

The relevance of integrating renewable energy is being felt in the face of a world transition towards sustainability. The greater the demand for clean energy, the more important it is to ensure the hassle-free integration of renewables into existing energy systems. Not only does this maximize the environmental benefits of renewables but also addresses concerns about energy security and reliability issues. Through building systems that draw on diverse sources of energy, such as sunlight and wind, communities are capable of de-linking themselves from traditional fossil fuels while becoming more resilient to supply and demand volatility in energy. The use of these resources put together results in an energy portfolio that can adapt to shifting conditions of supply and demand, eventually developing into a more sustainable and balanced system of energy. Energy storage devices provide the foundation for effective implementation of renewable energy systems. They are essential components in managing the natural fluctuation in renewable sources since they are founded on natural conditions. For instance, advances in alternative energy storage technologies such as electro-chemical batteries and pumped hydroelectric systems enable buffer capacity to accumulate surplus energy generated when production is at peak levels to be utilized in the future when demand is higher than supply. Unified energy storage solutions, as adopted by communities in recent research, can optimize energy utilization and enhance autonomy, enabling greater even matching of local energy consumption and production cycles. Effective energy storage methods are hence not just technological advancements; they are groundbreaking innovations that have the potential to drive the successful embedding of renewables in the energy mix. Renewable energy sources and energy storage integration also introduce other social values by raising the security of energy infrastructure from exogenous shocks and social justice through access. For instance, multi-energy hub modeling stresses the challenge of managing the stochasticity of renewable power, illustrating the necessity of intelligent systems that are able to optimize use of resources among various sources of energy. Energy integration in various places, like the Swiss village of *Cartigny*, stresses the necessity of daily and seasonal storage to balance the irregular supply of energy. This integrated approach of energy management is an indication of commitment to not just achieving the sustainability goals but also creating equal access to clean energy for all residents of the community, thereby ensuring environmental and social development in one integrated renewable energy system (Coccolo et al., 2017; Curto et al., 2024).

9.1. TYPES OF ENERGY STORAGE TECHNOLOGIES

The transition to renewable resources necessitates the development of efficient energy storage technologies to offset the intermittent nature of these sources. Two of the most prominent types are pumped hydro storage and compressed air energy storage (CAES), which employ gravitational and pressure energy, respectively. Pumped hydro involves pumping water into a reservoir during off-peak periods, which can be released to generate electricity in times of demand. Similarly, CAES compresses air with surplus electricity that gets stored in underground storage caverns awaiting the time when energy recovery can be executed. These technologies are renowned for their scalability and capability but are governed by geographical and environmental constraints. The further exploration of alternatives such as transcritical carbon dioxide cycles coupled with thermal energy storage is an appealing alternative that is both competitively efficient and cost-effective and holds the potential to revolutionize the system for energy storage for integrating renewables (Villanueva et al., 2019). Hydrogen fuel cell systems and hydrogen storage represent another new technology in energy storage. As renewable energy systems become increasingly common, they generate electricity whose production varies with the weather and time of day, emphasizing flexibility as a key feature of energy management. The production of hydrogen through electrolysis from surplus renewable energy allows for energy storage beyond traditional boundaries. The energy is then converted back into electricity through fuel cells or even utilized in cogeneration and tri-generation systems, further boosting efficiency. There has been great advancement in the production and adoption of hydrogen technology. Institutions such as the National Centre of Hydrogen and Fuel Cell Technology Experimentation in Spain have undertaken this work, pioneering the central position of hydrogen as an energy carrier to optimize renewable energy supply chain (Jul et al., 2011). Other than traditional and hydrogen-storage technologies, batteries have become a major component in the technology of energy storage, particularly lithium-ion batteries. Their rapid deployment in electric vehicles and grid storage demonstrate their capability to meet both short-term as well as long-term energy storage needs. Innovations in battery chemistry, such as solid-state batteries and flow batteries, focus on delivering greater energy density, cycle life, and cost. As the need for grid stability and electric vehicle infrastructure boosts up, battery technology must improve to facilitate easier integration for renewable energy sources. They do not only help with on-demand power support but also act as load-balancing entities in dynamic power markets, facilitating access to a cleaner energy future with the enhanced use of renewables (Figure 9.3).

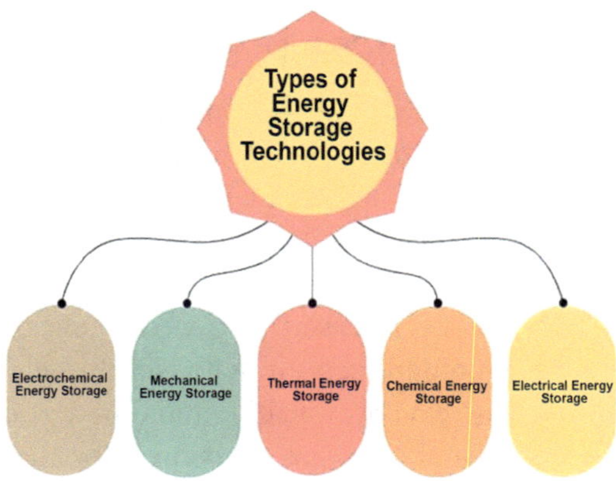

Figure 9.3. *Types of energy storage technologies.*

Source: https://b1879915.smushcdn.com/1879915/wp-content/uploads/2025/04/Types-of-Energy-Storage-Technologies.jpg?lossy=2&strip=1&webp=1.

9.1.1. Mechanical Energy Storage

Mechanical energy storage systems are especially crucial to facilitate the integration of renewable energy sources onto the grid. Through application of various mechanical means, such as pumped hydro and compressed air energy storage, they can smooth out the fluctuations of renewable energy sources such as solar and wind. Gravitational potential energy is utilized through pumped hydro energy storage, storing excess electricity during peak production times by pumping water into an upper reservoir. This stored energy is released during peak periods, thereby smoothing energy supply and demand (Ault et al., 2005). Compressed Air Energy Storage compresses air in underground storage caverns and releases it subsequently to drive turbines. These two systems fully qualify as a cushion for surplus power, making the grid stable and facilitating a more reliable supply of renewable energy. Other than traditional means, innovative techniques for mechanical energy storage are being developed to further increase their efficiency and applicability. For instance, flywheel energy storage systems store energy in the rotational kinetic state for rapid discharge and recharge operations.

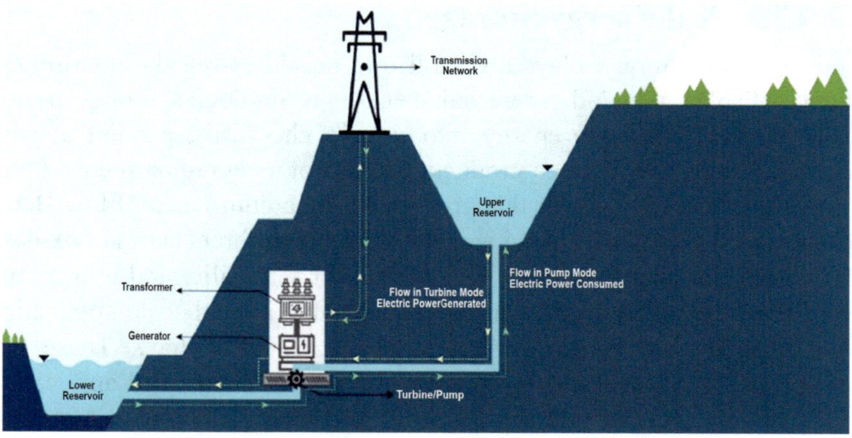

Figure 9.4. *A pumped-hydro energy storage (PHES) system.*

Source: https://www.researchgate.net/profile/Ammu-Jacob/publication/357900492/figure/fig1/AS:1113526802350080@1642496886358/A-pumped-hydro-energy-storage-PHES-system.png.

Flywheels are able to respond rapidly to grid demand fluctuations and are thus well-suited to smooth the output of renewable sources during sudden changes in wind or sunlight (Al-Kayiem et al., 2015). In addition, materials research has made it possible to create advanced flywheels with increased speed capabilities and energy densities, further improving the potential of flywheels as a mass-scale storage technology. The ongoing innovation of such technologies underscores the transforming potential of mechanical energy storage to create a stable and sustainable energy infrastructure. The application of mechanical energy storage systems, however, is not free from challenges. The geographical limitations of pumped hydro systems restrict their use to sites where the topography is favorable (Ault et al., 2005). Furthermore, compressed air systems are developed in a variety of geological formations but usually entail significant front-end costs and operational costs. Flywheel systems, although with high efficiency, suffer from disadvantages with respect to maintenance and longevity in the long run. Therefore, one needs to consider the economic, environmental, and regulatory frameworks that surround these technologies as they become part of an integrated future energy landscape that emphasizes integration of renewables. The complementarity between the mechanical energy storage technologies and other renewable sources will ultimately decide their efficiency in providing a sustainable energy transition (Figure 9.4).

9.1.2. Chemical Energy Storage

Chemical energy storage plays a central role in addressing the intermittency challenge of solar and wind as renewable energy technologies. Storage systems convert surplus renewable energy into storage chemicals, a point at which energy can be consumed when production does not meet requirements. One of the prevailing technologies in this area is the Vanadium Redox Flow Battery (VRFB), which operates by using vanadium ions in different states of oxidation as the vehicle for energy storage. Its capacity for scalability and utilization of a long lifetime cycle makes it particularly valuable in grid-scale applications, as identified by the extensive review of VRFB technology. These traits position VRFBs as ideal energy storage devices for smart grid applications, whereby supply and demand must be matched while harvesting intermittent renewable sources to provide system dependability (Guarnieri et al., 2016). Secondly, the need to incorporate chemical energy storage within current power infrastructure is also highlighted by the increasing need for distributed generation. The ability to store energy chemically at high efficiencies not only facilitates smoothing of peak load demands but also helps to soften the volatility linked with renewable energy generation. Current work explores the merger of chemical storage technologies with traditional power systems, with a focus on building a more flexible and resilient grid for electricity. There exist a number of integration methods, pointing to the necessity of innovative solutions that can scale in order to keep pace with increasing energy storage requirements. Examination of these storage methods reveals that while electrical storage can be prohibitively expensive, chemical storage methods often present feasible alternatives, particularly where scale and longevity are of greatest importance (Ault et al., 2005). Lastly, the future energy system scenario will definitely be shaped by the continued development of chemical energy storage systems. The research efforts continue to gear towards the advancement of such systems for higher efficiency and cost-effectiveness, with mechanisms realized to enhance further integration of the systems into energy infrastructures. Through better chemical process optimization and storage materials, gains in efficiency can lead to dramatic decreases in operating costs and lifecycle efficiency. Moreover, as the competition to integrate renewables intensifies, collaboration between researchers, industry participants, and regulatory agencies will become increasingly important. These partnerships have the potential to generate conditions that enable deploying chemical energy storage technology that promotes adapting energy landscapes and enhances overall grid resilience, ultimately leading to a sustainable energy future (Guarnieri et al., 2016) (Figure 9.5).

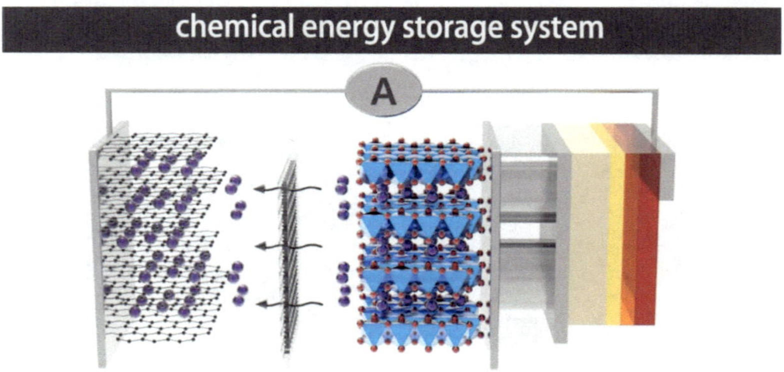

Figure 9.5. *Chemical energy storage system.*

Source: https://www.huntkeyenergystorage.com/wp-content/uploads/2023/08/chemical-energy-storage-system.webp.

9.1.3. Electrical Energy Storage

The integration of electrical energy storage systems is critical in order to enhance the operational reliability of power grids, especially with the growing use of renewable energy sources like wind and solar. These energy storage systems serve as useful intermediaries, compensating for the inherent variability and intermittency that characterizes renewable energy generation. In absorbing excess electricity during high generation, energy storage systems can release stored capacity during times of peak demand or low generation. This aspect not only provides a buffer against supply-demand discrepancies but also supports grid frequency stability, ultimately making the energy system stronger. In order to achieve these targets, it is essential to evaluate the current grid flexibility status since, as in the ASEAN nations' case, storage availability, and access for improving operation are most essential to achieve renewable energy targets (Huang et al., 2019). The development of a more responsive and adaptive electric grid is greatly facilitated through emerging energy storage technologies. Various energy storage technologies, such as lithium-ion batteries, pumped hydro storage, and flywheels, pose unique opportunities and challenges within the framework of particular energy market applications. These technologies can more efficiently store and deliver energy, aligning generation and matching consumption patterns.

Figure 9.6. *Electrical energy storage systems.*

Source: https://www.renhotecev.com/wp-content/uploads/2023/07/Electrical-Energy-Storage-System.webp.

Moreover, utilizing smart grid principles makes it easier to incorporate these storage systems into current infrastructures. Technological innovations in Information and Communication Technologies (ICT) make this transition possible by enabling operators to manage energy flows dynamically with security and optimality in grid operations (Borghetti et al., 2013). With additional innovation in energy markets, solving regulatory issues and establishing cooperation will only enhance the potential for efficient energy storage solutions. In examining the broader implications of electricity storage, it is evident that these systems are not passive stopgap arrangements but active components of a long-term energy policy. The possibility of storing electricity has paved the way for higher penetration of renewables, reducing fossil fuel dependence and mitigating greenhouse gas emissions. Successful integration of energy storage systems can promote a decentralized energy system, enabling consumers to participate in energy production and consumption. As the ASEAN experience indicates, increased grid flexibility can lead to more efficient overall systems, ultimately lowering the price for consumers and more sustainable energy systems. Thus, provoking national and regional policies to aid energy storage programs becomes important to ensure sustainable economic growth together with energy independence (Huang et al., 2019) (Figure 9.6).

9.2. ROLE OF ENERGY STORAGE IN RENEWABLE ENERGY SYSTEMS

The integration of renewable energy systems into the grid is increasingly recognized as a condition for long-term sustainability goals. However, the intermittent nature of renewable energy sources such as solar and wind power is a crucial issue for grid stability and reliability. It is here that energy storage systems play a central role. They provide backup power during low-production intervals while storing excess power generated during high-generation intervals. By offering a sure supply of power, energy storage systems not only make the grid more stable but also enable the broader deployment of renewable technology. The concepts and arguments discussed in energy storage and grid balancing conferences highlight the necessity of putting these systems in place because they complement renewable generation and can significantly lower the risk associated with volatility in energy supply and demand (Mishra et al., 2023). Moreover, the role of energy storage extends beyond mere backup mode; it adds to the overall grid flexibility as well. With the Association of Southeast Asian Nations (ASEAN) forecasting a rise in the proportion of renewable energy by 2025, the role of advanced energy storage systems becomes essential. Energy storage can help balance variable demand profiles and can be integrated with other emerging technologies such as smart grids and advanced forecasting devices. Maximum grid flexibility allows for the optimization of renewable energy generation and the highest capacity to accommodate sudden changes in energy supply. Thus, the use of energy storage devices can lower the costs of the system as a whole and increase access to reliable electricity, and this becomes a necessary investment as ASEAN strives to meet its goals for renewables (Huang et al., 2019). In addition to technological advancements, the socio-economic effect of energy storage cannot be overemphasized. Storage technologies enable greater penetration of renewable energy and encourage the generation of employment in renewable energy and stimulate the local economies. With the advancement of energy storage technologies, costs decline even further, and this makes integration with renewables cheaper. Such conferences as the National Conference on Renewable Energy will continue to bring into focus these developments, with major focus on the necessity of policy frameworks and regional coordination. Overcoming challenges through collective action will ensure that energy storage remains an enabler of the overall energy transition, aligning with global goals on sustainability and resilience of energy systems (Mishra et al., 2023) (Figure 9.7).

Figure 9.7. *The role of energy storage in renewable energy systems.*

Source: https://www.mitarsh.com/wp-content/uploads/2025/01/02-The-Role-of-Energy-Storage-in-Renewable-Energy.jpg.webp.

9.2.1. Balancing Supply and Demand

Harmonizing renewable energy resources into the existing energy grid is a huge task as far as supply and demand balancing is concerned. As generation from renewable energy sources, such as solar and wind, is naturally un-predictable, energy systems must acclimatize to fluctuations to guarantee stability. This calls for advanced energy storage facilities, which can save excess energy when there is high generation and replenish it during low-generation hours. By optimizing supply and demand management with the use of storage facilities, energy providers can reduce the consumption of fossil fuels and enhance grid stability. The integration will be successful if it is possible to devise intelligent energy policies that consider the interaction between supply security and climate change, as per discussions by experts (Rious, 2025). The policies will provide a platform conducive to innovation and efficiency in the consumption of renewable energy. To utilize energy resources most effectively, the concept of smart microgrids has developed as an answer. Microgrids are localized systems that facilitate local control of energy production and consumption, with greater flexibility in managing the supply and demand side imbalances. Microgrids can also exploit different types of energy production sources, including small renewable generators, to make it more accessible and reliable. Nonetheless, deployment of intelligent microgrids is beset with load balancing, pricing policy, and participation of the customer (Elmenreich et al., 2013). Overcoming these challenges will be important to facilitate efficient operation of microgrids, which can be elements of a complete smart energy infrastructure. As technology evolves and technologies improve, microgrid potential in facilitating storage of energy as well as coordination of consumption will become a bigger factor in overall energy systems. Lastly, the relationship between supply/demand balancing and energy storage cements the necessity of combined methodologies considering not only technological advancements but regulatory schemes. With changing energy markets towards increased dependence on renewable sources, effective demand-side management is an integral component of overall grid stability. This entails not only promoting awareness and engagement by consumers but also incentivizing energy-efficient actions associated with corresponding actual time-of-day energy availability. Coupling energy storage technologies with regulatory measures can create the gateway to a more dynamic and responsive energy system, which can respond to the requirements of a developing context (Rious, 2025). With stakeholder cooperation and an application of technological innovation, energy systems can better solve the intricacies of supply and demand management, enabling the move toward a sustainable energy future (Figure 9.8).

Figure 9.8. *How energy storage helps balance supply and demand on the grid during peak periods.*

Source:https://bioenergytimes.com/wp-content/uploads/2024/06/solar-panels-8593759_1280.jpg.

9.2.2. Enhancing Grid Stability

The integration of the grid with renewable energy sources is a heavy burden, primarily concerning grid stability. As dependence on intermittent sources such as wind and solar power increases, there is a need to maintain a stable supply of electricity. Solutions for energy storage have become prominent as key factors in helping to alleviate these challenges because they enable operators of the grid to effectively manage supply fluctuations. These systems store excess energy generated at times of peak generation and release it during times of high demand or low generation, thereby enhancing overall grid stability. Policy support for the installation of energy storage technology is vital in ensuring an environment that can support transition towards more reliance on renewable energy sources. Microgrid development is also another available approach to enhancing grid stability, especially in decentralized setups. Microgrids can be isolated or synchronized with the primary grid to provide improved resilience to outages. By integrating various energy storage systems, microgrids are able to balance supply and demand more effectively and hence improve reliability. Policies by the federal and local governments assist in promoting microgrid projects and hence enhancing energy stability within communities. As advanced in recent studies, microgrid development in the United States demonstrates how technology innovation and supporting policies can facilitate innovative solutions for integrating renewables and overcoming some of the inherent issues surrounding grid stability (Bao et al., 2018). For broader regional objectives, say for the ASEAN countries, grid flexibility becomes an imperative for effective integration of renewables. As these nations seek to increase their percentage of renewable energy, grid flexibility is the key to ensuring the variability of these resources is accommodated. A tool employed in measuring the grid flexibility of ASEAN is key in areas of improvement such as the quality of forecasting systems and improved access to electricity markets. With these deficiencies filled, member countries can improve their ability to absorb renewable sources stability and economically. The collective efforts of nations in increasing the flexibility of grids ultimately underscore the role of regional collaboration in meeting the challenges presented by high penetrations of renewables, both in ensuring grid stability and meeting renewable energy targets (Huang et al., 2019).

9.2.3. Facilitating Renewable Energy Deployment

Facilitating deployment of renewable energy is critical in creating a sustainable energy future since it makes the energy system more efficient and reliable. Incorporation of energy storage systems is one of the cornerstones in this transformation. Due to the prevalent use of distributed generation, a greater demand for numerous storage technologies is created, and the ensuing future power system will be one with ubiquitous energy storage (Ault et al., 2005). The system enables supply and demand to be balanced, hence ensuring equal energy availability while combating the intermittency that comes along with renewable sources of solar and wind power. Having the ability to store energy in a preferable way not only enhances the flexibility of the grid but also spurs further investments in renewable technologies, hence enabling conditions for sustainable use of energy practices.

Figure 9.9. *In China, generation-side and grid-side energy storage dominate, making up 97% of newly deployed energy storage capacity in 2023.*

Source: https://assets.weforum.org/article/image/responsive_big_webp_ATI-wjQpaYV-kGsXuFzRfOAI--I9aSDUTgS2FCTnWAtM.webp.

Further, the development of smart grid technology is a key factor in facilitating the ease of renewable energy deployment. The traditional grid electricity network infrastructure has been increasingly inadequate in addressing the new problems such as variable electricity load demand and integration of alternative energy sources (Borghetti et al., 2013). Smart grids employ advanced Information and Communication Technologies (ICT) to enhance the reliability and flexibility of energy management systems. Through enabling real-time monitoring and automated controls, these systems optimize usage and utilization of renewable energy and hence increase the overall grid resilience. Transition to a smart grid not only supports integration of energy storage solutions but also tackles technical as well as regulatory complexities, allowing an integrated approach towards integrating renewable sources into the prevailing energy system. Lastly, dismantling regulatory barriers is necessary to facilitate the use of renewables and allied storage options. Current policies are typically lagging behind new technologies, creating hurdles for private sector investments and public adoption of renewables. Policymakers need to create facilitating regulatory frameworks that encourage energy storage development and the interconnection of renewables to markets. This involves reshaping incentive schemes, streamlining the permitting process, and fostering harmonization among regulatory entities and stakeholders. As discussed in the case studies presented in present literature, enhancing the economic viability of storage technologies with supportive regulation can similarly significantly enhance their adoption. By coordinated planning of economic interests and requirements for sustainability, governments can make deployment of renewable energy not merely viable but also economic (Ault et al., 2005). Creating an environment of concordance for energy policy innovation will be critical in ensuring the achievement of goals stipulated in transitioning towards sustainable energy systems (Figure 9.9).

9.3. BENEFITS OF ENERGY STORAGE FOR RENEWABLE INTEGRATION

The incorporation of energy storage systems in renewable energy systems has numerous advantages that enhance the efficiency of the electric grid. With solar and wind, sources of renewable energy, dominating the scene, variability in their output makes efficient solution management necessary to stabilize supply and demand. Energy storage technologies, i.e., batteries and pumped hydro storage, provide the grid the freedom to deal with intermittency so that utilities can provide energy demands reliably. Furthermore, energy storage has the potential to alleviate stress on grid operation as excess generation during periods of peak production can be stored and dispatched when demand is high or generation is low. This capability significantly enhances the whole grid reliability and stability and provides means of achieving regional renewable energy generation targets by initiatives, such as ASEANs intends to increase the percentages of renewables by 2025 (Huang et al., 2019). Energy storage systems also achieve economic benefits in the renewable energy industry. With the effective utilization of generated power, such systems can potentially lower the reliance on peaking fossil-fueled power plants, which are more expensive to operate. Economic studies indicate that energy storage can lead to lower system costs in the aggregate from more effective energy dispatching and less waste from excess renewable generation. In addition, energy storage technologies can improve energy market participation by allowing efficient achievement of grid needs from distributed generation sources such as rooftop solar. For a city like Los Angeles, integrating energy storage with current solar capacity can unlock large quantities of excess energy to improve the overall performance of the grid while benefiting low-income populations through additional renewable generation opportunities (Cheng et al., 2020). Finally, grid flexibility is enhanced through energy storage to enable the shift to a cleaner power system. Greater renewable integration into the grid comes with challenges, particularly in its ability to dynamically respond to fluctuations in power production and consumption. As highlighted in present analyses of grid flexibility in grids like ASEAN, the ability to store surplus energy and release it when required is a critical component for optimizing grid efficiency (Huang et al., 2019). Furthermore, effective energy storage systems help prevent environmental impacts through the use of clean energy while avoiding greenhouse gases emissions from traditional energy consumption. Lastly, with advances in energy storage technologies and declining costs, their use in renewable energy integration will be at the forefront of a sustainable and resilient energy future.

9.3.1. Reducing Greenhouse Gas Emissions

The sheer need to reduce the emission of greenhouse gases is, in a natural way, related to the successful integration of renewable energy sources into the present power infrastructure. With the reduction in fossil fuel reliance, the role of energy storage becomes crucial in preventing carbon emissions and ensuring a safe provision of energy. Various studies highlight that enhanced energy storage solutions not only complement the variable characteristics of renewables but also facilitate the transition towards a low-carbon energy system. Storage technologies optimized and expanded infrastructure can assist nations to cut greenhouse gas emissions to a great extent, envisioning a cleaner power sector. To illustrate, Brazil is at a crossroads in its energy policy, choosing to invest in energy that can either power carbon-driven growth or position it as a clean energy leader. Renewable integration through next-generation energy storage systems represents a significant opportunity for Brazil to deliver on its climate goals while enhancing economic competitiveness (Lucon et al., 2015). As efforts to balance energy production with environmental objectives are made, the circular economy is found to be a support strategy to integration schemes of renewables. A critical component of the success of climate neutrality, the circular economy promotes resource efficiency and lowers overall greenhouse gas emissions through sustainable production. This revamp of material reuse helps in reducing waste, thereby lessening the carbon footprint of production and disposal cycles. In addition, the creation of lead markets for green circular economy products requires creative regulation policies that encourage low-carbon substitutes and spur the use of greener technology. In the words of the European Union further, effective circular economy practices through the avenue of regulations can spur investment opportunities in addition to efforts in emissions reduction (Egenhofer et al., 2019). This inclusion is advantageous in two aspects: spurring economic resilience and fashioning sustainable environmental strategies. Lastly, the joint effort in reducing greenhouse gas emissions calls for coordinated policy and technology deployments using energy storage to facilitate the integration of renewables. All players must make joint decisions and implement measures that address energy efficiency as well as climate ends. Newly emerging mechanisms such as carbon contracts for difference and green finance can offer a pro-environmental space in which firms can invest in energy storage technologies compatible with emissions reduction targets. In addition, efficient monitoring systems play an imperative role in monitoring the progress and making necessary changes in strategies to raise ambition for climate goals. This multifaceted strategy not only improves the viability of renewable energy but also contributes towards pursuing a comprehensive approach to combat climate change while ensuring sustainable economic growth (Egenhofer et al., 2019). In these situations, energy storage not only seems a technological necessity but also a keystone to systemic change.

9.3.2. Increasing Energy Efficiency

There is a need to enhance energy efficiency in the context of integration of renewable energy sources with energy storage technology. Increased energy consumption could undermine the sustainability of renewable technologies that are biased towards dependency on variable energy generation. With a rise in the use of energy efficiency measures across various sectors, the total energy demand can be significantly diminished. This reduction not only complements the capability of energy storage systems but also decreases fossil fuel reliance, thus integrating global sustainability objectives. Regulations, such as those in Brazil potentially at the disposal for policy-making decisions regarding energy investment, also help bring about this transition. Investments of the order of 0.5 trillion U.S. dollars can either pave the way for carbon-intensive infrastructure or drive progress towards a low-carbon economy, depending on how much weight is accorded to energy efficiency measures and renewable energy integration in future planning (Lucon et al., 2015). Renewable energy integration into the energy system also needs to consider technological advancements that enable efficiency. Data centers, owing to their traditionally energy-hungry nature due to their power consumption, are a natural choice for applying principles of energy efficiency. Through the application of advanced technologies such as low exergy air-conditioning technologies and optimization of IT load management, data centers can significantly reduce energy consumption before turning to renewable energy sources. The emphasis on efficient electrical supply and energy storage integration in smart grids can de-stress existing energy infrastructures, facilitating a more robust energy system. The emphasis on the implementation of advanced technical principles can thus increase the proportion of renewable energy while ensuring that energy consumption is sustainable and manageable (N/A, 2022). Moreover, the relationship between storage technologies and energy efficiency highlights the intricacy of modern energy systems. Waste heat reuse mechanisms and the utilization of free cooling methods not only enhance efficiency but also enhance the utilization of renewable energy. Through optimization of the general energy profile of systems, such strategies can create synergies that optimize both supply and storage times. Lastly, energy-efficiency-focused interventions have the ability to lower operating and capital expenses, thus enhancing economic sustainability and environmental ratings. Measuring energy efficiency through metrics such as Power Usage Effectiveness (PUE) assists in obtaining practical knowledge regarding areas that can be optimized, confirming that more energy efficiency is not only an operational matter but also a crucial component of providing productive energy storage alongside renewables integration (N/A, 2022).

9.3.3. Lowering Energy Costs

Lowering the energy cost is paramount in the effort of properly integrating renewable energy sources. One of the most significant means of achieving this is using energy storage systems. These systems enhance the efficiency of using renewable energy by storing extra energy generated at times of surplus production and releasing it during high demand or when there is low production. For instance, thermal energy storage is a feasible alternative, especially in the construction of buildings, which can reduce energy consumption by as much as 30% or more. Evidence indicates that such systems can guarantee more than 30% energy savings, thereby not only reduce the cost of operations but also encourage the utilization of renewable energy sources in the constructed environment (Padura et al., 2017). The compatibility between energy storage and renewable integration thus becomes an essential element in minimizing monetary expenditure and facilitating the shift towards a clean energy market. As cities evolve leaps and bounds to becoming carbon neutral, the application of energy efficiency solutions becomes crucial in minimizing energy cost. Major cities, like Boston, have recognized the need to adopt clean technologies in their transportation systems and buildings. The anticipated increase in power demand necessitates a shift away from fossil fuels to other sources of energy to ensure future energy requirements are met in a sustainable manner. Improved energy efficiency and local energy generation from sources like rooftop solar photovoltaics are essential in reducing energy use cost-related expenditures. Moreover, as urban areas attempt to reduce the consumption of natural gas and petroleum-based fuels, affordable energy sources remain intertwined with their broader social justice and environmental sustainability goals (Cleveland et al., 2019). Therefore, a strategic focus on energy efficiency and renewable integration forms the basis of the long-term goal of maintaining low energy prices in urban energy planning. Emerging energy storage technologies not only facilitate cost savings but also pave the way for building more sustainable and resilient infrastructures. Thermal energy storage materials technology holds great promise, especially in building systems striving towards zero energy consumption. Utilization of latent and sensible heat storage technology has been shown to attain impressive volumetric thermal energy storage capacities, hence promoting the implementation of renewable resources into the grid. The innovation equalizes the playing field of the energy bills through the ability of buildings to properly handle their thermal loads, paving the way for less peaky demand-side management methods and reduced peak energy bills (Padura et al., 2017). Thus, by optimizing energy use within buildings and leveraging renewable integrations, cities can harness significant economic benefits while simultaneously addressing climate change challenges.

9.4. CHALLENGES IN ENERGY STORAGE IMPLEMENTATION

The use of energy storage systems has numerous challenges to be addressed in a bid to facilitate the use of renewable energy sources in the grid. One of the major challenges is the technological limitations with respect to current energy storage media, such as batteries and hydrogen energy storage. While there is advancement, there are still limitations with respect to energy density, cycle lifespan, and operational efficiency to be addressed. For instance, the performance of batteries in large-scale integration is often tempered by aging and variable performance based on the environment. It is thus necessary for new innovations, for instance, hydrogen energy storage systems, to be integrated; the systems can enhance reliability by storing surplus energy as hydrogen with power-to-hydrogen technology, but they also have their own constraints, such as conversion efficiency and cost constraints (Mishra et al., 2023). Besides, regulatory and market forces are central drivers in shaping the energy storage environment. Existing energy policies cannot efficiently capitalize on the value provided by energy storage systems to the grid. Poor supportive regulatory environments can deter investment in storage technologies and restrict the advancement of important grid balancing services. Whereas utilities and policymakers are still grappling with how to incorporate renewable energy, it is becoming more apparent that innovative storage solutions for energy must be incorporated into policy and market designs so that they may be implemented. Containing energy storage within market structures will make available new revenue streams, prompting energy producers to invest in storage solutions that increase grid stability, lower operational costs, and ultimately make renewable energy integration projects achievable (Marzband et al., 2019). Finally, public perception and acceptance are also barriers to the deployment of energy storage. While the environmental benefits of renewable energy are as well appreciated as they are, safety, aesthetics, and land issues can come in the way of public acceptance of energy storage facilities. For example, large battery facilities can be suspicious in the eyes of local residents, because of fears over chemical spills or adverse environmental impacts. To respond to such apprehensions, aggressive outreach and education are required. Engaging communities early and openly discussing benefits and impacts of energy storage systems can garner public trust and pave the way for less controversial project approvals. Addressing these issues will be crucial to building infrastructure that will enable integration of renewable energy to allow effective use of energy storage systems as a crucial component of a clean energy future (Mishra et al., 2023).

9.4.1. Technological Limitations

In the quest for effective energy storage technologies to facilitate the use of renewable sources, technical limitations pose significant challenges. Any of the current storage technologies, such as lithium-ion batteries, are constrained by their capacity, lifespan, and material degradation over time. These batteries, required to balance intermittent sources like solar and wind, have short lifespans with frequent replacement, contributing not only to increased cost but also environmental impact from disposal and recycling. While innovations are being driven in battery management systems and chemistry, current technologies are not adequate for the amount of efficiency and capacity needed for large-scale deployment to contain excess energy generated during off-peak production hours. It is an application requiring innovative solutions that not only enhance existing technologies but also find novel sources of energy storage. The intersection of distributed energy resources, particularly in cities, also amplifies the technical complexities associated with energy storage. Rooftop solar installations, for instance, have vast potential to stabilize energy demand; however, they are generally underutilized due to grid capacity constraints and policy regulations. The research demonstrates that, even with the huge potential to create net exports to the electricity grid, most of the buildings consume more electricity than they can produce, which implies inefficiency in energy distribution (Cheng et al., 2020). Accordingly, transcending these technological limitations involves not only enhanced storage systems but an entire overhaul of the current energy infrastructure. Advanced policy and intelligent grid technologies can effectively improve distributed generation integration while energy storage equipment may be maximized to utilize their full potential. With the exception of solar photovoltaic challenges, hybrid energy systems are emerging as a critical solution to fill technology gaps in power production and storage. The setup of trans-critical carbon dioxide heat pumps with photovoltaic arrays and combined heat and power (CHP) is a solution towards multi-angled energy refurbishment and storage (Santoli et al., 2019). The systems are designed to be efficient and reduce the use of traditional fossil fuels, moving toward Near Zero Energy Building (NZEB) levels. Yet, realizing the deployment of these hybrid systems depends on their capacity to deliver stable, consistent energy output while addressing issues of high upfront investment and integration with current infrastructure. Thus, more research is needed in order to maximize these hybrid solutions, ensuring they can achieve the twin objectives of increasing energy storage capacity as well as encouraging sustainable energy applications across different settings.

9.4.2. Economic Barriers

Integration of renewable resources into power grids is a serious economic challenge, particularly when considering energy storage technologies. The intermittent nature of renewable sources such as solar and wind causes variability in electricity supply that complicates grid management. As highlighted in recent assessments, achieving sufficient grid flexibility is essential to facilitate the utilization of such intermittent energy resources (Huang et al., 2019). The price tag for upgrading the infrastructure and investing in future-ready energy storage can prove to be a significant challenge for most utilities, particularly for developing regions. Financial incentives and effective funding mechanisms are needed to spur investment in technologies that make the grid more flexible, reducing reliance on fossil fuels and constructing the way toward a more sustainable future for energy. Investment in renewable energy infrastructure is slowed down by money issues, with long-term implications for a nation's energy policy. Brazil, for instance, is faced with the urgent challenge of investing approximately 0.5 trillion U.S. dollars on energy infrastructure over the next decade. The direction of that investment will result in a reliance on carbon technologies or building a low-carbon economy (Lucon et al., 2015). This economic turning point necessitates that policy initiatives promote clean energy technologies, lest neglecting the shift to renewable investments has the risk of locking nations in unsustainable energy strategies. In addition, without robust economic policies to support energy storage initiatives, nations may fail to adequately integrate renewable resources, thus leaving the pursuit of aggressive climate and energy goals stagnant. In addition, differences in economic endowments between regions compound these issues, presenting a compounded situation for energy storage deployment. In places such as ASEAN, differences in readiness of the grid for flexibility reflect the disparate efforts in upgrading energy infrastructure (Huang et al., 2019). Economic impediments, in terms of restricted access to funding and absence of technological infrastructure, especially impact developing countries. This dilemma calls for regional collaboration to aggregate resources and knowledge, further illustrating how economic implications encroach on the development of energy storage technology. Encouraging investment through international pacts and development aid can help sidestep such challenges, enabling states to implement their energy integration policies effectively. Ultimately, conquering such economic barriers enables not just the integration of renewables but also economic stability and environmental well-being.

9.4.3. Regulatory and Policy Issues

The policy framework for energy storage for renewable integration is central to building a secure energy future. Pools of varied policies must be harmonized to encourage the installation of energy storage technologies, which are central to dealing with the intermittent nature of renewables such as solar and wind. In Los Angeles, for example, current policy regarding solar array interconnection refers to the significant disconnect while those systems have the capacity to produce vast quantities of energy, current policy actually leads to levels of wasted power output that are extremely high. More precisely, approximately 1700 MW of solar power is left untapped due to oppressive interconnection policies, which say that regulatory systems need to be reviewed in order to capture the true potential of renewable energy facilities. Streamlining these regulations enables jurisdictions to create an environment that not only supports energy self-sufficiency but also grid reliability. Further, international best practices say that the merit of enhancing grid flexibility in terms of regulatory systems. The ASEAN strategy to increase the share of renewable energy to up to 23% by 2025 indicates a shared commitment to address the need for clean energy and the management issues of grids. Most notable is the differential capacity of countries to modernize their grid structures, which indicates an actual imperative for having tailored regulation measures that consider the unique contexts and capacities of each nation. Since they have already been defined, forecasting equipment quality and availability of electricity markets are crucial determinants of readiness in every country toward renewable energy. By indirectly combating their own weaknesses in their regulation systems, ASEAN countries can better position themselves toward meeting renewable energy targets as well as improving grid resilience as a whole. Lastly, addressing the policy and regulatory issues of energy storage is not merely a matter of enhancing the infrastructure but also giving all communities equal access and opportunities. Vulnerable and low-income communities, such as those being discussed in the Los Angeles study, have a unique potential for clean technology investments to yield social returns in terms of energy equity. Policymakers must consider these dynamics in crafting inclusive structures that enable poorer communities to benefit from and contribute to the integration of renewables. Investment and policy innovations that take into account such neighborhoods' solar generation potential can translate into more engagement in the energy economy and contribute to enhanced energy resilience. Therefore, an intermultifaceted approach combining robust regulatory regimes with community-focused initiatives shall be imperative in achieving an integrated and just energy future.

9.5. CASE STUDIES OF SUCCESSFUL ENERGY STORAGE PROJECTS

To enjoy the possibility of energy storage in facilitating integration of alternative resources, careful review of successful case studies is essential. A great instance is the application of a hybrid energy storage project that combines a mixture of batteries and supercapacitors, with demonstration of the merits of integrated systems. This project presented useful insights into how interaction among diverse storage technologies can not only optimize performance but also mitigate the intermittency problem of renewable energy technologies. The ability to utilize smart control and management schemes, as highlighted in recent literature, also boosts the efficiency of these systems in accessing stable energy storage solutions. Such case studies point out that new strategies can bridge the gap between the generation and utilization of renewable energy and, consequently, drive the agenda for sustainable energy integration forward (Venkataramana et al., 2024). Another eminent case study is the one in the case of microgrid systems, which are key elements in the transition towards more intelligent energy networks. Microgrids facilitate the integration of distributed energy resources and storage facilities, hence enhancing the system reliability. A detailed analysis of a specific microgrid project showed the significant increase in the reliability indices, especially for critical loads necessitating stable power supply. The stochastic nature of renewable energy sources causes problems; however, advanced analysis methods, including Monte-Carlo simulations, are capable of effectively compensating for these issues. This widespread implementation of microgrids not only addresses reliability issues but also enables various stakeholders to explore low-cost energy solutions with less distribution expansion requirements (Ansari et al., 2017). Analysis of the successful energy storage projects identifies several recurring patterns that enable them to be successful. The necessity to install integrated storage systems is the first recurring pattern that is a major performance and reliability improvement driver. Case studies emphasize using advanced energy management systems with the use of new-generation algorithms to optimize operations. Besides that, sustainability is emphasized through all these projects since economic and environmental considerations are fundamental to ensuring sustainable long-term success. By embracing a synergistic strategy that combines technology, strategy, and sustainability, these successful cases provide a model for subsequent projects in energy storage. Ultimately, such projects offer lessons in policy directions and research priorities through lessons that strengthen energy storages understanding of its role in enabling renewable energy goals and building resilient systems (Venkataramana et al., 2024).

9.5.1. Tesla's Hornsdale Power Reserve

Tesla's Hornsdale Power Reserve is a pioneering instance of how energy storage systems can enable the integration of renewable generation into existing power systems. Located in South Australia, the plant uses advanced lithium-ion battery technology to manage the supply of electricity, particularly during peak demand or when renewable generation, such as wind and solar, is insufficient. By storing surplus power produced during peak generation hours, Hornsdale Power Reserve can provide electricity quickly back into the grid and hence enhance the stability and resilience of the grid. This aspect is important in a changing energy landscape, where the working problems related to frequency management and supply-demand equilibrium are becoming more advanced. Using such energy storage systems is the way ahead in the requirement of new power generation, towards a sustainable and stable energy future as suggested in recent studies of frequency control policies for Australia (Aziz et al., 2018). The operating process of the Hornsdale Power Reserve also leads towards an insight into the far-reaching implications of energy storage in the integration of renewable energy. This plant works harmoniously with traditional power stations, producing ancillary services that contribute significantly to grid stability. Tesla's battery technology has a fast response rate, allowing the reserve to cater to supply changes by renewable means, basically plugging into condition blackouts that can happen because of surplus demand or generation shortfalls. Recent studies indicate the manner by which the inclusion of high-capacity energy storages like Hornsdale can balance out the deficiencies in the intermittent nature of renewable energy supply, highlighting the importance of advanced energy management systems in contemporary electricity markets (Ding et al., 2022). Not only does its addition make power systems more reliable but also offers a blueprint for such initiatives across the world desirous of expanding their renewable energy capacities. Moreover, the Hornsdale

Power Reserve has set itself up as a model for future energy storage programs, proving the potential of hybridizing multiple energy storage technology. With both electrochemical systems and novel solutions combined, the plant aims to optimize efficiency without the environmental drawbacks of conventional energy production methods. Literature has established that such hybrid systems possess the capability to improve the general performance and profitability of energy storage systems and address problems with regard to energy restoration from renewable sources after disturbances (Ding et al., 2022). With the world's energy systems shifting more and more, Tesla's Hornsdale Power Reserve is a flagship benchmark that shows how effective energy storage makes intermittent renewable energy not an obstacle but a leap towards cleaner and sustainable energy (Figure 9.10).

Figure 9.10. *Hornsdale Wind Farm – world's largest Tesla battery storage.*

Source: https://ghd-p-001.sitecorecontenthub.cloud/api/public/content/974aa2c7f78d4ecca17ae3c0bc26c3e1?v=cf1d41c9&t=w1744h1090.

9.5.2. The Kauai Island Utility Cooperative (KIUC)

Kauai Island Utility Cooperative (KIUC) is the prime example of forward-looking integration of energy storage systems into its renewable energy policy, demonstrating a firm transition towards distributed power generation and increased grid resilience. KIUC, as a utility owned by the community, has focused investment on renewable energy resources, being able to leverage the islands' abundant solar and wind resources. The cooperatives' focus on energy storage technologies, such as battery systems, mitigates the intermittency of such renewable resources. With a more reliable power supply, KIUC enhances its capacity to serve customers during peak demand hours and lessens reliance on traditional fossil fuel-based generation. This aligns with broader industry trends that advocate for a more decentralized grid that is more resistant to disturbances, as highlighted by previous criticisms of centralized electric grids, particularly in the wake of Hurricane Sandy (Jones et al., 2015). KIUC's strategy also highlights the role of business model innovation in the deployment of energy storage technology. The cooperative has explored diversified applications of battery storage, and the worth of developing business models to address specific operational contexts, such as utility-scale structures and community-led initiatives.

Figure 9.11. *Kaua'i Island Utility Cooperative achieves 57.9% renewable generation for 2023.*

Source: https://ncbaclusa.coop/content/uploads/2024/07/Kauai-island-electric-coop-solar-500x333-1.png.

The fact that there is no one-size-fits-all battery storage solution implies that an accurate understanding of local energy needs, regulatory schemes, and market designs is essential, as discussed by Chalvatzis et al. (2019). KIUC's forward-thinking strategy manifests commitment to raising energy efficiency in addition to addressing the economic and environmental concerns inherent in energy storage deployment. To do so, the cooperative not only enhances its delivery service but also contributes to broader discourse on clean energy transition, which serves as a model for other utilities that are in the same position. Further, the installation of energy storage at KIUC demonstrates the cooperatives commitment to enhancing energy security and sustainability for its service territory. By investing in a range of storage technologies, KIUC is able to better balance electricity supply and demand swings, and to provide a stable power supply. This approach becomes increasingly necessary as climate change increases the effects of extreme weather events, necessitating that utilities start adopting resilient infrastructure approaches. KIUC's forward-looking strategy serves as a benchmark for how utilities can be proactive and innovative when meeting the challenges of evolving energy systems, toward a more resilient and sustainable power grid. As the focus on renewables integration continues, KIUC's activities are instructive about how to effectively regulate energy storage technologies, toward a future of greater energy autonomy and ecological stewardship (Figure 9.11).

9.5.3. The Hornsdale Power Reserve in Australia

The South Australian Hornsdale Power Reserve is among the most prominent energy storage technology advancements and a major impetus to the transition of renewable energy sources into the grid. The facility, constructed in 2017, is composed of large-scale lithium-ion battery systems whose aim is to stockpile excess energy generated through renewable energy sources such as solar power and wind power. This is a valuable skill since the variability of renewable energy has the tendency to produce supply-demand volatility. Harvesting excess energy during periods of peak production, the Hornsdale Power Reserve does a great deal to stabilize the grid, offering a stronger and more sustainable power infrastructure. The project highlights the pivotal role of strategic energy storage solutions in reshaping the energy map towards a sustainable future, as they not only enhance energy security but also facilitate the shift to low-carbon economies. Apart from its technical significance, Hornsdale Power Reserve is an exemplary case of the economic benefits of energy storage systems (ESS). The performance of this plant has proved that mass battery storage can bring down the price of electricity for consumers. By injecting stored power into the network during periods of peak usage, the reserve relieves pressure on the grid and minimizes the requirement for expensive peaking power plants. The project is also a guinea pig in terms of efficiency of operation: it can easily respond to grid commands, optimizing total energy management. This responsiveness is more important in order to offset the growing share of renewable energy in the grid. As energy markets transition to becoming decarbonized, initiatives like Hornsdale are significant in showing that economic viability and environmental responsibility can intersect, establishing a template where other parts of the world can learn a lesson in energy planning. The Hornsdale Power Reserve also educates more broad discourses surrounding the future of energy systems, particularly through its engagement with demand response technologies and regulatory frameworks. As regulators and the utilities work to address requirements for making the grid more resilient, ESS integration demonstrates a concrete pathway to flexibility of operation. Combining energy storage and demand-side management is capable of leading to effective energy utilization, as demonstrated by the study that documents the benefit of combining ESS and demand response programs. The Hornsdale project also tests scalability considerations and implementing the system on a different geography. As documented by contemporary research, the adaptability of battery technologies in various uses will be at the heart of developing infrastructure that can manage energy demands of the future and achieving sustainability targets, ultimately influencing policy and investment choices in the energy sector (Agbanusi et al., 2025; Grov et al., 2022) (Figure 9.12).

Figure 9.12. *Hornsdale Power Reserve, South Australia.*

Source: https://www.aurecongroup.com/-/media/images/aurecon/content/expertise/power-generation/power-generation_1280x768.jpg.

9.6. FUTURE TRENDS IN ENERGY STORAGE

The energy storage horizon is changing at a rapid rate, driven by the compelling need for reliable integration of renewable energy sources (RES). One key trend in this case is the growing application of digital twin (DT) technology for energy systems. Through this new concept, monitoring and forecasting of energy storage solutions' maintenance are made possible in real-time, increasing the efficiency and reliability of their operations. With more developed DT technology, its applications in grid management optimization will become increasingly significant. By surpassing current capacity limitations of energy storage, DT has the potential to achieve a more streamlined integration of variable RES like solar and wind, ultimately resulting in a cleaner energy infrastructure. Further, the potential for real-time data analysis and dynamism highlights the imperative of ongoing research and development in the domain, so that energy storage systems can dynamically adjust according to variable energy demand and supply uncertainty (Al-Shetwi et al., 2025). Concurrent with technological advancements, the economic environment surrounding energy storage solutions is undergoing a revolution. Investment in emerging battery technologies like lithium-ion and yet-to-be-developed alternatives like solid-state batteries will revolutionize energy storage capacity. Such advancements do not only promise increased energy density and longer lifespan but also play a role in cost savings that enhance their economic viability for wide-scale adoption. With governments and private sectors further emphasizing sustainable energy initiatives, the role of energy storage becomes more critical. The declining cost of batteries, coupled with supportive policy enablers, will likely drive growth in installations of energy storage. The trend not only enhances the stability and reliability of the grid but also plays a central role in facilitating the shift towards a more sustainable model of renewable energy that lessens environmental degradation (Mahmoud et al., 2024). Furthermore, energy storage technologies are integrated more into smart grid innovation, enabling improved management of energy resources. Advanced metering infrastructure (AMI), distributed control systems (DCS), and machine learning algorithms integrated together enhance the capabilities of energy storage systems to operate efficiently alongside RES. The technologies provide automatic management of energy supply according to real-time consumer demand and generation patterns to guarantee efficient dispatch of energy. Trending developments indicate that breakthroughs in artificial intelligence will continue to enhance energy management systems by tapping into massive data inputs to enhance decision-making models. These developments are vital in achieving resilience and sustainability in modern power grids and their capacity to accommodate future expansion in renewable energy integration and to counteract complications arising from climate change and energy security (Mahmoud et al., 2024) (Figure 9.13).

Figure 9.13. *The future of energy storage.*

Source: https://sustainability-news.net/wp-content/uploads/2023/07/shutterstock_1075034981.jpg.

9.6.1. Advancements in Battery Technology

Advancements in battery technology are a key determinant to overcoming the energy storage challenges for renewable integration. Recent advances have been focused on enhancing the energy density, durability, and recharge rate of batteries to make them increasingly viable options for the support of intermittent power supplies such as wind and solar. For instance, lithium-ion batteries, currently widespread across a set of applications ranging from electric cars to grid storage, deliver improved performance metrics compared to the previous technologies. Even more critical, research has indicated towards optimized battery systems with high tolerance to renewable energy penetration levels. These types of systems must have the capacity to handle immense supply as well as demand fluctuations, thus underscoring the importance of new battery technology in providing grid reliability and efficiency, as demonstrated by recent research on Volt/VAR Optimization and Control (VVOC) techniques (Arghandeh et al., 2013). Coupled with operational efficiency, battery technology advances have far-reaching implications for electric vehicle (EV) adoption that, in turn, support the overall philosophy of integrating renewable energy. With growing popularity of EVs, particularly those connected to ordinary household power outlets, there is an urgent requirement to evaluate their impact on low voltage power distribution networks (PDN). With more and more EVs, so does the loading on these infrastructures, which can lead to overloading.

Figure 9.14. *Advances in battery technology for energy storage.*

Source: https://beasfa.com/wp-content/uploads/2024/08/Picture1-1.png.

That advanced battery technologies are able to achieve coordinated charging strategies can mitigate such issues, enabling better load management on the power grid. Specifically, studies point out that with the use of such technologies, the adverse effects of uncontrolled EV charging may be significantly reduced, thereby the stability of the distribution grid ensured while ensuring a seamless shift to cleaner transport (Butans et al., 2016). Furthermore, the interlink between battery technology and the integration of renewable energy extends to policy and regulatory spheres. Strong research identifies critical weaknesses in existing methodologies that inhibit the use of advanced battery technologies in the distribution grid. As states transition toward aggressive renewable energy targets, such as Californias target of achieving 33% renewable penetration, integrating advanced energy storage solutions is vital. Closing the VVOC technology research gaps is essential to realizing the full potential of emerging battery technologies that can enhance voltage control and grid efficiency overall. In the future, facilitating innovation in battery technology has the potential to not only revolutionize energy storage models but also enable a stronger, sustainable energy culture as the world transitions more heavily to renewable resources (Arghandeh et al., 2013) (Figure 9.14).

9.6.2. Integration with Smart Grid Technologies

The integration of smart grid technologies with energy storage systems is crucial to the management of the inherent intermittency of renewable sources of energy such as wind and solar power. The traditional grid for electricity is increasingly faced with modern demands, necessitating a shift towards smart grid infrastructure that is able to handle both renewable energy consumption and generation effectively. The application of Information and Communication Technologies (ICT) has enhanced the reliability and flexibility of such networks, offering an adaptive infrastructure required to facilitate the utilization of renewable energy. Smart grids, in addition to optimizing energy flow and distribution, also facilitate active customers' participation in energy management, ensuring a sustainable energy environment. The implementation of these future technology's places energy storage at the forefront to maximize grid stability so that excess energy produced at peak production periods is stored properly to be used in the future, maximizing efficiency as a whole (Borghetti et al., 2013). Also, the expansion of microgrids, particularly in America, raises the revolutionary potential of decentralized energy solutions with the enabling power of smart grid technologies. These microgrids are stand-by networks that are not dependent on the main grid, and they supply energy resilience and stability to a single project site or community. Microgrids' combined nature facilitates real-time energy management, which is critical when combined with energy storage and renewable energy sources. Federal and local policies that have fostered the adoption of microgrids have added to the phenomenal technological and control mechanism advancements in these systems. As the professional literature demonstrates, effective pilot projects focus not only on technical viability but also on the general economic value of microgrid implementation, paving the way for lower-fossil-fuel-dependency energy systems and greater compliance with sustainability goals (Bao et al., 2018). With ongoing advances in energy storage and smart grid technologies, there are a range of issues still present relating to regulatory environments and technical specifications. Technological advancements at a faster pace demand end-to-end amendments to existing regulations that are able to encapsulate the smart grid characteristics and the use of energy storage in this perspective effectively. Efficient policies must adopt the multidimensional array of stakeholders beginning from utility companies to customers, in a manner so that everyone's interests remain intact. Another concern which has emerged is that of cybersecurity as there's increased usage of digital communication in handling the grid. Hence, creating a cohesive strategy that considers both regulatory and security issues will be of utmost importance in realizing the full potential of smart grids in providing renewable energy integration. This comprehensive strategy will ultimately provide a strong framework capable of addressing the transition to a sustainable energy future (Borghetti et al., 2013).

9.6.3. Potential for Decentralized Energy Systems

The idea of decentralized energy systems has been highly discussed in recent years, particularly as the complexity of incorporating renewable energy into the main grid heightens. It is possible for such systems to offer a tangible alternative to the traditional centralized energy generation, where energy is both generated and utilized locally. This shift not only enhances the complexity of the grid but also allows for active consumer participation in the energy market. Thanks to the utilization of innovative technologies such as smart grids and energy storage, decentralized systems can stabilize supply and demand shifts in a harmoniously balanced energy system. The National Conference on Renewable Energy conference highlights the importance of bringing these innovations up for discussion, such as microgrids and decentralized energy systems that pave the way towards a more sustainable energy system (Mishra et al., 2023). The advent of prosumers, or energy consumers who also generate and consume energy, further increases the trend towards decentralized power grids. These consumers tend to embrace renewable technologies, like solar cells, in order to generate their own electricity. The inclusion of prosumers gives birth to a peer-to-peer (P2P) electricity market model with direct energy exchange between peers. This economic system makes consumers independent in their decisions regarding power sources and fosters collaboration in energy communities. When prosumers engage in this emerging market, they are not only becoming more energy autonomous but also building the stability of the overall grid. The actualization of these P2P markets is a revolutionary opportunity for distributed energy systems to build new energy management and resource sharing paradigms (Baroche et al., 2018). As decentralized energy systems mature, they hold the promise of great gains in efficiency and sustainability. Decentralized systems make it possible to integrate a wide variety of renewable energy sources, presenting an answer for smoothing the intermittency generally inherent in such technologies. Energy storage technologies combined with decentralized systems have the ability to harvest excess energy produced during periods of peak output to use at another time. This interaction not only optimizes the use of energy but also minimizes waste and reliance on fossil fuels. Furthermore, decentralized system development inspires technological innovations and investment in local energy resources, enabling communities to be self-sufficient in generating the energy they need. Therefore, distributed energy systems offer a dual solution for a more sustainable and resilient energy landscape in conjunction with global action to mitigate climate change and promote the consumption of renewable energy (Mishra et al., 2023).

SUMMARY

Briefly, the addition of energy storage systems represents a milestone in the transition to an increasingly sustainable energy system that is renewables-driven. As indicated, there will be an escalating need for storage with the rise of distributed generation sources such as wind and solar power. Technological advancement in energy storage technologies like lithium-ion batteries and thermal storage can enhance the reliability and stability of grids as well as facilitate efficient handling of variable power generation. This not only reduces the intermittent nature of renewable energies but also helps in reducing reliance on fossil fuels, thereby minimizing greenhouse gas emissions. The importance of developing an integrated strategy for introducing various storage technologies into the power grid cannot be overstated, as it will have a central role to play in the long-term energy efficiency and sustainability (Ault et al., 2005). The challenges in financial and regulatory structures also illustrate the intricacies which are to be overcome while selling energy storage. Policymaking needs to change to accommodate groundbreaking business models for storage deployment, especially during continuous technological evolution. It can block the way to progress if the policymaking is fractured, undermining the potential of energy storage systems to provide economic solutions for supply and demand balancing. As countries such as Brazil take significant energy policy decisions, emphasis on the integration of energy storage in their infrastructure will influence not only economic stability but also global climate objectives. By investing in low-carbon technologies addressing energy concerns, such as thermal storage, countries can secure their future energy security and meet aspiring climate objectives (Lucon et al., 2015). Short of that, effective deployment of energy storage technologies is critical to actualize the maximum benefit of renewable energy sources. Careful deployment and application of storage technologies will allow utilities to address an increasingly changing energy landscape. Moreover, sufficient care needs to be taken when considering regulatory policies and investment plans to render these technologies economically viable and widely practiced. The journey to a cleaner, longer-lasting energy system will require government, industry, and academia to collaborate. As the globe shifts toward a more electrified future, the advancement of energy storage is a critical pillar in achieving a sustainable low-carbon economy for generations to come (Ault et al., 2005; Lucon et al., 2015).

The integration of alternative energy sources into existing electricity grids comes with a set of challenges that require new energy storage devices. The traditional energy infrastructure is no longer sufficient with the added alternative sources of energy and fluctuating electricity demand. As contended in the ongoing debate regarding smart grids, there is an urgent need to design new electricity grids with the ability to accommodate different energy inputs and provide reliability and flexibility. The application of Information and Communication Technologies (ICT) provides for a more robust management framework for such smart grids, thereby facilitating the deployment of energy storage systems that help make the transition to a low-carbon power sector (Borghetti et al., 2013). Not only does the deployment of such systems assist in mitigating grid instability caused by fluctuations in renewable energy, but it also facilitates effective practices of energy use, which are crucial for a low-carbon energy economy (Borghetti et al., 2013). In a consideration of the policy and financial structures for energy storage, it is revealed that regulatory frameworks play an important part in promoting the incorporation of renewable energy. A meticulous consideration reveals that differentiated subsidies for renewable energy are needed in a bid to optimize investments in energy storage. The subsidies can assist in lowering the amount of curtailment of excess production as well as address the need for large-scale storage infrastructure. Through the adoption of a public policy approach that considers the unique characteristics of various renewable technologies, stakeholders are in a position to optimize the overall performance of the electricity market. Through the implementation of this niche approach, not just the value of the renewable energy sources in operation is optimized but also supply and demand shifts are optimally balanced, thus bolstering the case that judiciously designed support policies have the potential to drastically lower reliance on energy storage systems (Abrell et al., 2019). The synergy of renewable energy and energy storage technologies is central to power sector decarbonization. Based on studies, the marginal value of the storage alternatives diminishes considerably in shorter durations, suggesting that an emphasis on short-term storage may not be the best response to addressing the energy supply-demand imbalance. Instead, long-term storage options, albeit with lower returns, remain crucial for ensuring national and regional energy security. This interplay between energy storage capacity guides the strategic direction necessary to steer policymakers, grid operators, and energy investors. With the acknowledgment of the character of operation of renewables as well as the evolving regulatory landscape, an effective strategy may be formulated, and successful transition to a more sustainable and decarbonized energy regime could be achieved (Abrell et al., 2019).

REFERENCES

1. Abrell, J., Rausch, S., & Streitberger, C. (2019). Buffering volatility: Storage investments and technology-specific renewable energy support. https://core.ac.uk/download/286703813.pdf (accessed on 30th September 2025).
2. Agbanusi, I. L., Ajiboye, J. A., Azubuike, C. H., Ojo, T., et al. (2025). Enhancing power grid resilience through energy storage and demand response. https://core.ac.uk/download/646468829.pdf (accessed on 30th September 2025).
3. Al-Kayiem, H. H. (2015). Hybrid techniques to enhance solar thermal: The way forward. https://elar.urfu.ru/bitstream/10995/57442/1/ijepm_2016_v1_1_05.pdf (accessed on 30th September 2025).
4. Al-Shetwi, A., Atawi, I., El-Hameed, M., & Abuelrub, A. (2025). Digital twin technology for renewable energy, smart grids, energy storage and vehicle-to-grid integration: Advancements, applications, key players, challenges and future perspectives in modernising sustainable grids. *IET Smart Grid.* https://www.semanticscholar.org/paper/f531924cffd551bd15db5c6de5ee5836cf5cfa22 (accessed on 30th September 2025).
5. Ansari, O. A., Chung, C. Y., & Safar, N. (2017). Reliability assessment of microgrid with renewable generation and prioritized loads. http://arxiv.org/abs/1709.07970 (accessed on 30th September 2025).
6. Arghandeh, R., Brown, M., Cibulka, L., Miller, R., et al. (2013). Distribution system voltage management and optimization for integration of renewables and electric vehicles: Research gap analysis. https://core.ac.uk/download/288432751.pdf (accessed on 30th September 2025).
7. Ault, G. W., Currie, R. A. F., Foote, C. E. T., McDonald, J. R., et al. (2005). Ubiquitous energy storage. https://core.ac.uk/download/9034038.pdf (accessed on 30th September 2025).
8. Axon, C., Balta-Ozkan, N., Cipcigan, L., Connor, P. M., et al. (2014). Scenarios for the development of smart grids in the UK: Literature review. https://core.ac.uk/download/155776124.pdf (accessed on 30th September 2025).
9. Aziz, A., Stojcevski, A., & Than Oo, A. M. (2018). Issues and mitigations of wind energy penetrated network: Australian network case study. https://core.ac.uk/download/395416549.pdf (accessed on 30th September 2025).
10. Bao, Y., Feng, W., Jin, M., Liu, Y., et al. (2018). A review of microgrid development in the United States–A decade of progress on policies,

demonstrations, controls, and software tools. https://core.ac.uk/download/323310623.pdf (accessed on 30th September 2025).

11. Baroche, T., Moret, F., Pinson, P., Soares, T., et al. (2018). Peer-to-peer and community-based markets: A comprehensive review. http://arxiv.org/abs/1810.09859 (accessed on 30th September 2025).
12. Barrios Padura, Á., Chacartegui, R., Lizana Moral, F. J., Valverde Millán, M., et al. (2017). Advances in thermal energy storage materials and their applications towards zero energy buildings: A critical review. https://core.ac.uk/download/157763138.pdf (accessed on 30th September 2025).
13. Becerra Villanueva, J. A., Calderón, B., Carvalho, M., Chacartegui, R., et al. (2019). Transcritical carbon dioxide charge-discharge energy storage with integration of solar energy. https://core.ac.uk/download/322845447.pdf (accessed on 30th September 2025).
14. Beluco, A., Canales, F. A., Guezgouz, M., Jurasz, J., et al. (2019). A review on the complementarity of renewable energy sources: Concept, metrics, application and future research directions. https://core.ac.uk/download/286642706.pdf (accessed on 30th September 2025).
15. Borghetti, G., Menga, G., Procaccianti, G., Kok, K., Ardito, L., Morisio, M., Mo, D., et al. (2013). Smart grid technologies in Europe: An overview. https://core.ac.uk/download/pdf/11431923.pdf (accessed on 30th September 2025).
16. Butans, E., Nieto-Martin, J., & Zafred, K. (2016). Electric vehicles–Effects on domestic low voltage networks. https://core.ac.uk/download/74409962.pdf (accessed on 30th September 2025).
17. Chalvatzis, K., Kaldellis, K., Kondyli, E., Li, X., et al. (2019). Bringing innovation to market: Business models for battery storage. https://core.ac.uk/download/196594876.pdf (accessed on 30th September 2025).
18. Cheng, D., Federico, F., Fournier, E., Gustafson, H., et al. (2020). Net solar generation potential from urban rooftops in Los Angeles. https://core.ac.uk/download/322493841.pdf (accessed on 30th September 2025).
19. Clemente Jul, M. del C., Funez Guerra, C., & Montes Ponce de León, M. (2011). Assessment of hydrogen as sustainable clean energy. https://core.ac.uk/download/148660705.pdf (accessed on 30th September 2025).
20. Cleveland, C. J., Fox-Penner, P., Pollack, A., Walsh, M., et al. (2019). *Carbon Free Boston: Energy Technical Report*. https://open.bu.edu/bitstream/2144/39216/2/CFB_Energy_Technical_Report_190514.pdf (accessed on 30th September 2025).

21. Coccolo, S., Kuehner, A. L., Mdeihli, N., Mohajeri, N., et al. (2017). Extending building integrated photovoltaics (BiPV) using distributed energy hubs: A case study in Cartigny, Switzerland. https://core.ac.uk/download/631287079.pdf (accessed on 30th September 2025).
22. Curto, D., Di Silvestre, M. L., Franzitta, V., Martorana, P., & Montana, F. (2024). Energy storage systems: A comparison of different technologies and possible application in Sicily. https://core.ac.uk/download/638994827.pdf (accessed on 30th September 2025).
23. De Santoli, L., Garcia, D. A., Lo Basso, G., Piras, G., et al. (2019). Dynamic simulation model of trans-critical carbon dioxide heat pump application for boosting low temperature distribution networks in dwellings. https://core.ac.uk/download/188832812.pdf (accessed on 30th September 2025).
24. Ding, J., Ding, Y., Sun, L., et al. (2022). Energy storage for black start services: A review. https://core.ac.uk/download/519561305.pdf (accessed on 30th September 2025).
25. Egenhofer, C., Elkerbout, M., & Rizos, V. (2019). *Circular Economy for Climate Neutrality: Setting the Priorities for the EU* (CEPS Policy Brief No. 2019/04, 22 November 2019). https://core.ac.uk/download/286731360.pdf (accessed on 30th September 2025).
26. Elmenreich, W., & Sobe, A. (2013). Smart microgrids: Overview and outlook. http://arxiv.org/abs/1304.3944 (accessed on 30th September 2025).
27. Grov, S. (2022). Electrochemical energy storage for an HVDC transmission system connected to an offshore wind farm. https://core.ac.uk/download/646127150.pdf (accessed on 30th September 2025).
28. Guarnieri, M., Mattavelli, P., Petrone, G., et al. (2016). Vanadium redox flow batteries: Potentials and challenges of an emerging storage technology. https://core.ac.uk/download/80155277.pdf (accessed on 30th September 2025).
29. Guntreddi, V., Yegireddy, N. K., Vasupalli, M., Ramakrishna, R., Reddy, R. M., & Aakula, S. (2024). Storage solutions for sustainable future: Integrating batteries, supercapacitors, and thermal storage. https://core.ac.uk/download/620307867.pdf (accessed on 30th September 2025).
30. Huang, Y. W., Kammen, D. M., & Kittner, N. (2019). ASEAN grid flexibility: Preparedness for grid integration of renewable energy. https://core.ac.uk/download/323310932.pdf (accessed on 30th September 2025).
31. Jones, K. B., Ruiz, S., & Thomas, K. R. (2015). Promoting clean reliable energy through smart technologies and policies: Lessons from three distributed energy case studies. https://core.ac.uk/download/154564805.pdf (accessed on 30th September 2025).

32. Kiasari, M. M., Ghaffari, M., & Aly, H. H. (2024). A comprehensive review of the current status of smart grid technologies for renewable energies integration and future trends: The role of machine learning and energy storage systems. *Energies*. https://www.semanticscholar.org/paper/cbe0ae36d1805023045628693b9d7e080b346c57 (accessed on 30th September 2025).

33. Lucon, O., Fransen, T., & Romeiro, V. (2015). Bridging the gap between energy and climate policies in Brazil: Policy options to reduce energy-related GHG emissions. https://core.ac.uk/download/75782972.pdf (accessed on 30th September 2025).

34. Marzband, M., Mirzaei, M. A., Mohammadi-Ivatloo, B., et al. (2019). Day-ahead network-constrained scheduling of CHP and wind-based energy systems integrated with hydrogen storage technology. https://core.ac.uk/download/199234955.pdf (accessed on 30th September 2025).

35. Mishra, S., Nayak, B. T. M., et al. (2023). National conference on 'Renewable Energy, Smart Grid and Telecommunication–2023. https://core.ac.uk/download/588430838.pdf (accessed on 30th September 2025).

36. N/A. (2022). Advanced concepts for renewable energy supply of data centres. https://core.ac.uk/download/548537672.pdf (accessed on 30th September 2025).

37. Rious, V. (2025). *The Design of The Internal Energy Market in Relation to Energy Supply Security and Climate Change*. https://core.ac.uk/download/pdf/6768188.pdf (accessed on 30th September 2025).WW

CHAPTER 10

SMART GRID AND DEMAND RESPONSE

Contents

INTRODUCTION

In the past years, the shift in the energy landscape has necessitated more efficient and sustainable electricity infrastructure. The concept of the smart grid is at the forefront of this revolution by integrating advanced communication, automation, and data analytics technologies into the traditional electrical grid. All these advances are crucial in being able to manage the increasing levels of complexity brought about by decentralized production sources like wind and solar energy. By enabling two-way communications between utilities and consumers, smart grids enhance the ability to track energy consumption and balance demand and supply. Demand response programs built into that initiative further bolster it by allowing consumers to adjust energy usage in response to price signals or grid needs, thereby enhancing system reliability and efficiency overall (Bucur et al., 2019). The translation of these dynamic behaviors into controllable systems is called for to address the pressing challenges of energy sustainability and security. The interaction of smart grid technology and demand response programs is a promising way to act against the problems brought by traditional energy management. With the spread of renewable energy sources, it becomes increasingly challenging to maintain an equilibrium balance between supply and demand.

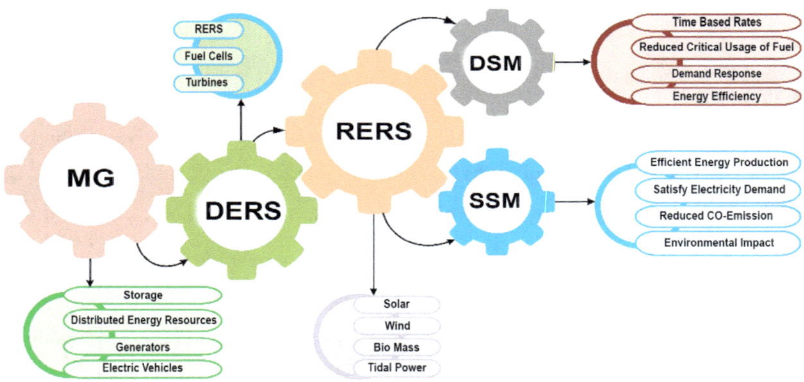

Figure 10.1. *Conceptual approach of smart grid (SG).*

Source: https://www.mdpi.com/energies/energies-15-02003/article_deploy/html/images/energies-15-02003-g001.png.

Demand response programs have an important role by inducing consumers to modify the consumption of their electricity at peak times or in response to fluctuating energy prices. In addition to providing relief to the grid, they encourage energy efficiency, and as a result, which saves greenhouse gas emissions. Besides that, the implementation of demand response programs provides worthwhile business prospects for corporations, particularly energy service companies (ESCOs), to promote their business and gain additional streams of revenue by undertaking energy efficiency projects (Bucur et al., 2019). The twin-targeted model is in support of the imperativeness of smart grid technology in the maximization of sustainable energy solutions. Realistic implementations of smart grid and demand response technologies unveil innovative strategies to increase energy management. As an example, research at UC Berkeley has demonstrated the value of creating Building-to-Grid testbeds, where building energy flows are closely monitored and analyzed. This project employed cutting-edge metering technologies and collaborative efforts on the part of several academic and industry partners to understand energy behavior and grid interactions. By examining the means of making energy use more grid-conformant, such projects illustrate the ability for greater efficiency on both the grid and the building. These findings illustrate the urgent necessity for adding smart grid technologies and demand response solutions to addressing the multifaceted issues presented by contemporary energy systems and in clearing the path for a more sustainable future (Figure 10.1).

Definition of Smart Grid

The concept of the smart grid is a revolutionary technological innovation in the management of electricity distribution, characterized by its application of cutting-edge technology to facilitate communication and efficiency in the electric structure. Essentially, a smart grid operates on two-way communication networks that provide for real-time interaction between electricity providers and consumers. It is this interactivity that enables improved monitoring and control of energy flow, enabling the management of intermittent renewable resources effectively. The application of smart meters and advanced communication networks is essential in establishing the dynamism, enabling users to communicate with their consumption of energy and be included in energy demand measures—a central element when one takes into account the role of demand response in shaping consumer behavior and patterns of use. Demand response is one of the most important utilities in the smart grid, allowing consumers to alter electricity consumption based on availability and price signals. Providers can impose time-dependent tariffs using Autonomous Demand Response (ADR) systems that encourage customers to change energy usage patterns so as to enhance overall grid performance. Despite the potential of ADR systems to save money and manage electricity more efficiently, it is not possible to accurately predict user energy demand. Shifts in consumption patterns among users can result in unforeseen peaks in demand, undermining the objectives of cost-effective energy management. In response to such problems, software such as the APD-Analyzer has been developed to assist Distribution System Operators in checking whether tariff structures satisfy demand response targets, so that the planned benefits of installing smart grids are realized even in the face of user uncertainty (Elmegaard et al., 2018). The development and implementation of smart grid technologies are influenced by a set of interconnected factors that involve supply and demand sides. Regardless of its robust political and financial backing, there still exists a necessity for enhanced theoretical and operational definitional precision. The most common key factors associated with smart grids include effective supply management, enhanced two-way communication, and the use of information technology to control responsive demand (Axon et al., 2014). These are aimed at addressing broader environmental goals, such as the reduction of carbon, while simultaneously responding to public needs driven by data security, privacy, and the affordability of new technology. Thus, effective smart grid deployment relies not just on technological innovation but also on creating a platform for stakeholder participation, regulatory issues, and public acceptance, all of which must be achieved in order to make a sustainable energy future a reality (Figure 10.2).

Figure 10.2. *Smart grid.*

Source: https://electricalacademia.com/wp-content/uploads/2017/11/smart-grid.jpg.

Importance of Demand Response (DR)

Demand response (DR) has never played a more central role in the smart grid, as it is an essential mechanism for real-time supply-demand equilibrium. With DR, utility companies can incentivize customers to reduce or re-tune their usage of electricity during peak periods and thereby alleviate grid pressure. This proactive approach not only minimizes the need for additional power-generating stations but also encourages consumers to take a more active role in energy consumption habits. With dynamic pricing systems, utilities can control peak load effectively while creating a more responsive energy system. Such initiatives are critical in a climate where renewable forms of energy are dominating the scene due to their inherent intermittency. Through this, the integration of DR in smart grid infrastructure supports a transition to a more sustainable and robust energy system that aids climate goals as well as reduces end-users' and utilities' operational expenditures. Furthermore, smart grid infrastructure deployments of DR solve various socio-technical challenges beyond mere energy management. The importance of establishing public acceptance and confidence with demand response technologies cannot be overemphasized, since consumers may be more resistant to alter their energy consumption habits. All stakeholders from the regulators to the consumers must take part in deliberations regarding the benefits of DR programs and attendant risks for ensuring proper deployment. As revealed in current research, the installation of smart grids is politically and financially supported with the expectation that it will enhance carbon reduction targets and maximize energy efficiency in general (Axon et al., 2014).

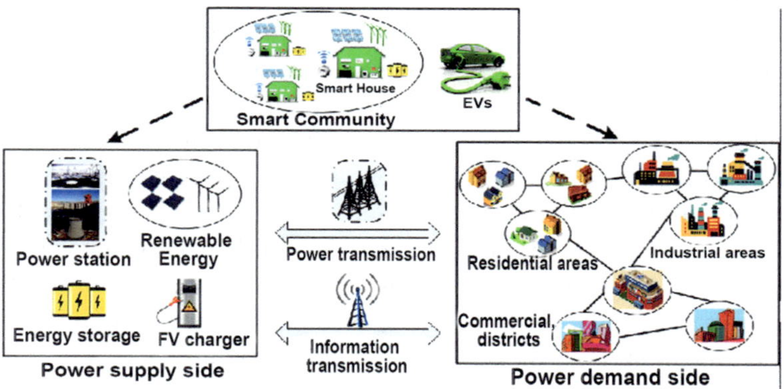

Figure 10.3. *Demand response in the smart grid.*

Source: https://www.researchgate.net/profile/Sabita-Maharjan-2/publication/298902712/figure/fig2/AS:347841848463361@1459943367431/Demand-response-in-the-smart-grid.png.

The advantages, nonetheless, come at the cost of having to cater to challenges such as data security, privacy, and susceptibility to cyber-attacks. Hence, the creation of a cohesive strategy encompassing regulatory mechanisms and consumer education is instrumental in optimizing the functioning of demand response in smart grids. Finally, the economic impacts of effective demand response programs illustrate the intricate relationship between utility price and consumer action. Research indicates utilities could utilize DR to charge revenue-maximizing prices alongside responding to peak load constraints (Afsar et al., 2016). This equilibrium helps utilities achieve maximum revenue while minimizing user cost, hence creating a win-win scenario for the stakeholders. Moreover, by including consumer choice and task scheduling in their energy pricing models, utilities can create a better energy experience that encourages participation in DR programs. As the energy market matures, embracing demand response is a critical component in ensuring a financially viable energy ecosystem. Briefly, demand response development not only enhances grid efficiency and reliability but also fuels economic and technological innovation within the backdrop of smart grid infrastructure, ultimately resulting in a greener and smarter future energy (Figure 10.3).

10.1. THE CONCEPT OF SMART GRID

The Smart Grid is a revolutionary plan for transforming the electric power delivery system with the integration of advanced technologies and communication networks. The new paradigm is focused not only on efficiency and reliability but also on allowing utilities and customers to engage in real-time interactions. By integrating digital communications technologies, the Smart Grid facilitates two-way communications, and customers can react to real-time prices while being empowered to adjust energy consumption accordingly. This responsiveness is instrumental in minimizing peak load demands and maximizing overall energy use efficiency, which is precisely what the aims of demand response programs are. Furthermore, as greater quantities of distributed energy resources (DERs), such as solar panels and electric vehicles, become more popular, the Smart Grid can simply accommodate these resources, maximizing their inputs into the electrical grid and reducing the demand for traditional energy supplies. Among the key elements in the Smart Grid is the integration of local flexibility markets, which enable the smarter way of providing and consuming energy. Research indicates that the various stakeholders, including consumers, aggregators, and distribution system operators, get the opportunity to engage in the sale and purchase of flexibility services through such markets. Such services comprise load management, distributed resource generation of power, and other demand response alternatives, ultimately boosting the grids operational efficiency. With participation allowing participants to compete in providing flexibility, local markets are capable of reducing massive infrastructure upgrades while concurrently slashing operational costs. This synergy is particularly critical in enabling the increased penetration of renewable sources of energy, as evidenced by the local flexibility framework, which can enhance the hosting capacity of distribution networks significantly (Sumper et al., 2018). Smart Meters are a critical part in the Smart Grid, enabling an enhanced robust and responsive system of energy management in homes. With the use of Smart Meters, customers actively observe and manage their energy consumption, and synchronize the energy usage of adjacent houses or multi-unit residencies. The proposed framework highlights this cooperative aspect by establishing paradigms for communication that enable sharing of consumption data among connected devices. Such collaboration not only optimizes the energy distribution but also employs alternative energy sources, like solar energy, to support energy efficiency in communities. Besides, performance analysis proves that the systems guarantee the longevity and functioning of Smart Meters without sacrificing, while enabling efficient demand-side management actions working towards the overall goals of the Smart Grid (Atzori et al., 2014) (Figure 10.4).

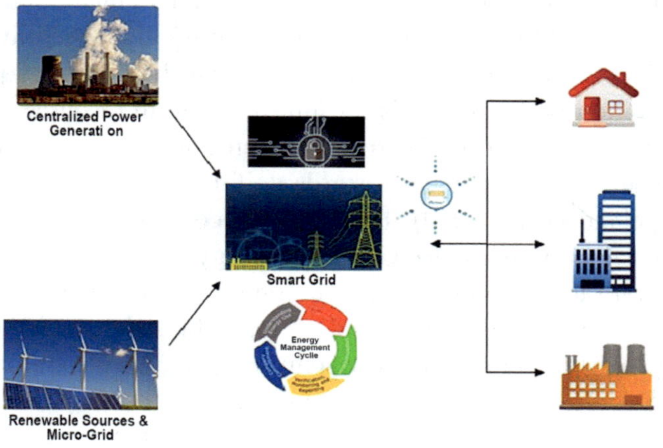

Figure 10.4. *Working concept of smart grid.*

Source: https://ars.els-cdn.com/content/image/1-s2.0-S2090447920301064-gr1_lrg.jpg.

10.1.1. Key Components of Smart Grid Technology

At the foundation of Smart Grid technology is a sophisticated integration of several components for the purpose of increasing energy efficiency, reliability, and consumer engagement. At the center of which the realization of these aims takes place are Smart Meters, providing real-time readings of electricity consumption, allowing consumers to make well-informed decisions about their energy usage. This real-time feedback has a key role to play in demand response programs, where customers modify their energy consumption patterns based on utility-provided signals, e.g., time-varying rates during peak hours. The application of Smart Meters enables a more engaged relationship between customers and energy companies, leading towards a more responsive energy system. With the heightened importance of involving consumers in managing energy, the Smart Meters remain a vital component within the overall system of Smart Grid technology that promotes efficiency and responsiveness in energy consumption. Advanced sensors and automation devices dispersed throughout the energy infrastructure are another core component of Smart Grid technology. These devices facilitate round-the-clock monitoring and control of the electricity grid to ensure that any issues such as power outages or inefficiencies are identified and corrected in real-time. With the Internet of Things (IoT), Smart Grids connect numerous appliances that are capable of sharing information and receiving commands, hence creating an adaptive and dynamic energy system. Such two-way communication of information not only enhances the operation efficiency of the grid but also enhances the security and reliability of energy supply. The inclusion of IoT devices in Smart Grid systems as proposed in existing research provides the basis for improvement in resource management for optimization and support of demand response initiatives, a demonstration of the significance of IoT devices towards unifying technology and consumer demand (Copeland et al., 2019). Furthermore, communication infrastructure is also a crucial part of Smart Grid technology that facilitates interaction between the various elements. Effective communication networks allow for free data exchange among utilities, end-users, and regulatory authorities, thus guaranteeing that the entire energy system is operated in harmony. Such an infrastructure comprises different communication technologies such as fiber optic, wireless networks, and advanced telemetry systems that enable real-time monitoring and control of power sources. Demand response program efficiency relies on robust communication infrastructure that can effectively convey knowledge of energy prices, system operation, and grid needs. As highlighted in recent reviews, upgrade of such communication networks is core to building a more responsive and secure energy future, owing to their core role in the Smart Grid era (McIlvennie et al., 2020) (Figure 10.5).

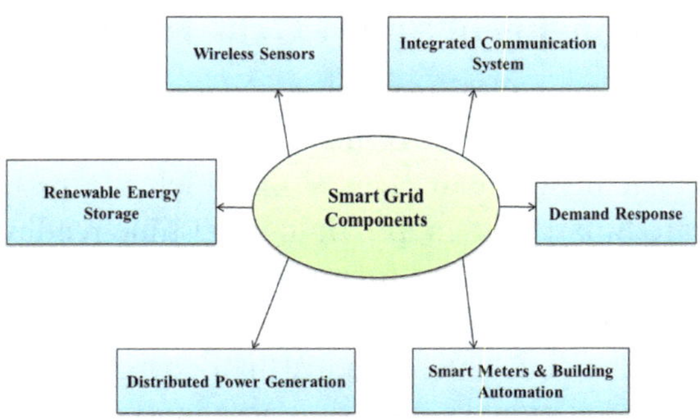

Figure 10.5. *Diagram of smart grid components.*

Source: https://www.stet-review.org/articles/stet/full_html/2024/01/stet20240131/stet20240131-fig2.jpg.

10.1.2. Benefits of Smart Grid Implementation

Smart grid implementation has numerous benefits which contribute towards ensuring stability and sustainability of energy systems. Undoubtedly, one of the most significant benefits is enhanced energy management through enhanced communication and information sharing between consumers and utilities. Smart grids are enabled by advanced information technology to make real-time monitoring and control of energy flows possible, thereby enabling utilities to react more dynamically to shifts in demand. This responsiveness not only optimizes energy supply but also facilitates the integration of intermittent renewable power sources. As discussed earlier, smart grids facilitate the transition to a low-carbon energy future through the promotion of effective management practices and carbon-reduction targets. The political and economic support these technologies receive further solidify their capacity to create new employment opportunities and reduce energy production and distribution costs (Axon et al., 2014).

The second significant aspect of deploying smart grids is enhancing energy resilience and reliability. With decentralized power generation and the use of microgrid technologies, communities can increase energy security. Microgrids, as indicated, are key to this mission by providing energy solutions that are dependable even when the main grid fails. The ability of communities to function independently in cases of disaster enhances energy resilience in general, demonstrating a trend towards decentralized energy systems with service continuity. Supportive policies at national and state levels have driven many microgrid projects, and these have served as reference examples that have pushed stability in community power systems forward. They not only showcase the actual benefits of smart grids but also set an example for other regions wanting to emulate their approach to energy management (Bao et al., 2018). Moreover, the smart grid feature of triggering demand response functions is a significant step towards energy efficiency. Demand response programs encourage customers to change their energy use at peak hours on the basis of incentives or time-of-use rates, altering the very nature of energy use. This shift not only reduces pressure on the energy grid during peak usage times but also allows consumers to play an active role in managing their energy use. By utilizing smart technologies, consumers can be informed and motivated to reduce their energy usage or reschedule it to off-peak hours, resulting in reduced energy expenditure and a lower demand for additional power generation facilities. The synthesis of these aspects exemplifies how implementation of the smart grid promotes an increased sustainability in the energy sector, ultimately proving advantageous for both consumers and utilities (Axon et al., 2014) (Figure 10.6).

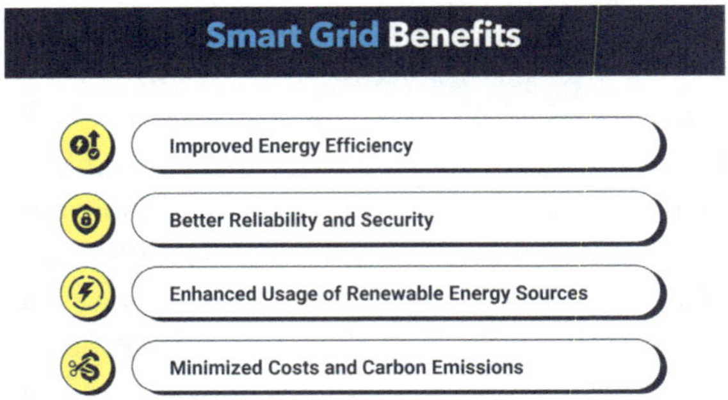

Figure 10.6. *Smart grids improve energy efficiency, reliability, and renewable energy use while minimizing costs and carbon emissions.*

Source: https://assets.eweek.com/uploads/2024/11/EW_SmartGrids_2024_MEL_rnd2-03-1024x683.png.

10.1.3. Comparison with Traditional Power Grids

The shift from traditional power grids to smart grids is a paradigm shift in terms of consuming and managing energy. Traditional power grids are predominantly characterized by one-way flow of electricity whereby electricity is generated in centralized systems and distributed to consumers with no instant feedback mechanisms. This kind of arrangement is prone to inefficiencies and weaknesses such as energy losses in transit and failure to respond to varying energy requirements. On the other hand, smart grids utilize advanced technologies, such as IoT, to facilitate two-way communication between utilities and consumers. With two-way interaction, enhanced monitoring, and control of energy delivery with real-time modifications that optimize efficiency and reduce waste are achieved. Thus, the paradigm of the smart grid facilitates a more robust and flexible power system capable of managing intermittent energy sources and fluctuating consumer loads, something traditional grids can barely perform with efficiency. A significant aspect of smart grids is the addition of demand response (DR), which is a big advantage over traditional power systems. In traditional grids, supply and demand are typically decoupled from one another, which makes it challenging to balance the two, especially during peak demand. Demand response management (DRM) enables real-time change of consumer power usage according to price signals or grid needs. Through the application of advanced algorithms and forecasting techniques, such as artificial neural network-based methods, smart grids can optimize demand-side management for ensuring a secure energy supply with lower cost to the consumer. This capability is essential to enhance operational efficiency and reduce the environmental impact of energy consumption. Traditional grids are not adaptive in this way, leading to inefficiencies and potentially increased operating costs and emissions through use of fossil fuels during peak demand. Embedding smart technologies into the grid enhances operations effectiveness while ushering in a new era of energy sustainability. Intelligent grids incorporate advanced technologies like predictive maintenance, renewable energy integration, and peer-to-peer energy trading via IoT, which collectively transform the energy system into a more efficient and greener one. For instance, predictive maintenance allows infrastructure to be proactively managed, resulting in reduced downtime and higher reliability, which is an area conventional grids fail because they follow a reactive maintenance approach. Secondly, such innovations also support dynamic pricing strategies which empower consumers to manage their energy use decisions, resulting in greener consumption behavior. The shift towards IoT-enabled smart grids is a reflection of the advancement towards cleaner energy consumption, ultimately leading to a greener and secure future for both consumers and utilities (Belhaiza et al., 2024; Jain et al., 2025).

10.2. DEMAND RESPONSE EXPLAINED

Demand response (DR) is a crucial component of contemporary energy systems, particularly in the case of smart grids. DR is a phrase used to denote the actions taken by utility companies to manage consumers' energy use in accordance with supply. By offering incentives to consumers for lowering or deferring electricity use during peak usage times, DR programs enhance grid stability and reduces dependence on additional power generation infrastructure. Not only does the use of demand response programs enhance energy reliability, but also environmental sustainability by limiting the use of fossil fuels that overwhelm the grid during peak times. The complement of new technologies, such as smart meters and automated control systems, also maintains the effectiveness of these programs, demonstrating the way in which consumer participation can be invaluable for both end-users and utilities. The process of load balancing is inherent to understanding the demands of demand response. By providing users, ranging from individual residential households to commercial organizations, greater control over their energy consumption, demand response programs are developing a new class of energy consumers who are known as prosumers. These are the consumers who take an active role in energy generation and consumption, often using renewable sources of energy, thereby creating a self-generative system of energy. Information and communication technology (ICT) integration takes center stage in this dynamic, facilitating real-time utility-consumer communication. In addition to supporting real-time decision-making, this supports the roll-out of innovative demand response paradigms that promote energy efficiency. The ultimate result is a more robust energy system that is better positioned to manage fluctuating demand and developing energy challenges. Big data analysis is also complemented by the implementation of demand response measures. Utilities can sort through vast amounts of consumption information to learn from trends, identify load management potential, and tailor incentive offers to consumers. Furthermore, education initiatives like serious games are able to actually induce energy-saving behaviors among users, changing the way consumers interact and experience their energy use. Such technology encourages greater engagement in demand response programs as energy savings become more appealing and within reach. With cities adopting such technology, the potential for demand response to improve the overall performance of the grid grows stronger. Through their focus on outreach to communities and technology integration, demand response projects, as demonstrated in the UC Berkeley and other university studies, represent a huge step towards the building of smart grid infrastructure for an energy-sustainable future (Bonetto et al., 2017; Peffer et al., 2010) (Figure 10.7).

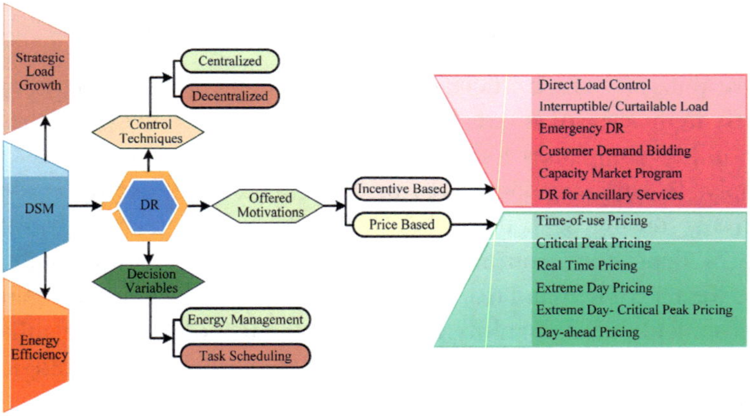

Figure 10.7. *DR infrastructure.*

Source: https://www.mdpi.com/energies/energies-15-02003/article_deploy/html/images/energies-15-02003-g004.png.

10.2.1. Definition and Purpose of Demand Response

Demand Response (DR) refers to a set of measures and mechanisms that are meant to regulate consumer electricity use in relation to supply. This can involve anything from incentivizing customers to reduce or postpone their consumption during peak times to providing flexible energy products that support grid management. The function of DR serves a two-pronged intention: one, it stabilizes the grid by alleviating the pressures of peak loads, which are normally at greater rates and at the risk of blacking out. Two, it really encourages energy efficiency by making consumers more conscious of the pattern of their usage. Thus, DR is an essential component in the context of modern energy management, enhancing reliability and promoting sustainability, particularly with the growing need for electricity amidst emerging renewable sources. Adding Demand Response to a Smart Grid network significantly enhances its operational capabilities.

Figure 10.8. *How does demand response work?*

Source: https://sympower.net/hs-fs/hubfs/Imported_Blog_Media/sympower-demand-response-.png?width=2400&height=1601&name=sympower-demand-response-.png.

Smart Grids utilize advanced technologies such as real-time communications and smart meters to facilitate the communication between the consumers and the utility providers. With this convergence of technology, access to demand-side resources is now achievable, allowing utilities to manage demand dynamically. Consequently, DR programs can provide real-time incentives or notifications, allowing consumption adaptation depending on price signals or grid conditions. As indicated in recent research, the movement toward Smart Grid technologies is a revolutionary approach to the management of electricity, one with a focus on flexibility and effectiveness toward not only the traditional energy supply challenges but also toward contemporary issues of sustainable energy practice and consumer participation (Burstein et al., 2011). The end purpose of Demand Response goes beyond its immediate monetary benefits or grid reliability improvements. It plays a vital role in shaping bigger energy policy goals, such as enhancing the resilience of the energy system and reducing greenhouse gas emissions. When utilities integrate DR into their operating platforms, they can directly engage consumers in energy saving measures, and consequently, contribute to making the energy system more sustainable. In addition, the smart and adaptive nature of Smart Grids allows for maximum integration of intermittent renewable sources of energy, such as solar and wind, which are supported by responsive demand management strategies to mitigate their natural intermittency. This convergence results in Demand Response as one of the main components of managing the complexity of modern energy systems, hence driving technological advancement as well as significant progress in environmental sustainability (Borghetti et al., 2013) (Figure 10.8).

10.2.2. Types of Demand Response Programs

Demand response (DR) schemes are an integral part of efficient operation of smart grids as tools that align consumer electricity consumption and supply availability. Demand response schemes could be categorized broadly into incentive-based and price-based schemes. Incentive-based demand response induces consumers to save energy consumption during peak demand times through payment for participating. This is particularly relevant for large commercial customers, who can schedule their loads according to the signals from utilities. Price-based demand response, by contrast, uses real-time pricing mechanisms to encourage customers to shift their loads to off-peak hours and flatten load curves and reduce the load on the grid. These classifications exhibit the flexibility and versatility of DR programs, which are required to enhance grid reliability and make the best use of renewable energy resources. Among the various classes of demand response programs, load management and ancillary services are unavoidable components in ensuring grid stability. Load management is aggressive managing or holding back electricity usage during peak periods, and it accounts for decreasing the demand for costly peaking power plants. Ancillary services, on the other hand, include demand response programs that assist in sustaining grid reliability by providing services like frequency regulation and reserves in response to unforeseen swings between demand and supply. These ancillary markets create new incentives for consumers as well as firms, such as electric vehicles and data centers, to encourage their engagement in demand response. The inclusion of these various services highlights the importance of an integrated strategy for energy management that addresses not only technology solutions but also consumer participation and engagement in demand-side management strategies (Chen et al., 2015). Emerging technologies may also have the capability to reshape demand response programs, especially that of smart home energy management systems (SHEMS). SHEMS applies advanced controls and communications technologies in order to reduce energy usage at the home level while allowing consumers to view real-time feedback of their consumption patterns. Interestingly, more sophisticated participation in demand response can be facilitated by SHEMS through integrating user-centric strategies with technical measures to maximize energy benefits. Research indicates that improved incorporation of user behavior into SHEMS can lead to greater effectiveness of demand response and overall energy saving. It is crucial that future studies look into how these technologies interact with consumption patterns by consumers to enhance the leverage of their likely impacts in support of demand-side policies (McIlvennie et al., 2020) (Figure 10.9).

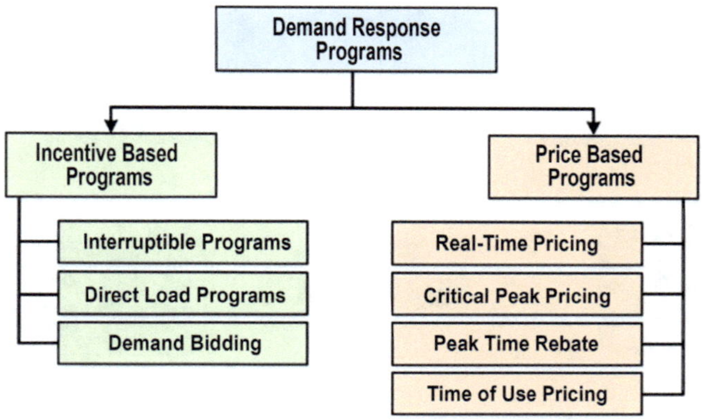

Figure 10.9. *Types of demand response programs.*

Source: https://www.researchgate.net/publication/342494330/figure/fig3/AS:907032873930754@1593264897493/Types-of-demand-response-programs.png.

10.2.3. Role of Consumers in Demand Response

Smart grid technology development has transformed the energy consumption paradigm in a basic sense by making consumers active players in Demand Response (DR) programs. Unlike traditional energy systems founded on passive customers, advanced grids promote consumer participation based on real-time prices and detailed information about energy consumption patterns. This adjustment encourages consumers to regulate their usage during peak hours, thereby ensuring better energy utilization. Utilization of advanced energy management systems (EMSs) causes consumers to utilize energy in the most efficient manner possible according to price signals and grid needs, illustrating how consumer behavior can directly influence energy efficiency and sustainability. Thus, active engagement of consumers not only induces load balancing but also enables a paradigm shift to an interactive responsive energy regime, with system reliability and cost-effectiveness overall highlighted by

Azarinejadian et al. (2017). Furthermore, consumer participation in Demand Response is essential to unleashing the full potential for integrating renewable energies in the smart grid paradigm. With the increasing reliance on intermittent renewable sources such as wind and solar power, flexible demand solutions are crucial.

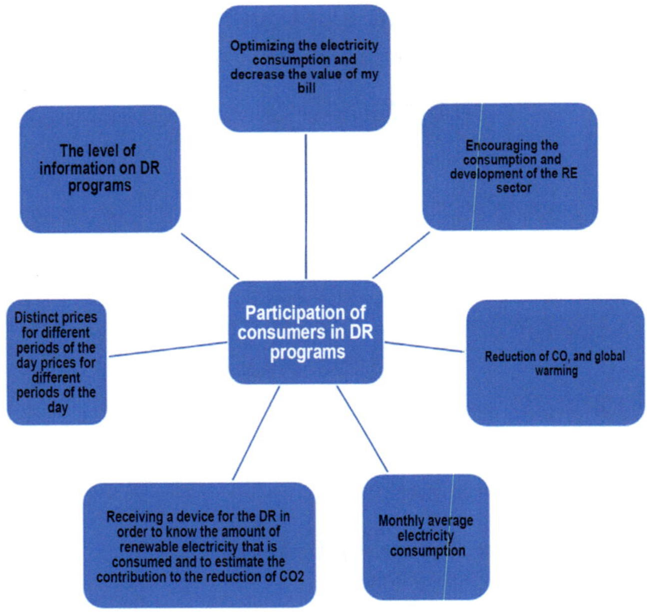

Figure 10.10. *Estimated factors which influence the participation of consumers in demand response programs.*

Source: https://www.mdpi.com/energies/energies-14-08273/article_deploy/html/images/energies-14-08273-g004.png.

Customers play a central role in this new paradigm by adjusting their energy use in harmony with the unpredictable nature of the renewable generation. It is an inclusive approach that aims at forging a symbiotic partnership between supply and demand, leading to grid stabilization and environmental sustainability. Moreover, insights developed from consumer participation in DR programs can help utilities gain insights into consumption habits, informing policies to encourage more participation and enhance the efficiency of the grids operation. In this regard, an in-depth examination of consumers informs DR modeling, thus contributing to the critical role consumers play towards the establishment of an efficient smart grid by Azarinejadian et al. (2017). An understanding of socio-technical dynamics is a key to the successful deployment of Demand Response programs since consumers' attitude towards adopting new technologies decides DR success to a significant extent. Various factors that shape DR, such as concerns about data privacy and perceptions of technology risks, decide smart grid evolution attitudes among consumers. Furthermore, legal frameworks and regulations that govern energy markets must consider consumer interests such as equitable access and protection of consumer rights. The interdependence of industry stakeholders, for example, regulators, utility companies, and consumers, necessitates the formulation of a coordinated approach to monitoring and encouraging consumer engagement. The interdependence emphasizes the need for coordinated approaches to educate consumers about the benefits of DR while soothing their concerns related to cost and reliability. By encouraging a synchronized environment, the smart grid can not only meet energy demands effectively but also assist broader social goals such as carbon reduction and economic efficiency in relation to the challenges and opportunities set out in (Figure 10.10).

10.3. TECHNOLOGIES ENABLING SMART GRIDS

The evolution of the smart grid is greatly facilitated by a broad set of smart grid enabling technologies that enhance its operational efficiency and consumer participation. At the center of this transformation is the convergence of communications technologies, which provide the backbone of smart grid infrastructure. The technologies facilitate real-time information exchange among various components, including home energy management systems, electric vehicles, and renewable resources. By merging the wired and wireless communication networks, smart grids can achieve interoperability and connectivity between scales that vary from Home Area Networks (HANs) to Wide Area Networks (WANs). Smart grids ensure the vital function of collecting and transmitting information regarding energy consumption, which allows for more adaptive and responsive energy management solutions, ultimately resulting in a more sustainable and resilient energy ecosystem (Erol-Kantarci et al., 2015). Another imperative aspect of the technologies powering smart grids is the use of sophisticated monitoring and sensing equipment. Wireless sensor networks play a pivotal role in enabling real-time monitoring of transmission lines, distribution plants, and energy generation plants. In addition to enhancing monitoring during operations, they also reduce man-to-machine involvement in remote locations, thereby enhancing efficiency. Furthermore, technologies such as cognitive radio networks are able to solve issues related to congestion within the spectrum, thereby optimizing the communication flow within the smart grid. With embedded sensing technologies, demand-response programs are enhanced within the smart grid in a way that loads are better managed and peak shaving methods for reliving pressure on traditional power sources become simplified (Erol-Kantarci et al., 2015). The technology's human dimension is also crucial, particularly with regard to user engagement and change in behavior. Smart home energy management systems (SHEMS) have been a centerpiece of this process by empowering consumers with smart tools to play an active role in controlling energy use. However, research indicates a resounding focus on SHEMS's technological dimension at the cost of user engagement and interaction. This technology-centered approach limits SHEMS's potential to impact end-use energy consumption. A coordinated model incorporating both technical and people-centered solutions could lead to substantial energy savings as well as enhancement of demand response measures. A facilitative approach to a complete understanding of SHEMS allows stakeholders to better reconcile technology innovation with consumer expectations, eventually to enhance the impact and deployment of smart grid systems in energy management initiatives.

10.3.1. Advanced Metering Infrastructure (AMI)

Advanced metering infrastructure (AMI) integration with the smart grid platform is key to enhancing demand response mechanisms and overall grid efficiency. AMI comprises smart meters, communications networks, and data management systems, which enable two-way communication between utilities and consumers. Two-way information exchange enables energy usage monitoring in real-time and promotes consumer participation based on dynamic pricing models. With AMI, utilities can more accurately predict load behavior and manage spikes in demand, leading to reduced operating costs and increased reliability. Such advancements are needed by the development of electricity grids, particularly together with rising alternate sources of energy and the imperative need for energy-saving policies (Borghetti et al., 2013). Therefore, AMI not only represents technological development but also represents a strategic reorientation of energy management that responds to the challenges of today's sustainability targets. With shifting energy policy and energy consumption profile, the stakes of AMI become even higher. It is the backbone on which smart demand response programs are based, whose function is to align consumer demand with the supply capacity of the grid. The capacity to record real-time data makes it possible for utility operators to formulate efficient demand-side management policies. This can involve compensating consumers to reduce or shift their power consumption during peak hours and thereby ease the load on the grid and prevent power outages. The infrastructure characteristics of AMI, with higher consumer participation, can lead to a more resilient energy system (Burstein et al., 2011). By establishing a culture that encourages customers to behave responsibly when controlling their energy usage, AMI becomes a leading catalyst for greener energy use and allows individuals to take an active role in maintaining the grid stable.

In addition to improving operational efficiencies and consumer engagement, AMI simplifies integration of renewable sources of energy into the grid. As the energy landscape shifts toward more decentralized forms, AMI plays a part in preserving the variability and uncertainty of sources like solar and wind. With its capacity to provide near-instantaneous reports on energy consumption and production, AMI systems enable utilities to accurately balance supply and demand.

This is especially important as utilities seek to reduce greenhouse gas emissions while they permit an increasing share of renewables. Moreover, with the enhanced data analytics capability provided by AMI, utilities can find patterns and trends and enhance their grid management processes (Borghetti et al., 2013). Thus, the deployment of Advanced Metering Infrastructure not only makes operations easier but also supports the overall goal of a more sustainable, robust, and efficient smart grid (Figure 10.11).

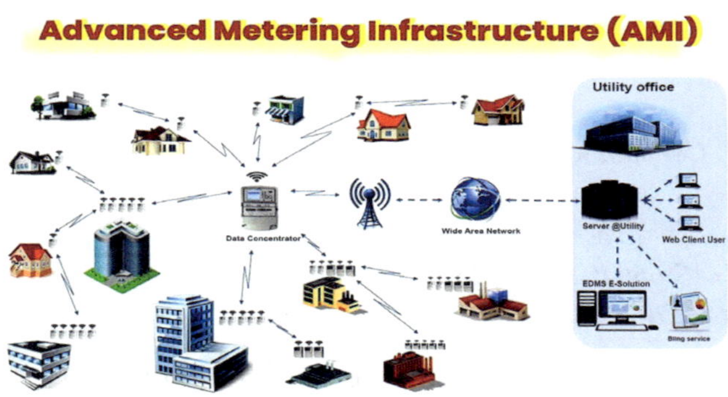

Figure 10.11. *Understanding advanced metering infrastructure (AMI).*

Source: https://i.ytimg.com/vi/Lby1sDHpXKo/maxresdefault.jpg.

10.3.2. Communication Technologies in Smart Grids

The evolution of communication technology has revolutionized the Smart Grids' scenario significantly, enabling better management of energy and active participation of consumers. The basis of this transformation lies in the merging of smart information and communication technology (ICT) allowing real-time data exchange between the consumers and the power grid. This dynamic interactivity allows buildings to function as prosumers, producing and using electricity, that is, from renewable sources, thus stabilizing the grid overall.

Successful demand response is a vital consideration with buildings balancing their energy usage against operational demand and, in doing so, maintaining thermal comfort, as well as indoor environmental quality, optimizing energy use. The call for interoperable systems highlights the importance of robust communication infrastructure that can reconcile disparate legacy systems with new-age technologies, a problem that remains very crucial for industry participants (Boudreau et al., 2016).

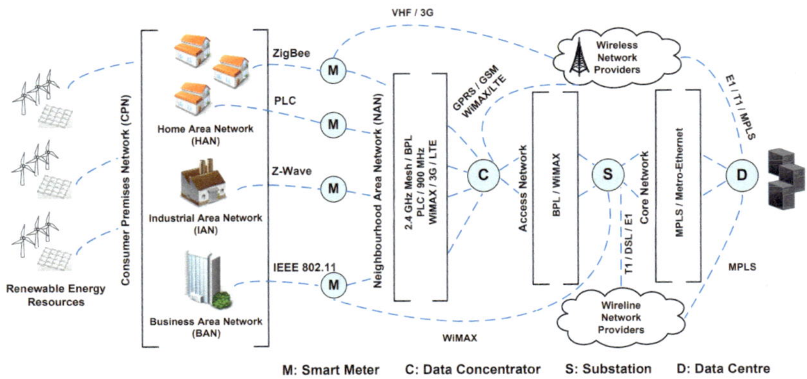

Figure 10.12. *Communication technologies in smart grid.*

Source: https://www.researchgate.net/profile/Mitja-Kolenc/publication/284159647/figure/fig1/AS:669498948452365@1536632396874/Communication-technologies-in-Smart-Grid-22.png.

As growth in Smart Grids accelerates, integrity of their communication infrastructure increasingly takes center stage. Smart devices numbering millions talk to one another in these grids and raise fears of making the whole power infrastructure rely on optimal communication efficiency. The integration of distributed communication intelligence allows utilities to manage energy demand more effectively, incorporating demand response methods that can stabilize the power grid. However, this connectivity is fraught with reliability susceptibilities, and thus it becomes imperative to address the architecture and security in neighborhood area networks (NAN). These problems necessitate ongoing research and innovation to ensure that communication systems are not only effective and fast but also fault-tolerant against failures that could compromise the reliability of power supply (Hu et al., 2015). Moreover, the development and deployment of communication technologies in Smart Grids require collaboration between public policy, economic policies, and technology advancement. It is the responsibility of policymakers to establish standards in order to facilitate interoperability between various systems, thereby enhancing integration of new technology into existing infrastructures. Economic incentives can also lead to deployment of innovative solutions that are energy-efficient but remain consumer-acceptable. As the grid continues to advance, these non-technical barriers must be overcome for a successful transition to a smarter and more responsive energy system. By working together as stakeholders including regulators, technology developers, and utilities, it is possible to establish an effective communication structure in Smart Grids, hence enabling a sustainable future (Boudreau et al., 2016) (Figure 10.12).

10.3.3. Integration of Renewable Energy Sources

Incorporation of renewable power sources in existing grid systems is an integral component in creating a smart grid with possible opportunities and issues. While changing from traditional energy systems to a more decentralized system, the concept of smart microgrids has emerged as a viable solution. Microgrids enable local level generation of energy using small renewable sources, enhancing energy security and load balancing. But such integrations come at the expense of some challenges, like effective consumer involvement, pricing models, and the integration of various technologies (smart and dumb appliances). These are challenges to be surpassed through creative architectural models that can connect fragments like generators and storage units, ultimately forming an integrated and sustainable energy network (Elmenreich et al., 2013). Besides, enhancement of real-time communication in smart grids plays a significant role in the efficient integration of renewable power. Advanced networking technologies enable near-real-time exchange of information among generation, transmission, distribution, and consumption points. Highly sophisticated communication system is key to dynamic energy load management and responding to variability of renewable generation, which can be highly uncertain. Thus, smart metering and grid communication must evolve to support interoperability and overall grid reliability. Standardization coordinated throughout the regions is important in consolidating these communication protocols, thereby simplifying much of the complexity involved in the integration of renewable energy (Chin et al., 2011). The benefits of renewable energy source integration within smart grids extend beyond mere environmental considerations; they encompass economic and social considerations as well. With a diversified energy supply, communities can reduce consumption of fossil fuels, hence less emission of greenhouse gases and improved air quality. Further, consumer involvement in demand response programs can stimulate conservation of energy during peak periods and support the consumption of renewable substitutes. The greater the number of homes brought into energy management, the more the aggregate shift towards renewable systems can redefine energy consumption patterns. Hence, successful integration is also contingent not only on technology but also on building a culture of sustainability and participation by the community (Chin et al., 2011) (Figure 10.13).

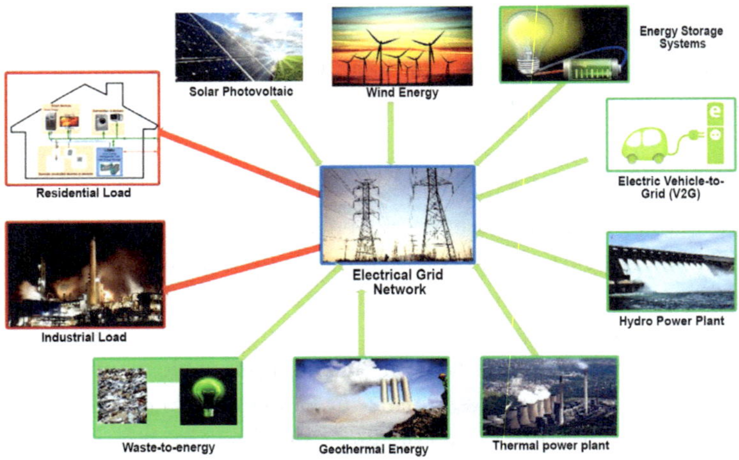

Figure 10.13. *Various types of energy sources in the smart grid.*

Source: https://www.mdpi.com/applsci/applsci-12-08914/article_deploy/html/images/applsci-12-08914-g004.png.

10.4. ECONOMIC IMPACTS OF SMART GRIDS

Smart grid deployment revolutionizes the financial picture of energy supply and usage. Through the integration of advanced technologies such as smart meters and automated demand response systems, the grids become more efficient in use and reduce wastage of power. The process yields end-user as well as utility company cost savings. The ability to monitor and control electricity consumption in real time makes maximum demand response strategies possible, which can potentially shift energy usage during off-peak periods, decreasing pressure on the network. Not only are such advancements critical to economic efficiency, but also environmentally because less use of peaking power plants can lead to lower emissions. Hence, the introduction of smart grid technology has a dual advantage of cost savings and environmental sustainability. Moreover, the economic benefit of smart grids extends beyond the direct cost savings. With increased interoperability and communication capacity, smart grids allow for higher integration of renewable energy sources, thus ensuring energy independence and long-term sustainability. By empowering consumers to participate in decentralized energy markets, such grids empower individual homes and businesses to make money selling energy. This transition to a more decentralized energy future can ignite local economies, creating installation work, maintenance work, and technology development work. Furthermore, greater energy security reduces exposure to the volatility of fossil fuel prices, hence stabilizing the overall economic conditions. Therefore, smart grids not only nurture individual economic benefits but also usher in a better, robust, and self-sufficient economy. Lastly, issues related to security and privacy typical of smart grid technology are paramount in defining the economic viability of such an infrastructure. With recent advances in privacy-preserving algorithms to tackle economic dispatch as well as demand response issues, ensuring data security is of topmost concern in establishing consumer confidence in smart grids. For instance, blockchain authentication enhances security in consumer-facing demand response management and reduces risk elements associated with unauthorized access to sensitive information. Utilization of these safeguards for protection not only protects consumer interests but also has immediate economic benefits by reducing the risk of costly breaches and system downtime. In addressing these challenges, utility firms are able to enable mass adoption of smart grids, thereby maximizing their economic value across various sectors. With these advances, the potential for increased economic resiliency in the energy sector arises.

10.4.1. Cost Savings for Utilities and Consumers

The arrival of smart grid technologies fundamentally shifts the energy consumption environment, enabling utilities and consumers to save significant amounts of money. Demand response systems are one of the key mechanisms by which savings are realized, with the ability to enable utilities to manage consumer usage during high-demand periods. Utilities can mitigate the need for expensive peak generation capability by incentivizing consumers to reduce or offset usage. This reduces the operational cost while also delaying the capital outlay in infrastructure needed to service boom-and-bust demand otherwise. Further, enhanced energy efficiency through smart grid technology reduces waste while lowering overall energy costs, resulting in significant savings for the customer. The net effect of these technologies is a more stable system that enables price stability for the benefit of consumers and utilities alike. In addition to short-run monetary savings, involvement in demand response schemes can lead to long-run savings for the consumer. The majority of smart grid implementations have real-time pricing, through which consumers make price-based decisions regarding varying their consumption patterns. Dynamic pricing, which is this approach, tends to make consumers consume energy when it is cheapest, thus lowering their monthly bills. Secondly, cost savings on reduced peak load can be equivalent to reduced electricity costs overall since utilities wear down the need for building new, costly plants. This union of short-run responsiveness and long-run planning catches the wave of the evolving consumer landscape that craves openness and convenience in energy consumption. As demand response continues to become a thread in the tapestry of electricity supply, consumers can expect to see even more economic benefit as their involvement changes energy consumption patterns towards efficiency (Rolnick et al., 2022). Additionally, the integration of the Internet of Things (IoT) into smart grid infrastructure contributes even further to these cost-saving efficiencies. IoT sensors provide for more accurate monitoring and analysis of energy usage, allowing both utilities and consumers to identify areas for optimization. New technologies, such as Federated Learning, enable devices to locally process data smartly without compromising anyone's privacy, generating value that can be directly mapped into reducing energy expenditure. For utilities, which means improved outage management and advanced load forecasting, reducing operational costs in the bargain. Customers, nonetheless, obtain access to advanced devices for managing their energy consumption, often achieving savings that extend beyond the actual utility bill. The combined advantages of introducing IoT into smart grids create a win-win condition where both parties can share in optimal energy usage, enhancing the economic value of embracing such state-of-the-art technologies (Dinh et al., 2021).

10.4.2. Job Creation in the Energy Sector

The transition to an energy future which is sustainable has generated tremendous controversy on the creation of employment within the energy sector. The implementation of smart grid technologies not only maximizes the efficiency of energy use but also generates innumerable numbers of job opportunities, even more so in areas such as system design, installation, and maintenance. Through investment in sophisticated infrastructure that incorporates demand response technologies, the electricity sector can create employment opportunities for both technical and operational needs. The change offers prospects for highly qualified experts in software development and engineering and electricians and technicians, thereby offering a wide range of employment opportunities for various segments of the labor force. By propelling innovation and transforming our electric grids, the energy sector can fuel economic growth through job creation while supporting environmental stewardship. Moreover, investment incentives and government policies are the major factors that help generate employment within the energy sector. The American Recovery and Reinvestment Act is a case in point and refers to the heavy investments made in clean energy technologies, and the way the acts potential for generating employment opportunities on a massive scale. The more than $90 billion investment in green energy projects is significant because it not only stimulates economic growth but also accelerates the rollout of low-carbon technology. The acknowledgment of such investments comes within an overall effort to develop a resilient workforce that can accommodate future energy needs. As these initiatives are brought to fruition, outcomes indicate that the transition to clean energy not only diminishes the environmental impact but also triggers job growth, hence balancing economic benefits with sustainability efforts (Aldy et al., 2025). The employment creation potential extends beyond existing employment levels because the establishment of the energy sector influences other supporting industries. Because the smart grid consolidates dispersed sources of energy, like renewable energy, a ripple effect is felt along technology, manufacturing, and service sectors. For instance, the increased demand for energy efficiency solutions spurs innovation in the smart technologies themselves, which further boosts research and development employment even higher. Employment in supporting sectors like customer services and energy consulting will also expect to rise as the energy sector evolves. Besides, such diversification encourages enhanced economic resilience and market shock tolerance because communities are less reliant on traditional energy sources. As highlighted by the green recovery program evidence, large investments in clean energy can generate up to 2 million jobs, highlighting the high employment generation potential and economic recovery of this sector (Garrett-Peltier et al., 2025).

10.4.3. Economic Incentives for Demand Response Participation

The concept of economic incentives is central to the instigation of Demand Response (DR) participation in the context of smart grids. By providing money rewards or cost savings, policymakers and utilities aim to motivate consumers to make energy consumption shifts based on levels of supply. When participants shift their consumption from peak periods to off-peak periods, they not only stabilize the grid but also save on tariffs or are given direct monetary benefits. Such direct incentives in this way can encourage mass participation by making it economically worthwhile for consumers to participate in DR programs. Moreover, as competition increases in the energy industry, the scope for earning with such economic incentives can lead to efficient resource allocation so that consumers and energy providers both achieve the most desirable outcomes from consumption. It is significant to understand how various appliances respond to economic incentives so that Demand Response measures can be implemented efficiently. Bundling of diverse consumer loads can generate significant flexibility in the grid, allowing a more stable supply of energy. Clustering different appliances based on their consumption patterns and service constraints allows for the identification of the optimum participation strategies. However, it is measuring and anticipating customers' response to these incentives, based on their individual behavioral characteristics and inclinations, which is posing the challenge. After the study, it remains difficult to reflect the group action of heterogeneous appliances and will require more elaborate modeling and analysis tools to achieve maximum DR participation. Addressing such complexity matters because more precise forecasting of consumers' reaction to incentives equates to improved program design and overall demand-side management effectiveness (Alizadeh et al., 2014). Besides participant incentives, the competition structure of energy producers also influences the success of Demand Response initiatives. The interaction between heat power generation and hydropower can decide the market condition under which DR plans operate. Research has indicated that in the case of incentive-based demand response, peak energy demand is minimized, and power demand is transferred to off-peak hours. This not only enhances consumer surplus due to the financial benefits earned but also compels generators to reshape their strategies in competitive markets. While DR programs reduce profits for generation companies due to lower energy trading during peak times, these mechanisms end up creating a more efficient and stable power market for consumers as well as suppliers. This underlines the utmost significance of economic incentives being tuned to balance all stakeholders' interests in the energy sector (Ruiz et al., 2018).

10.5. ENVIRONMENTAL BENEFITS

The transition to smart grids marks a paradigm shift in the search for sustainable energy sources, offering a large environmental dividend. One of its principal advantages is in the enhanced ability to incorporate solar and wind energy supplies, which are notoriously variable. Smart grids facilitate efficient management of such unstable inputs with two-way communications systems that allow real-time monitoring and adjustment. This ability not only limits energy wastage but also decreases dependence on fossil fuels, which results in less greenhouse gas emission. The potential of smart grids comes from how they have the ability to optimize energy delivery while enabling the overall aim of carbon reduction, a fundamental aspect identified in existing research and policy-making debates on green energy strategies (Axon et al., 2014). The other significant environmental benefit of smart grids is heightened energy efficiency, both supply side and demand side. Through the implementation of demand response practices, customers are incentivized to drop or shift their electricity load during peak times, thus easing pressure on the grid. This dynamic process not only makes customers frugal, but it also reduces the demand for building additional power plants that otherwise could have adverse environmental impacts. Moreover, efficient demand response programs reduce the losses of electricity during transmission and distribution, where a significant percentage of energy is wasted. Through enhanced supply and utilization of electricity, smart grids enhance a cleaner energy usage pattern that directly impacts environmental conservation efforts (Cheriet et al., 2014). Besides, the integration of smart grid technologies allows for an enhanced understanding of energy consumption patterns, which is required for attaining green practices among consumers. Thanks to advanced analytics and smart metering, households and companies can monitor their energy use more closely, allowing them to make well-considered decisions favoring lower overall consumption. Rising awareness has a tendency to induce changes of behavior that work to sustainability, essentially generating a feedback process in which consumers actively help conserve energy. With the growth of smart grids, the collective impact of these numerous independent initiatives can realize significant savings in energy usage and emissions, underpinning the value of mass participation in the transition to a low-carbon economy (Axon et al., 2014).

10.5.1. Reduction in Greenhouse Gas Emissions

One key path to greenhouse gas emissions mitigation is the convergence of smart grid technologies and demand response initiatives. Through the use of real-time data analytics and communications technologies, smart grids enable efficient management of electricity usage. Demand response programs incentivize customers to make changes in energy usage during peak hours, thus easing the load on the grid while optimizing energy efficiency. This shift in behavior is especially significant in the reduction of fossil fuel-based energy dependencies, a primary source of emissions. When consumers engage in demand response, the general trend in consumption of energy shifts towards a more sustainable trend, which allows greenhouse gas emissions associated with electricity generation to be reduced. Consequently, the deployment of smart grid systems not only improves grid reliability but also has a categorical contribution to carbon level reduction in various industries. Furthermore, electrification of the transportation system enhances the efficiency of smart grid and demand response technologies in emission control efforts. The transition from gasoline-powered vehicles to electric vehicles involves a re-wiring of the energy supply chain; here, smart grids can optimize charging. In an entirely decarbonized electricity grid, electrification of transport is projected to be able to reduce greenhouse gas emissions from light-duty vehicles up to 80% if efficiently implemented. This emphasizes the imperative for consistent policies that promote electrification alongside renewable integration. As the transport sector consists of a significant share of emissions, smart grid technologies in conjunction with EV adoption represent a firm plan for fulfilling domestic as well as international obligations concerning emissions cuts. With these technologies, not only will targets be achievable, but they will also play a significant role in determining the overall energy path. Evidence from global case studies, such as China's ambitious energy policies, shows the potential for smart grid and demand response solutions to make a game-changing contribution towards emissions reduction. China aims to peak its carbon dioxide emissions by 2030 and raise the share of non-fossil energy supplies to 20% by then. Deployed correctly, technologies for reducing carbon emissions and higher energy efficiency will be able to bring 2050 CO_2 levels down to 42% below 2010 levels. Smart grid capabilities included would enable efficiencies that are compatible with these bold objectives and enable a structure for managing energy use as it transitions away from coal and oil-based fuels. The co-benefits of these transitions—improved local air quality, improved human health, and more decisive global action on climate change mitigation—underscore the value of smart grids and demand response not just in alleviating emission problems, but also in facilitating greater overall benefits to society, thus initiating a holistic transition toward climate action.

10.5.2. Promotion of Renewable Energy Usage

The promotion of the use of renewable energy is one of the main drivers for the realization of a sustainable energy future, particularly in the case of smart grids and demand response schemes. The integration of distributed renewable energy sources such as solar and wind power into the traditional energy infrastructure demands sophisticated energy management practices that facilitate two-way communications between consumers and producers. By incorporation of technologies facilitating such interaction, home-homes equipped with renewable energy generators can become integral to energy production. Such a model offers a platform by which electricity consumers can become more involved in their energy consumption patterns, building a culture of conservation and efficiency that ultimately leads to lowered total net energy use, as demonstrated in the design by Turck et al. (2010). Moreover, the deployment of smart grids is significantly reinforced by the requirement to reduce carbon output and combat climate change. Smart grid systems can provide best use in effective management of volatile energy supplies created through renewable energy resources, which are unpredictable in nature. By incorporating highly advanced IT systems that respond based on real-time information, smart grids have the capability of maximizing energy delivery and improving grid reliability. Such a capability not only ensures the reduction in the use of fossil fuels but also lays the groundwork for the development of decentralized energy networks where houses are able to pump excess energy back into the network and thus become participants in a communal energy system. The political and financial backing of smart grid initiatives also signifies the importance of transitioning to low-carbon energy sources, as highlighted by Axon et al. (2014). Finally, in order to make full use of renewable energy in smart grids, it is essential to tackle public acceptance and attitudes towards connected technologies. Fears regarding data privacy, security, and the reliability of smart meters can hinder bulk installation of these systems. Policy, competitive rates, and knowledge-based public debate are crucial in alleviating concerns while making implementation of smart grid technologies accommodating to community needs. A proactive stakeholder approach can help consumers participate more in demand response programs, as it encourages the consumers to embrace the culture of energy efficiency. Therefore, encouraging the utilization of renewable energy is more of a socially inclusive strategy that entails empowering the consumers, enhancing the stability of the grid, and enabling the overall sustainability of energy systems.

10.5.3. Impact on Energy Efficiency

The advancement in smart grid technology has completely revolutionized the energy efficiency sector. With the integration of sophisticated computing and communication technologies, the smart grid enhances the efficiency and reliability of power systems. It has a profound impact on consumption patterns of energy, where there is room for real-time demand response policies able to react to the user's consumption pattern. For instance, with the use of smart meters, consumers can monitor the consumption of energy in detail, leading to more effective choices on the use of energy. The systems facilitate easy incorporation of renewable forms of energy in an efficient manner, which is essential for energy management in a sustainable way. Thereafter, not only do these advancements fulfill present energy demands, but they also minimize the total environmental effect of electricity generation, thus maximizing energy efficiency overall (Hu et al., 2015). In addition, utilization of demand response techniques through intelligent grids maximizes energy efficiency via load balancing and peak shaving. Households involved in demand response approaches—whereby prosumers generate some fraction of their electricity load—play a critical role in energy demand management. Through shifting utilization to off-peak periods, such methods effectively decrease peak load on the grid during peak demand periods, resulting in smoother patterns of energy supply (Bonetto et al., 2017). This is made possible by state-of-the-art information and communication technologies that allow efficient monitoring and regulation of energy flow throughout the grid. As societies embrace new demand response paradigms, they not only reduce overall energy use but also realize optimized adaptive energy use strategies that achieve sustainability goals. Finally, the integration of big data analytics in the smart grid system transformed the energy efficiency idea once again by promoting active energy management behaviors. With the analysis of large data generated from customer activities with smart devices, utilities can better forecast demand and create tailored energy efficiency programs. Such intelligence allows customers to manage their energy behavior knowingly, making peaking demand reduction and total energy consumption higher. Incorporation of serious games as a learning instrument also results in an engaged consumer base, which promotes more energy-efficient practices in everyday life (Bonetto et al., 2017). With advancing smart grid technologies, their combined approaches to data analysis not only optimize energy flows but also develop an energy stewardship culture among consumers, making energy efficiency across the board all the more effective.

10.6. CHALLENGES AND BARRIERS

The transition to smart grid systems has a number of challenges and hurdles that prevent the full realization of demand response potential. One actual challenge lies with integrating renewable sources of power, which, while being crucial in ensuring a future with sustainability, are prone to intermittent and unreliable patterns of generation. This unreliability has the tendency of introducing volatility in the power grid, causing the control of energy supply and demand more complex by the day. As noted in landscape storage smart grid analysis, technologies must be developed to enhance energy absorption capacity, but the inherent volatility of sources like solar and wind energy requires advanced forecasting and grid management abilities to minimize these issues. Despite smart grid and energy storage technology development, the integrations are still a complicated issue that continues to present high operation hurdles to be addressed in order to facilitate smart grids' trustworthy demand response (Li et al., 2025). Compounding the complexity is the technologies gap in communications infrastructure needed in putting the smart grid into effect. Wireless communication technology is crucial to facilitate real-time data exchange and seamless interconnectivity of grid devices. Adoption of these technologies is uneven in all areas, leading to differences in smart grid operation. As indicated, communications technology like 5G and LoRa significantly enhance monitoring and management capability; however, access differentials in these technologies prevent uniform smart grid progress. Security matters regarding data transmission should also be addressed to guarantee the integrity and reliability of smart grid operations. Without secure and dependable wireless communication infrastructure, the full potential of demand response can be severely limited since consumers might not be ready to engage in plans that expose their data and privacy (Zhou, 2025). Finally, organizational and regulatory problems are key barriers to mass deployment of smart grid and demand response technology. Regulative practices today are often lagging behind technological development, which leads to regulative environments that can serve as disincentives for investment in smart grid projects. Moreover, calls for cooperation between different stakeholders like utility companies, regulative authorities, and consumers complicate decision-making. Effective demand response schemes must include collaborative efforts towards coordinating incentives and making all stakeholders capable and willing to adapt to new paradigms of operation. As the landscape storage smart grid analysis suggests, addressing these organizational challenges involves not just technological innovation but also comprehensive policy reform. A comprehensive approach is most crucial to overcome these hurdles and enable the development of a robust and efficient smart grid system capable of addressing future energy needs aptly (Li et al., 2025) (Figure 10.14).

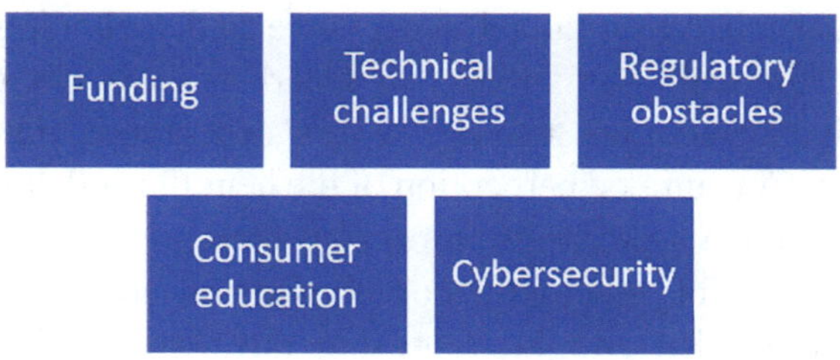

Figure 10.14. *The challenges of implementing the smart grid.*

Source: https://www.researchgate.net/publication/375175638/figure/fig2/AS:1143 1281202483659@1698859032526/The-challenges-of-implantation-of-smart-grid. png.

10.6.1. Technological Challenges in Implementation

The employment of smart grid technologies by default has to contend with a series of technological challenges that may hinder its success and widespread application. One significant impediment is the integration of various communication technologies that are able to facilitate two-way data exchange between the distributed energy resources and the grid operators. Heterogeneity of sensor and meter types presents it with difficulties to create an integrated network enabling real-time monitoring and control in accuracy. For instance, the UC Berkeley group's Building-to-Grid testbed reflected the importance of rigorous testing and assessment of metering subsystems required for accurate data communication and the grids' performance (Peffer et al., 2010). In the absence of interoperable and standards-based technology, utilities can be prevented from efficiently harnessing demand response mechanisms, and hence, the potential of the smart grid is reduced to inefficient optimization of energy consumption among users. Moreover, data management and cybersecurity concerns present stringent challenges in the deployment of smart grid systems. Since smart grids rely upon data analytics in forecasting and responding to user demand, high security and capable infrastructures for data become a top priority. The interconnectedness between smart grid stakeholders increases the risks to cyber-attacks, which can not only breach consumer privacy but also the stability of the power supply. Research shows evidence of consensus in effective management practices and security measures; however, the feasibility of such interventions has been variable (Axon et al., 2014). This mismatch has the effect of making stakeholders wary, thereby delaying the development of smart grid projects, and undermining public trust. Therefore, closing these weak spots must therefore be a priority for everyone involved in implementing smart grid. Finally, the balance between social acceptance and technological feasibility cannot be overestimated. The success of smart grids is not only reliant on technological advancement but also consumers' acceptance of new technology, e.g., smart meters. Consumers' perception of the benefits and drawbacks of such innovations might have a profound impact on their success. Current research emphasizes the need for successful communication strategies and effective educational programs that enlighten consumers regarding the advantages of smart grid technologies and allay their privacy and safety concerns (Axon et al., 2014). Consumers will be more inclined to engage in demand response programs if they are empowered and educated, hence making the smart grid operations more efficient. Thus, integration of wide stakeholder participation initiatives into the process of technological integration is important to achieving an effective adoption of smart grids.

10.6.2. Regulatory and Policy Hurdles

The development of smart grid technology and demand response programs is faced with tremendous regulatory and policy hurdles which can delay their development and deployment. The integration of the advanced technologies such as microgrids necessitates a reexamination of existing regulatory structures, especially in deregulated states like Pennsylvania. Such models have no real statutory provisions upon which microgrids would be categorized and their operating boundaries, thus not easily facilitating the division of responsibilities between private players, local authorities, and state utilities. As quoted by Parks et al. (2019), the absence of solid guidelines can create ambiguities concerning regulation, investor confidence, and participation from the public, thus influencing the objectives of resilience and efficiency incorporated in smart grid solutions. Therefore, it is necessary that such loopholes in the regulations are plugged to allow a conducive environment for deployment and innovation in the energy space. The regulatory landscape also becomes intricate due to the large number of stakeholders present within smart grid projects. Each stakeholder, from public utilities to independent power producers, typically has varying interests and degrees of influence, which can result in disputes during the execution of the project. In addition to this, the absence of an overarching law can lead to inconsistent interpretations of regulations in various jurisdictions and hence complicate the planning operations of smart grid technologies. As an example, Parks et al. (2019) illustrates how microgrid deployments in Pittsburgh require coordination among various organizations, issues regarding public utility control converting to stiff logistical hurdles. Therefore, it becomes imperative to develop harmonized policies that enable stakeholder engagement and establish roles in a way that makes it possible to have efficient smart grid projects and maximize demand response potential. In addition, the move towards electric vehicles (EVs) also brings a new regulatory complication that could impede advancement of smart grid systems in general. Widespread charging infrastructure will have to be fulfilled along with incentives for battery electric vehicles (BEVs) in order to make them available and part of the energy system. However, ineffective policy guidelines and weak policy architecture can serve as barriers to investors willing to invest in EV infrastructure. Moreover, existing regulatory documents may not adequately address the environmental pressure generated from conventional transportation, which will again become a hurdle to shift towards sustainable ones. As discussed in, the regulatory bodies must lead in formulating dynamic policies that will be able to adapt to changing technologies and market trends in the transport sector. This is vital not only for the development of smart grid initiatives but also for promoting a complete shift to sustainable energy consumption and demand response capabilities.

10.6.3. Consumer Acceptance and Engagement Issues

The deployment of smart grid technologies creates gigantic hurdles in terms of consumer acceptability and involvement. How the public views the scenario determines the success or failure of smart grids as the energy landscape evolves. Consumers are normally not willing to adopt new technologies, particularly where they are not certain about the advantages and how the new technologies would work. Concerns surrounding data privacy, security, and the potential for increased costs play a crucial role in shaping public attitudes. As noted, public acceptability of smart grid technologies hinges not only on perceived benefits but also on the management of associated risks, highlighting the importance of transparent communication between energy providers and consumers (Axon et al., 2014). Being proactive about such challenges can foster heightened trust, as well as drive consumer buy-in. Consumer participation methods involving active consumer involvement in the adoption of smart grids also ensure higher acceptance. Education and awareness programs about smart grid technologies play a key role in breaking the mystique that surrounds them. By providing open, comprehensible information on how demand response initiatives can lead to efficiencies and potential savings, utility companies can allow customers to become more active participants in their energy consumption. In addition, incorporating the input of customers into the creation and utilization of these technologies can assist with ensuring that they serve their needs and desires. Such co-creation strategy not only enhances user experience but also contributes to a sense of shared responsibility of energy tasks (Axon et al., 2014). Therefore, the integration of consumer intelligence in the design of smart grids is indispensable for building robust acceptance and facilitating authentic engagement within the energy market. In addition, overcoming consumer acceptance issues requires adequate knowledge of the socio-technical landscape that encompasses energy consumption routines. The terminology has to often switch between viewing consumers as end-users to recognizing them as part of the energy continuum. Socioeconomic condition, geography, and cultural habits may influence consumer sentiments and acceptance of demand response programs. Customized outreach strategies taking these nuances into account are therefore required. By utilizing social media websites, public forums, and educational workshops, energy utilities can craft dialogue that addresses different consumer personas. Not only does this help consumers learn about smart grid technologies, but it also facilitates a participatory atmosphere that inspires individuals around energy management. Ultimately, building such an atmosphere is central to realizing widespread consumer adoption along with the complete utilization of smart grids and demand response initiatives.

SUMMARY

In summary, the combination of demand response systems and smart grid solutions presents a paradigm platform for enhancing the energy efficiency and reliability on today's electric grids. As illustrated through various projects, such as the UC Berkeley Building-to-Grid testbed, studies have tackled the rich mapping of energy flows as a way of better understanding and optimizing building-level consumption patterns. The partnership approach between educational institutions and public utilities has not only raised technical concerns but also identified the role of holistic approaches that have the ability to bridge the technology growth-practical application gap in energy management. With this partnership knowledge, stakeholders are able to effectively contribute towards maximizing energy usage while encouraging greater environmental sustainability in the longer term (Peffer et al., 2010). Moreover, the potential of smart home energy management systems (SHEMS) to seriously influence household energy consumption cannot be overlooked. Literature demonstrates a prevalent focus on the technical with common neglect of user engagement and behavior modification necessary for effective demand response strategies. This lag presents an immediate need for interdisciplinarity that not only makes the use of advanced technological solutions a priority but also involves user-centered approaches that ensure active participation in energy management. Through this, it may pave the way towards an integrative understanding of energy consumption within home environments that would directly lead to greater effectiveness in demand-side management and increased savings in energy (McIlvennie et al., 2020). Finally, the future integration of demand response and smart grid will depend on the interaction between consumer involvement and technological innovation. The experience acquired from projects such as the UC Berkeley project provides a blueprint for future projects, demonstrating the successful interaction which exists between researchers, utility companies, and consumers. By considering both the technical and societal dimensions of energy use, the path ahead becomes clearer, with a focus on the requirement for innovative technology that takes into consideration diverse user experiences and requirements. Thus, ongoing innovation in smart grids and demand response is a critical steppingstone to a future with sustainable energy, where collective effort can yield great benefits for consumers as well as the planet. This dual focus on technology and user interface will be the key to realizing the full potential of smart grids and their accompanying demand response functions (McIlvennie et al., 2020; Peffer et al., 2010).

The presence of smart grid technologies has significantly changed the dynamics of supply and consumption of energy to such an extent that it emphasizes the need for an agile response to changing demand. By developing a solid base that supports two-way communication between utilities and consumers, smart grids effectively enable the application of demand response (DR) mechanisms. As the rapid advancement of the Internet of Things (IoT) goes, such technologies have been pivotal in maximizing the efficiency and reliability of power grids. However, along with the advent of IoT devices come security threats that must be mitigated in order not to suffer potential cyber-attacks and still have a reliable system. Therefore, it is critical to know the interplay of DR and smart grid technologies for developing a master energy plan that is responsive to existing demands while safeguarding the system against dangers (Khan et al.,

2024). The other point of interest in the debate is the pivotal function battery electric vehicles (BEVs) play in the evolving smart grid environment. As BEVs develop, they are increasingly being viewed as multi-faceted assets capable of stabilizing the grid and adding to grid efficiency. BEV participation in demand response programs adds value by providing not only charging flexibility but also opportunities for energy storage and distribution. Smart charging technology and developments in grid integration offer a platform for effective utilization of BEVs in DR programs. However, their proper utilization also requires the active participation of their owners, something that remains to be a problem. Gamification-based incentive programs are some of the new solutions suggested to boost smart grid usage among BEV drivers and get more of them registered in demand response programs (Lee et al., 2024). In summary, the discussion of the smart grid and demand response indicates the interdependence of technology, usage patterns, and energy control. Coordination among grid operators, consumers, and car owners is required to tap the full potential of smart grid technologies. As security concerns are dealt with and collaboration is fostered by innovative mechanisms, stakeholders can create a more resilient and effective power system. Further development in such systems has the potential to not only ensure better delivery of energy but the loftier virtues of carbon neutrality and sustainability. Thus, the implementation of smart grids and demand response schemes is a must to revolutionize the energy consumption and resource management in a world continuously electrified (Khan et al., 2024; Lee et al., 2024).

REFERENCES

1. Afsar, S., Brotcorne, L., Marcotte, P., et al. (2016). Achieving an optimal trade-off between revenue and energy peak within a smart grid environment. https://core.ac.uk/download/49456434.pdf (accessed on 30th September 2025).
2. Agenbroad, J., Bai, Q., Chrisman, K., et al. (2019). A roadmap for China to peak carbon dioxide emissions and achieve a 20% share of non-fossil fuels in primary energy by 2030. https://core.ac.uk/download/323308755.pdf (accessed on 30th September 2025).
3. Aldy, J. E. (2025). A preliminary review of the American Recovery and Reinvestment Act's clean energy package. https://core.ac.uk/download/pdf/9305642.pdf (accessed on 30th September 2025).
4. Alizadeh, M., Goldsmith, A., Kesidis, G., et al. (2014). Capturing aggregate flexibility in demand response. http://arxiv.org/abs/1404.1955 (accessed on 30th September 2025).
5. Atzori, L., Cherchi, R., Colistra, G., & Pilloni, V. (2014). Energy consumption management in smart homes: An M-Bus communication system. https://core.ac.uk/download/pdf/54602568.pdf (accessed on 30th September 2025).
6. Axon, C. J., Balta-Ozkan, N., Cipcigan, L., et al. (2015). Scenarios for the development of smart grids in the UK: Literature review. https://core.ac.uk/download/43094319.pdf (accessed on 30th September 2025).
7. Ayub, M. F., Li, X., Mahmood, K., Shamshad, S., Saleem, M. A., & Omar, M. (2024). Secure consumer-centric demand response management in resilient smart grid as Industry 5.0 application with blockchain-based authentication. *IEEE Transactions on Consumer Electronics, 70*, 1370–1379. https://www.semanticscholar.org/paper/0a02c65a020f601c62b6d62b9df0f41723b513b4 (accessed on 30th September 2025).
8. Azarinejadian, F., Guerrero, J. M., Marzband, M., et al. (2017). An optimal energy management system for islanded microgrids based on multi-period artificial bee colony combined with Markov chain. https://core.ac.uk/download/60613271.pdf (accessed on 30th September 2025).
9. Bao, Y., Feng, W., Jin, M., et al. (2018). A review of microgrid development in the United States–A decade of progress on policies, demonstrations, controls, and software tools. https://core.ac.uk/download/323310623.pdf (accessed on 30th September 2025).
10. Belhaiza, S., & Al-Abdallah, S. (2024). A neural network forecasting approach for the smart grid demand response management problem. *Energies*. https://www.semanticscholar.org/paper/421a7d891a25328a4e3ec821c59de15799473310 (accessed on 30th September 2025).

11. Bonetto, R., & Rossi, M. (2017). Smart grid for the smart city. https://core.ac.uk/download/84922332.pdf (accessed on 30th September 2025).
12. Borghetti, G., Menga, G., Procaccianti, K., Ardito, L., Morisio, M., et al. (2013). Smart grid technologies in Europe: An overview. https://core.ac.uk/download/pdf/11431923.pdf (accessed on 30th September 2025).
13. Boudreau, M. C., Helsen, L., Henze, G. P., et al. (2016). Ten questions concerning integrating smart buildings into the smart grid. https://core.ac.uk/download/129148665.pdf (accessed on 30th September 2025).
14. Bucur, M., de Agustin-Camacho, P., Macías, O., et al. (2019). Analysis of building energy upgrade technologies for implementing the dual energy efficiency and demand response scheme for non-residential buildings. https://core.ac.uk/download/228298459.pdf (accessed on 30th September 2025).
15. Burstein, A. J., Mulligan, D. K., & Wang, L. (2011). Privacy in the smart grid: An information flow analysis. https://core.ac.uk/download/288432753.pdf (accessed on 30th September 2025).
16. Chen, H., Coskun, A. K., Liu, Z., et al. (2015). Optimizing energy storage participation in emerging power markets. https://core.ac.uk/download/216247516.pdf (accessed on 30th September 2025).
17. Cheriet, M., Dandres, T., Lemieux, Y., et al. (2014). Challenges and complexities in application of LCA approaches in the case of ICT for a sustainable future. https://core.ac.uk/download/556226324.pdf (accessed on 30th September 2025).
18. Chin, W. H., Efthymiou, C., Fan, Z., et al. (2011). Smart grid communications: Overview of research challenges, solutions, and standardization activities. http://arxiv.org/abs/1112.3516 (accessed on 30th September 2025).
19. Copeland, R., Crespi, N., Rehmani, M. H., et al. (2019). Internet of Things-aided smart grid: Technologies, architectures, applications, prototypes, and future research directions. http://arxiv.org/abs/1704.08977 (accessed on 30th September 2025).
20. Dalton, R., ElBanhawy, E., & Thompson, E. M. (2013). Eras of electric vehicles: Electric mobility on the verge. Focus attention scale. https://core.ac.uk/download/17299059.pdf (accessed on 30th September 2025).
21. De Turck, F., Develder, C., Haerick, W., et al. (2010). Architectures for smart end-user services in the power grid. https://core.ac.uk/download/55730622.pdf (accessed on 30th September 2025).

22. Elmegaard, L., Hayes, B., Gruber, K., et al. (2018). Parallel statistical model checking for safety verification in smart grids. https://core.ac.uk/download/188826869.pdf (accessed on 30th September 2025).
23. Elmenreich, W., & Sobe, A. (2013). Smart microgrids: Overview and outlook. http://arxiv.org/abs/1304.3944 (accessed on 30th September 2025).
24. Erol-Kantarci, M., Rehmani, M. H., Radenkovic, M., et al. (2015). Special section on smart grids: A hub of interdisciplinary research: IEEE Access special section editorial smart grids: A hub of interdisciplinary research. https://core.ac.uk/download/48318388.pdf (accessed on 30th September 2025).
25. Fox-Penner, P., Gorman, W., & Hatch, J. (2018). Long-term U.S. transportation electricity use considering the effect of autonomous vehicles: Estimates & policy observations. https://core.ac.uk/download/286679086.pdf (accessed on 30th September 2025).
26. Garrett-Peltier, H., Scharber, H., Heintz, J., & Pollin, R. (2025). Green recovery: A program to create good jobs & start building a low-carbon economy. https://core.ac.uk/download/pdf/6310499.pdf (accessed on 30th September 2025).
27. Hu, R. Q., Qian, Y., & Xu, S. (2015). On reliability of smart grid neighborhood area networks. https://core.ac.uk/download/478907055.pdf (accessed on 30th September 2025).
28. Jain, N., Pillai, B., & Singh, N. (2025). Green IoT-based smart grid: Implementing intelligent demand-response systems for energy sustainability. *2025 2nd International Conference on Research Methodologies in Knowledge Management, Artificial Intelligence and Telecommunication Engineering (RMKMATE).* https://www.semanticscholar.org/paper/56921411eac251e767babd5737f6649c61ead5df (accessed on 30th September 2025).
29. Khan, M. A., Khan, T., Waseem, M., Saleh, A. M., Qamar, N., & Muqeet, H. A. (2024). Investigation and analysis of demand response approaches, bottlenecks, and future potential capabilities for IoT-enabled smart grid. *IET Renewable Power Generation.* https://www.semanticscholar.org/paper/d63da0099e6bdb3c8dd9019a293f01785d10611a (accessed on 30th September 2025).
30. Lee, J. W., Park, S. W., & Son, S. Y. (2024). Gamification-based vehicle-to-grid service for demand response: A pilot project in Jeju Island. *IEEE Access, 12*, 30209–30219. https://www.semanticscholar.org/paper/ec9cc52a98698cb501cf5b277ec3aef93613dd74 (accessed on 30th September 2025).

31. Li, J., & Wang, Z. (2025). Analysis of the characteristics and applications of wind-solar-storage systems in the smart grid. *Science and Technology of Engineering, Chemistry and Environmental Protection.* https://www.semanticscholar.org/paper/f9e90b84922063e52b15fd16f188fb0c98259308 (accessed on 30th September 2025).

32. Liu, B., Wu, J., & Chai, L. (2025). Distributed privacy-preserving algorithm for economic dispatch and demand response of smart grid with homomorphic encryption. *IEEE Transactions on Smart Grid, 16,* 173–182. https://www.semanticscholar.org/paper/b2bb4df282436c8baea9aff069280161ced3706a (accessed on 30th September 2025).

33. McIlvennie, C., Pritoni, M., & Sanguinetti, A. (2020). Of impacts, agents, and functions: An interdisciplinary meta-review of smart home energy management systems research. https://core.ac.uk/download/323062198.pdf (accessed on 30th September 2025).

34. Parks, R. (2019). Microgrids: Legal and regulatory hurdles for a more resilient energy infrastructure. https://core.ac.uk/download/212892728.pdf (accessed on 30th September 2025).

35. Peffer, T. (2010). UC Berkeley's Cory Hall: Evaluation of challenges and potential applications of building-to-grid implementation. https://core.ac.uk/download/288432727.pdf (accessed on 30th September 2025).

36. Rolnick, D., Donti, P. L., Kaack, L. H., Kochanski, K., Lacoste, A., Sankaran, K., Ross, A. S., et al. (2022). Tackling climate change with machine learning. *ACM Computing Surveys, 55,* 1–96. https://doi.org/10.1145/3485128 (accessed on 30th September 2025).

37. Ruiz, F., & Vuelvas, J. (2018). A novel incentive-based demand response model for Cournot competition in electricity markets. https://core.ac.uk/download/529514214.pdf (accessed on 30th September 2025).

38. Sumper, A., Bremdal, B. A., Rajasekharan, J., Lloret-Gallego, P., Olivella-Rosell, P., Villafafila-Robles, R., Ottessen, S. Ø., et al. (2018). Local flexibility market design for aggregators providing multiple flexibility services at distribution network level. https://core.ac.uk/download/157811721.pdf (accessed on 30th September 2025).

INDEX

P

R

S

T

U

V

W